The Evolution of Automotive Technology: A Handbook

Other SAE books of interest

Chrysler Engines 1922–1998
By Willem L. Weertman
(Product Code: R-365)

Hall-Scott: The Untold Story of a Great American Engine Maker
By Ric A. Dias and Francis H. Bradford
(Product Code: R-368)

World History of the Automobile
By Erik Eckermann
(Product Code: R-272)

For more information or to order a book, contact

SAE INTERNATIONAL
400 Commonwealth Drive,
Warrendale, PA 15096-0001, USA;

phone +1-877-606-7323 (U.S. and Canada only)
or +1-724-776-4970 (outside U.S. and Canada);
fax +1-724-776-0790;
e-mail CustomerService@sae.org;
website http://books.sae.org.

The Evolution of Automotive Technology: A Handbook

By Gijs Mom

Warrendale, Pennsylvania
USA

	400 Commonwealth Drive
	Warrendale, PA 15096
	E-mail: CustomerService@sae.org
	Phone: +1.877.606.7323 (inside USA and Canada)
	+1.724.776.4970 (outside USA)
	Fax: +1.724.776.0790

SAE Order Number R-435
DOI 10.4271/R-435

Library of Congress Cataloging-in-Publication Data
Mom, Gijs, 1949- author.
    The evolution of automotive technology : a handbook / by Gijs Mom.
       pages cm
    Summary: "This book covers one and a quarter century of the automobile, conceived as a cultural history of its technology, aimed at engineering students and all those who wish to have a concise introduction into the basics of automotive technology and its long-term development. Its approach is systemic and includes the behavior of drivers, producers, nonusers, victims, and other "stakeholders" as well as the discourse around mobility. Nowadays, students of innovation prefer the term co-evolution, emphasizing the parallel and mutually dependent development of technology and society. This acknowledges the importance of contingency and of the impact of the past upon the present, the very reason why The Evolution of Automotive Technology: A Handbook looks at car technology from a long-term perspective. Often we will conclude that the innovation was in the (re)arrangement of existing technologies. Since its beginnings, car manufacturers have brought a total of 1 billion automobiles to the market. We are currently witnessing an explosion toward the second billion. Looking back, we can see this history evolve through five distinctive phases: Emergence (1880–1917), Persistence (1917–1940), Exuberance (1945–1973), Doom (1973–2000), Confusion (2001–present). The Evolution of Automotive Technology: A Handbook helps us understand how these phases impacted society and, in turn, shows us how car technology was influenced by car users themselves." — Provided by publisher.
    Includes bibliographical references.
    ISBN 978-0-7680-8027-8
    1. Automobiles—Design and construction—History. 2. Automobiles—Social aspects. 3. Automobiles—Technological innovations. I. Title.
    TL15.M635 2015
    629.2'309--dc23
                                                                2014025046

ISBN-Print 978-0-7680-8027-8   **ISBN-epub 978-0-7680-8147-3**
ISBN-PDF 978-0-7680-8145-9   **ISBN-prc 978-0-7680-8146-6**

To purchase bulk quantities, please contact:
SAE Customer Service
E-mail: CustomerService@sae.org
Phone: +1-877-606-7323 (inside USA and Canada)
              +1-724-776-4970 (outside USA)
Fax: +1-724-776-0790

Visit the SAE International Bookstore
HTTP://BOOKS.SAE.ORG

to Maarten, and his great-grandfather

Contents

Preface

The Evolution of Automotive Technology: A Handbook is a quarter century-old dream come true.

Back in the 1980s, a Dutch publisher asked me to become the editor of a three-volume technical handbook conceived in the 1920s, by George Frederik Steinbuch, one of the pioneers of Dutch automotive engineering. I had just graduated from HTS Autotechniek (HTS standing for Hogere Technische School, or Polytechnic, focusing on Automotive Technology), and had also finished a Master's degree in literary history. Ambitiously, I proposed a ten-volume successor instead. Although this plan got stranded between the busyness of my subsequent career as a lecturer at the Polytechnic, and multiple reorganizations of the publishing house, five volumes were eventually finalized, with three of them translated into German. For every volume, I studied the history of the topic and wrote two introductory chapters myself (on history and basic theory), while editing the contributions of the experts in the remainder of each volume.

At the end of the 1990s, I received a PhD in history of technology, and transferred to Eindhoven University of Technology (TU/e) to take up a position in research and teaching, focusing on mobility research, traffic engineering, and the automotive culture.

Coincidentally, some years ago, the great-grandson of George Frederik Steinbuch, TU/e Mechanical Engineering professor Maarten Steinbuch, asked me if I would like to help set up a new Master's program in automotive technology. In addition, the board of the university decided to attract a new, more socially oriented type of engineering students by offering a brand-new Bachelor of Automotive Technology program at the Electrical Engineering Department, prioritizing the study of control and systems engineering, the electric and hybrid propulsion as well as the smart car. My answer should not be difficult to guess.

Suddenly, car technology proper, a closed book to many students of the history of the automobile, came alive again for me. I started to dig into my own history, combining this with the new educational and theoretical insights I had gathered during the last decade or so. The result is the book before you, tested extensively during the first three years of the Bachelor's program, and far more years at the Master's level. My audience was a continuously changing international group of students, whom I would like to thank here for helping me put the content to the test. Some of them produced reports that made it into the pages of this book.

Of these students, some have to be mentioned by name: student assistants Jorrit Bakker, Valerian Meijering and, especially, Wilco Pesselse and Texas van Leeuwenstein. They helped set up the courses, gather and scan illustrative material, develop the quiz questions, and go through numerous versions of texts, with their source references and requests for permission to use illustrations.

I thank Bart Smolders, Director of Education of the Electrical Engineering Department, who made the initial investment into the student assistant support; Johan Schot, Director of the Technology in Society group of the IE&IS Department for following up in the final phase, as well as Jos Hermus, Managing Director of the same department, for providing the basis for the continuation of this project beyond my retirement.

I was also able to make prolific use of the Netherlands Center for Automotive Documentation (NCAD), which I had co-founded at the end of the 1990s, and which keeps several thousands of printed material, including my own collection of old handbooks, as well as my complete set of all Steinbuch editions since 1922.

I would like to thank director Jan Wouters, the many NCAD volunteers who helped search for illustrations as well as its board of directors, chaired by Peter van der Koogh, for their cooperation during the past years.

I also would like to thank Jan Schukking, long-since retired from Kluwer Technische Boeken BV, who hired me in the 1980s as an editor of "the Steinbuch series", and who never lost his confidence in me. I am very pleased that I can show him the results of his confidence in the form of this book. I am also grateful to the current copyright holder of the Steinbuch series, mybusinessmedia, and to intermediary Henk Heuvelman, for granting me permission to use many of the original illustrations. The same gratitude extends to the many copyright holders of the other illustrations necessary in a handbook like this: they are duly credited in the caption of the figures. Every effort has been made to identify and contact copyright holders of illustrations. The author requests notification of any corrections that should be incorporated in future editions.

The preparation of the printed book was crucially advanced by Georgine Clarsen (Wollongong University), Ann Johnson (University of South Carolina), and Lee Vinsel (Stevens Institute of Technology), who read earlier versions of the manuscript, and acted as reviewers. I also thank the anonymous reviewer, who reminded me of the differences in language and study style between engineering students and other types of students. I hope that his influence on the writing style of this text will convince other teachers and professors, who are giving courses in automotive engineering, to make use of this book in their classes. It is also my hope that engineers and others interested in the long-term development of the car will find this book educational and fun to read.

My thanks also go to the staff at SAE International for guiding me through the labyrinth of book production: Martha Swiss and Monica Nogueira.

Finally, I would like to acknowledge the breadth of vision of Maarten Steinbuch, who, supported by many colleagues at TU/e, made this project possible in the first place. In less than a decade, automotive technology and "smart mobility" have been made into one of the strategic areas of the university, quite a feat after a long period of automotive lull at this institution. I was very lucky to help shape this program during the last phase of my professional career.

I dedicate this book to Maarten, and his great-grandfather. They stand at the beginning and the end of my version of automotive technology history.

Gijs Mom
Shanghai, 18 August 2014

Chapter 1
The Evolution of Automotive Technology

1.1   Introduction: Why This Book, and How?

This book covers one and a quarter century of the automobile. It has been conceived as a cultural history of technology proper. It differs from most academic histories of technology in that it emphasizes the latter in a particular sense: technology is interpreted and presented as constituted through a structured set of artifacts. These are made to "speak" through the study of handbooks and journals as part of a broader automotive discourse. The book has been expressly written for novices in automotive technology, students at the start of their engineering study, with the aim to convey a basic set of terminology, formulae, and graphs in their historical development. However, historians and social scientists, and especially those scholars from the burgeoning field of science and technology studies (STS), who wish to look over these students' shoulders to the way the car's structure changed over more than a century, are cordially invited to join us.

Histories of technology either emphasize the process or the product, production, or consumption and use. Traditionally, car histories with an emphatic technology content are production stories. Since professional social historians have discovered the car and its remarkable society-pervading history, the attention has shifted to the "car culture" writ large. This resulted in an explosion of sociocultural histories of the car while the car's technology proper has largely been black-boxed.

In contrast, the artifact-driven history developed here follows a perspective in the middle. It acknowledges that there is a distant, but crucial and often delicate relationship between the producer and the consumer, the engineer and the user, the manufacturer and the customer. Although we intend to emphasize the user side in this balance, the nature of our major source, automotive handbooks, compels us to observe the user's influence largely in an indirect way, through the lens of the producer.

The choice for this user perspective can be motivated from the insight that the automobile has evolved from an early stage of its history into a commodity, an item of consumption. Not many cultural histories of the car's technology from a user perspective exist, but recently some innovative studies have seen the light [1-1], [1-2], [1-3].

In this book, we will first, in Chapters 1 to 8, focus on technology proper, only incidentally referring to a wider context. In this we follow aeronautical engineer-turned-historian Walter Vincenti, who once said that the deeper one descends into the bowels of technology, the harder it is to connect to societal developments ([1-4], p.11). The further one reads in the following pages, however, the more context is offered, parallel to the increasing attention for dynamic over static phenomena. In Chapters 9 to 16, then, we have selected some clear trends where the context is so important that it should take center stage.

For this history, we used automotive handbooks rather than manufacturers' archives, as we intend to develop an evolutionary history of the car's technology. For the same reason, we will focus on the passenger automobile and neglect motorized two-wheelers, trucks and buses, and special vehicles such as wheeled cranes, shovels, or street cleaning vehicles. We will also focus on the "normal" car and hence will only occasionally refer to racing cars, taxicabs, and other examples of the myriad of automotive variants.[1]

In this chapter, then, we will first explain our systemic approach of the car and its context (Section 1.2). The next section will be dedicated to some basics of evolutionary history of automobile technology (Section 1.3), followed by a section on one of the central concepts of our history: the dual nature of technology (Section 1.4). We will then go on to sketch, in a very preliminary way, an evolutionary history of automotive technology (Section 1.5), and we will close this history with conclusions (Section 1.6).

1.2   Structure and System

How can we approach the history of the automobile as a cultural history of technological artifacts? To answer this question we need to specify what we mean by "a car." If I speak of "the car of the 1920s in the United States" I cannot be sure that you, as a reader, understand what I mean. In most colloquial cases this phrase is clear enough, but this is not so if we want to study the change in automotive technology in a scientific way. Therefore, we need to agree on the way we wish to approach the artifacts under investigation.

Structure

First of all, we approach the car and its culture as a system. For the internal, systemic setup of the artifact itself we reserve the term structure, indicating the system-like layout of the car's components such that if you change one element it has repercussions

---

1. We use the term "normal" in a Kuhnian way (normal car, normal change). Thomas Kuhn, in his seminal analysis of paradigm shifts in science, speaks of "normal science" [1-5].

for many other elements within the structure. The car's structure is a historical term: it was not established from the beginning. Of course, in order to call a certain vehicle (as proposed by a historical actor, such as an inventor or a company) a car, some basic systemic traits had to be in place from the very start, such as wheels, a place to sit, a contraption to attach the propulsion system to, and so on. We are not going into the scholastic issue of whether "a car" should have only four wheels or whether it may have three. We define automobile as a (rail-less) self-propelled land vehicle of any size and layout, and we use the term car to indicate the private passenger version (but often implicitly include the other vehicles within the automotive family in the definition as well).

Also, this structure is hierarchical, implying that some substructures (groups of components that perform a certain function, such as the suspension system) and components (a substructure that is delineated by its own housing and generally carries a proper name, such as a damper) and parts of a lower order (such as a piston, a housing, a mounting rubber) are more related to automotive technology, while others are more related to general mechanical engineering. The lower one descends in this hierarchy, the more the parts are generic, such as a bolt or a gear wheel. The higher in the hierarchy, the higher the complexity of the part and the more it has been constructed with the special purpose of the automotive context in mind.[2] In this book, we will generally not descend much deeper in this structure than three or four layers (Figure 1.1).

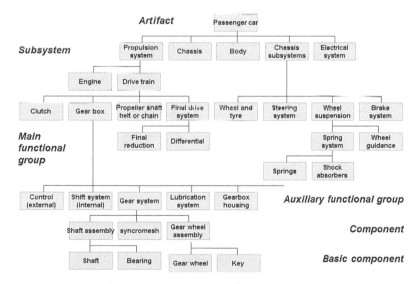

Figure 1.1 Structural diagram of the automobile: The car is a "nested hierarchy" of diminishing complexity the deeper one descends into the "bowels" of the technology. We distinguish between six levels, from the artifact until the basic component, but in this book we will mostly not go deeper than the auxiliary functional group. In this example, only the deeper structure of the gear box has been given.

---

2. There are exceptions to this rule: noncomplex parts visible to the consumer also develop as auto-specific parts, such as a steering wheel or a dashboard push button.

System

Next to this internal systemic setup, the wider automobile culture is also systemically structured. For this "car system" we use a less-hierarchical representation in the form of a flower, its petals arranged around a core formed by the car. This implies that we intend to keep the car central in our story even if we zoom out to the wider context, as we will do extensively in the second half of this book (Figure 1.2). It should be emphasized that this car-centric flower model is only a selection of the wider "car society" we have come to live in, as alternative technologies of mobilities (such as the railways or public transport in general) are only represented through the "petal" of the "mobility culture." Had we included other mobilities (as we will do in the last chapters of this book), the car would have been decentered, despite its ubiquity in quantitative terms. In fact, the flower petals represent only the inner ring of factors playing a decisive role in the societal context, factors needed to explain certain technological changes within the car's structure.

Car structure and car system are a part of a wider "car culture," which also includes the behavior of drivers, producers, nonusers, victims, and other "stakeholders" as well as the discourse (in the press, in government, in a novel or song, or in a video clip) around mobility.

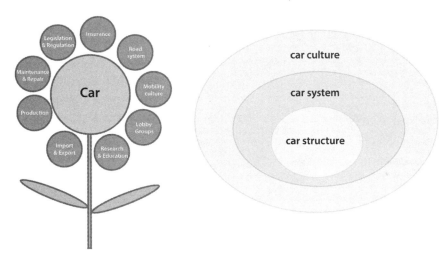

Figure 1.2 Flower model and nested hierarchical model of the car system and its societal environment.

1.3   A Quasi-Evolutionary Approach

In our history we make an eclectic, metaphorical use of Darwinian evolutionary insights. We acknowledge, of course, that artifacts do not mate and get offspring: it is people who invent, produce, distribute, and buy artifacts, thus setting in motion the process of variation, selection, and retention, although some theorists claim that artifacts have agency just like people [1-6], [1-7]. But even those theories will not claim that artifacts have intentionality, which human beings have and which sets their development (and the development of their technology) apart from the biological world.

Giving the artifact agency is not as strange as it may sound, as the development of the car has sometimes been described as a process of prosthetization. According to this perspective, the car and the driver merged into a cyborg, in which the distinction between the mechanical and the organic elements got blurred [1-8]. No wonder, then, that modern automotive engineering departments dedicate their attention also to the "animal" side of this cyborg, from ergonomics to the psychology of perception.

As a second caveat, we cannot claim that the evolutionary process of technological change is purely random, as Darwinian theory of change seems to imply, hence the use of the term pseudo- or quasi-evolutionary for those theories that try to apply biological concepts on the evolution of technology. This is often also done to avoid being accused by colleagues that evolutionary forms of history tend to foster a teleological, determinist conception of change.

According to the latter approach, very popular among engineers and the general public alike, technology develops, via a rather linear path, toward ever increasing perfection and sophistication, thus determining society (hence the term determinism). Nowadays, students of innovation prefer the term co-evolution, emphasizing the parallel and mutually dependent development of technology and society [1-9], [1-10]. Evolution is not teleological: it acknowledges the importance of contingency, of the coexistence of multiple paths of technical development and of the impact of the past upon the present, the very reason why we approach car technology from a long-term perspective ([1-11], p. 149).[3]

Incremental and Radical Change

The application of evolutionary insights on technical change goes beyond the metaphorical: if we define the set of extant artifacts (cars) in a certain period of time and at a certain locale as a Darwinian "population" (an amorphous set of specimens) in which more structured "species" (sets of identical specimens) exist, we can use a part of the evolutionary vocabulary to analyze and understand the basic principles of technical change. We will, however, not address the question in how far (technological) culture can in itself be described as a true evolutionary process [1-12].

In applying Darwinian insights to the history of automotive technology, we can distinguish between continuous and discrete change, and in the latter case we further distinguish between incremental and radical change and acknowledge that the distinction is not absolute, but that it depends on the level, the scale from which one observes it. Evolutionary theory favors the population as a unit of analysis, but in this book our

---

3. The second reason is that students tend to understand engineering problems easier when explained from a historical standpoint, as they see the historical actors struggle with such problems in the same way as they do. The invention of the diesel engine told in Chapter 12 is a case in point.

analysis will focus upon the typified species or car model. In other words, we use the "typical car" (singular) as a representation of a certain part of the car population.

We tend to say that the gradual growth of a species (expressed, for instance, in its weight, and shown in a gradually increasing curve) is a continuous change (Figure 1.3). Viewed in a more detailed way, however, it appears that the species grows through its individual, discrete artifacts, in tiny steps. In a similar vein, for an engineer working on engine mixture formation, the shift from a carburetor to an injection system represents a revolutionary jump, but seen from the population of millions of cars in the world, the change is not only tiny (especially, from a user's perspective, if it entails no major change in her behavior), but its introduction on the market is a very gradual, quasi-continuous process, which innovation specialists have called diffusion (as we will see in Chapter 13).

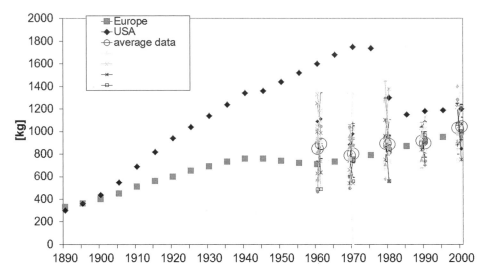

Figure 1.3 Continuous growth of the average car weight, divergence followed by a convergence between American and European values during the 1970s and 1980s.[4]

No surprise, then, that nontechnically informed histories often emphasize radical shifts, where a more technical approach would discern many intermediate or preceding steps [1-14]. It was economic historian William Abernathy who argued convincingly that incremental changes are cumulative. From a long-term perspective, these can lead to

---

4. This graph and related graphs in this book are based upon data provided to this author by the late Mr. A. Dekker. The bands of measurements in 1960 and 1961 (and those in 1970 and 1971, and every decade until 2000 and 2001) have been gathered by hundreds of first-year students of the Polytechnic for Automotive Technology (HTS-Autotechniek) in Arnhem, The Netherlands, during specially organized introductory classes in 2001 and 2002. The circles represent the arithmetic average of the bandwidth of data, taken from annual overviews of test accounts of the most sold cars in the Netherlands in the respective years. These series confirm the more shaky sequences of Dekker's European data, while the deviation in 1960/1961 (and possibly before those years) of the Dutch data perhaps can be explained by the need for a large car of Dutch families, which were significantly larger than the European average [1-13].

changes as radical as so-called paradigmatic changes. The real place to look for these changes, as historian of technology Jeffrey Yost stated, is the suppliers to the automotive industry, certainly in the United States [1-15], [1-16].

Normal Change

Whatever the preference, however, most students of technical change acknowledge that incremental change is normal change and that radical change is the exception rather than the rule. Radicalism seems to belong to the realm of technological fantasies and utopias (and in a patent description that needs a break with the past if it wants to be recognized as an innovation): they function as expectations that drive producers and users alike in their practices that bring about technological change. The introduction of the motor car is such a radical event, although there are good arguments to relativize this: it took place over a period of 10 to 15 years (perhaps even longer, if one includes the struggle between steam, electricity, and internal combustion) at the end of the 19th century, and canonized inventors such as Karl Benz did their utmost to present it as not much more than a horseless carriage, emphasizing the continuity (rather than the breach) with the past. A closer reading of the layout of the earliest motorized vehicles also reveals the influence of bicycle technology and the tradition of industrial, stationary engine technology. The car was new, as a new set of components and substructures, but its basic elements were not (Figure 1.4). In the following chapters we will encounter several examples of the car's heritage to earlier vehicle and other technologies. Often we will conclude that the innovation was in the (re)arrangement of existing technologies.

Figure 1.4 How new was the car when it first was proposed? This is a four-wheeled motorcycle developed in 1896 by Andrew Riker from Brooklyn, a later president of the Society of Automotive Engineers (SAE) ([1-17], p. 29) (*courtesy of Transportation Collections, Division of Work and Industry, National Museum of American History, Smithsonian Institution*).

Modern innovation theories acknowledge meanwhile that technical change is not a linear process, starting with invention in some genius, production in some factory, and diffusion to an eager public that passively waits to be provided with commodities. Not only is the inventive process very variegated [1-18], but users themselves are often involved in the inventive process, so much so that some students of innovation speak of co-construction. We will give several examples of this in the course of this book (see for instance the case of the tire in Chapter 7). Also, buying is an active, creative act involving expectations and fantasies of future use, as much as engineers generate fantasies while developing a product [1-19], [1-20], [1-21].

How, then, does technology change, seen from an evolutionary perspective? To get a grip on this process we use the concept of the dual nature of technology.

1.4   The Dual Nature of Technology

The concept of the dual nature of technology rests on the assumption that an artifact can be seen as consisting of two sets of characteristics (Figure 1.5): (technical) properties that describe its inner workings and other intrinsic traits (such as weight, color, tensile strength) on the one hand, and (relational) functions realized through its use on the other.[5] From an evolutionary point of view, artifacts exist in variants as diverse and hierarchically related as do functions. Early tire treads, for instance, came in all kinds of versions, some perhaps the result of a certain application philosophy, others simply proposed because they "looked cool", and only a very detailed (up to now not undertaken) historical analysis would be able to unearth the details of this variation.

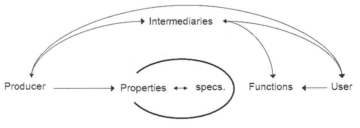

Figure 1.5 The dual nature of technology.

Selection and retention of one or a few variants take place through purchase and subsequent use, or by producers who take variants back and replace them by others anticipating perceived future changes in use. In the rare case of full substitution, the substituting solution may be locked-in and thus forms the start of a "path dependency", in which this solution represents a constituting element. Others speak of a dominant design [1-26], [1-27], [1-28]. Intermediaries (such as consumer organizations, or marketing

---

5. See a brief introduction by the two Dutch philosophers of technology who coined the phrase and developed the concept: [1-22]. See also [1-23]. For a critique see [1-24]. The text of this section is based on [1-25].

offices in car factories) emerge when the relation between producer and user become more complex.

Substitution and Coexistence

However, competition between artifacts does not necessarily result in full substitution; more often, alternative technologies coexist, covering different nuances in application (such as four- and six-cylinder engines) or geographical differences in car culture (such as the competition between battery and magneto ignition; see Chapter 3) (Figure 1.6).

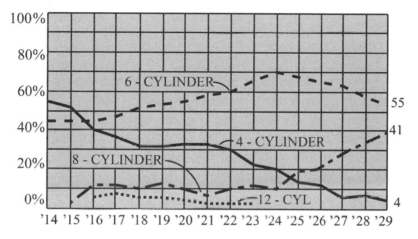

Figure 1.6 Substitution or coexistence? Engine types in American cars 1914–1929 [1-29].

Two basic insights govern the analysis undertaken with technology's dual nature in mind. First, the relationship between properties and functions is not straightforward but open for interference by the user, although some practices are more difficult to perform than others, and some are simply impossible. One could wish to drive 200 km/h during the first decade of the 20th century, but nobody did, for good reason: the engine's properties would not have allowed such extreme speeds.

Conversely, certain functions can be enabled by different sets of properties: both carburetors and injection systems can realize mixture formation, and driving around 1900 could be done on hollow, massive rubber tires or on pneumatic tires. This fuzziness between properties and functions enables the process of change of technology, because it allows for small changes in both technology and user behavior that subsequently can be hardened in a design or in cultural traits.

This is not to say that technology somehow prevents certain forms of use in an absolute way. Rather, technical constraints are historical: early manufacturers could perhaps have developed engines that allowed a car to drive at 200 km/h, but they did not, and it is the then-existing technology that allows us to conclude that such speeds were not possible. One should, in other words, be careful in saying that it was the primitive state of the material that prevented certain artifacts from being invented, as one sometimes hears.

This, indeed, sounds like full-blown determinism: most often there existed other solutions to get the artifact to function, if one really wanted. A nice example is the electric vehicle that has been haunted by assumptions ushered by later engineers that the energy content of its batteries or their longevity was simply not yet ready for road application. But early engineers were much more creative than is often thought: they implemented managerial and organizational innovations (by developing sophisticated recharging and maintenance systems) to circumvent current technical constraints in such a way that fleets of electric taxicabs functioned much more reliably and profitably than fleets of combustion-engine cabs [1-17].

The second insight derived from the dual nature of technology is that artifact populations evolve, split into subspecies, die out, and are (partially) replaced by substitutes, often as a response to changing user practices and sometimes just because the producers consider it worthwhile to change.

Taxonomies and Types

One way to get a grip on this process is by creating taxonomies, such as is often done in technical handbooks. As handbooks can impossibly cover all existing variants of a certain cross section in time (let alone cover all changes over time), they use one of the following typifying techniques: they present their examples through the construction of a (Jungian) archetype (a historically existing artifact representing all artifacts of a certain period); as a (Weberian) ideal type (a nonexisting, idealized artifact consisting of properties—not necessarily components—of existing or fantasized artifacts); or as an average type.[6] The latter is a nonexisting artifact resulting from a statistical assessment of the frequency of certain properties (such as constructed for Figure 1.3).

This typological approach to the existing and constantly evolving population of artifacts allows us to speak of "the car of the 1920s in the United States" in a more precise manner. This distinction between "types" is important: whereas Ford's Model T no doubt functioned in many handbooks (and in many heads of engineers and users alike) as the archetype of the American car in the 1910s and 1920s, its technology was far from typical in a statistical sense: the Model T had a magneto ignition built in the flywheel of the engine, combined with an equally atypical planetary transmission. It did not even have an electric starter motor as a standard item until after the First World War. Some journals in those days made statistics, for instance of the models proposed on car exhibitions, and Ford's idiosyncratic ignition system did not even deserve a mention in such statistics to determine an average type, as long as the models were not weighted according to their quantitative occurrence. The majority of the car *models* used alternative types of ignition (Figure 1.7).

---

6. According to sociologist Max Weber, ideal types "belong to the world of ideas" and are "inductively abstracted from reality." Quoted by [1-30] Hård distinguishes between ideal types and archetypes.

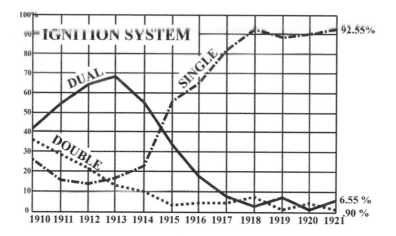

Figure 1.7 Construction of average types of ignition systems by calculating the frequency of the subspecies as occurring on automobile exhibitions in the United States. In this case, single refers to the battery ignition dominant in the United States at the time, double refers to a combined battery and magneto ignition system, whereas dual systems have two spark plugs per cylinder (see Chapter 3) [1-29].

Affordance

Even if we acknowledge that the relationship between properties and functions is fuzzy, this does not mean that we cannot get a better grip on the process of variation, selection, and retention. It was American psychologist James Gibson who developed a way to understand how people are able to use the distinction between properties and functions in their practices of use. Remarkably, Gibson developed his "ecological psychology" during the Second World War, working for the U.S. Air Force by analyzing aviation pilots' orientation while landing and throwing bombs, after he had performed (in 1938) a case study on automobile driving ([1-31], p.75, 82).

For Gibson, himself an ardent motorist, locomotion is "a special kind of space perception." Perception is an opportunity offered by the artifact to enable an activity by the user and perceived by this user through a process of acquisition of stimulation, as a result of purposeful action. He calls this opportunity *affordance*. Skills and affordances are closely related, through practices of use. Applying Gibson's insights to the dual nature of technology, we can conceive those combinations of technical properties as affordances that can be perceived by users as invitations to act. For instance, the steering wheel of a car offers steerability to the motorist, especially when she has acquired the skills to steer. According to Gibson, we perceive the affordance without having to use the artifact: the information is not added in our heads as meaning to a neutral stimulus (as cognitive psychology would have it); it is out there to be perceived by a purposeful user, even if she decided not to use the artifact.

The handbooks we will use throughout this book as main sources function as an intermediary in-between producers and users (see Figure 1.5) but do not always account for a

Gibson-like usability of the technologies described. Although some handbooks do include the skills necessary to handle the car's systems, in their depiction of the systems themselves they reflect the designers' bias rather than the viewpoint of the user. User reactions, and their influence on the evolutionary process of selection and retention, are better studied in journals of the car and touring clubs, especially in the letters sent to the editors. This type of sources has not been used for this book, as we intend to give a first overview rather than a bottom-up approach based on case studies. It is, however, possible, using Gibson's insights, to infer from the technologies' structure some information about its possible use, despite the fuzziness of the relation between the two, and we intend to do so in the coming chapters [1-32].

Equipped with these five tools (dual nature, taxonomies, typification, affordance, and intermediaries) we can, like Charles Darwin did when he entered the Galápagos Islands, approach the population of passenger cars as depicted in the handbooks with the aim to connect their evolution to the diversification and changes of automotive cultures in Europe and the United States (and later to other cultures such as Japan, Brazil, India, or China) during the 20th century. In other words, the handbooks are our Galápagos Islands, but unlike what Darwin found on these islands, we will find that most of the cars and their subsystems have already been dissected (in the handbooks), described in their properties, and categorized, enabling us to mobilize them for our overall evolutionary overview.

1.5   One and a Quarter Century of Cars

It seems that since its beginnings one and a quarter century ago, car manufacturers have brought a total of 1 billion specimens called "automobile" to the market. We seem to be on the brink of an explosion toward the second billion [1-33]. It is much too early to attempt a periodization of these 125 years on the basis of the car's properties, but our current intuitive and anecdotal grip on this history allows us to add the following technical characteristics to a periodization largely based on a cultural analysis of car use (Table 1.1) [1-34].

Table 1.1 The history of automotive technology can be structured through the construction of five phases	
Emergence (1880–1917)	pioneering phase, variation in propulsion types and lock-in of universal automobile configuration: heavy, large family car
Persistence (1917–1940)	basic configuration of components of combustion-engine car is locked-in: closed, small family car
Exuberance (1945–1973)	heydays of (Western, comfort-driven) mass motorization in "affordable family car"; emphasis on comfort
Doom (1973–2000)	energy- and pollution-consciousness in a phase of global motorization; heydays of automation; small cars vs. SUVs
Confusion (2001–)	climate change, BRIC motorization, revival of railways, metropolitanization; emphasis on fuel consumption; the car as an iPad on wheels; alternative propulsion systems: paradigm change to EVs?

According to this overview, the car emerged as a German invention in France in an enormous variety of setups, including propulsion types (electric, steam, petrol), body, and driveline configurations (engine in front, in the rear; passengers opposite each other or looking in the same direction), let alone alternative component layouts (solid rubber wheels and pneumatics; drive chains and shafts; cone and plate clutches). We know that by the First World War the heavy car for the big aristocratic family had wiped out most alternatives (such as the French light *voiturettes* from around 1900) from the market, although here and there (and especially in the United States) some lighter types were proposed for the less wealthy motorists (Figure 1.8).

Figure 1.8 A French *voiturette* (Decauville) [1-35].

The latter type then became the normal setup during the second phase, in which most components acquired a stabilized layout, such as the battery ignition (instead of the magneto ignition), the downdraft carburetor, and the low-pressure pneumatic tire (instead of the high-pressure pneumatic or solid rubber tire). For contemporary observers, the car in the second phase witnessed its first massive diffusion among the middle classes.

Although from a technical point of view the second phase may well represent the most interesting and exciting period (because many substitutes got locked-in on all levels of the car's structure, often after a fierce struggle), the third phase saw the true celebration of the car as commodity diffused beyond its middle class constraints, but still mainly as a Western vehicle for the nuclear family that sought "comfort" in its car. Again, a wave of innovation accompanied this phase.

The energy crises of the 1970s, preceded by some serious complaints about the car's unsafe behavior, brought an end to the unbridled fantasies about the freedom offered by the car and generated doubts about its environmental properties at a moment that its diffusion went far beyond the Western realm. Manufacturers tried to keep comfort at the same previous levels, at the same time starting a process of automation through electronization and diversifying the car population, with the tiny city car and the sport utility vehicle as the two extremes.

In the current, fifth phase, electronization of the previous period would allow for a "smartification" of the car, embedded in intelligent systems of mobility, perhaps even propelled by electric and hybrid-electric propulsion systems. We will study these trends in more detail in Chapters 9 to 11, chapters that prompt questions such as this: Are we on the brink of a new automotive era? Looking back over a period of 125 years may help us be prepared for what is about to come in this current phase of confusion, and beyond.

1.6   Conclusions

We have sought in this chapter to define automobile and car and to explain our quasi-evolutionary approach of technological change used in this book. We emphasized the importance of the dual nature of technology, a concept that allows engineering and user characteristics to flow into the constant shaping and reshaping of automotive artifacts. The fuzzy relation between technical properties and relational functions enables incremental changes to drive technological development.

We also gave three techniques of typology (archetype, ideal type, average type) meant to get a firmer grip on the enormously variegated and amorphous set of artifacts produced and used in the course of more than a century. We periodized the evolution of this artifact population into five consecutive phases.

Well equipped, we can now start to study the car's main subsystems and their evolution. We will do so in the following seven chapters (Chapters 2 through 8), after which we will move one level higher in the automotive system and investigate some trends, such as automation (Chapter 9), safety (Chapter 10), and the environment (Chapter 11). We will then take another step back and zoom out toward the system's context of engineering knowledge production and its scientification (Chapter 12). This will be followed by an investigation of the history and future possibilities of alternative solutions to the internal-combustion engine car, most particularly the electric and hybrid car (Chapter 13). Chapter 14 will focus on the innovation process and investigate production and diffusion; Chapter 15 will zoom out toward mobility on a global scale. We will attempt a final overview of the main trends of the previous four phases, as well as give a wrap-up of our main arguments, in the final chapter, Chapter 16. By then, we will be able to fine-tune our periodization along a more technical pattern and better characterize the major trends that in the course of the coming chapters should help us to explain the trajectories that the car and its engineers and users took during the past one and a quarter century.

References

1-1. Ann Johnson, *Hitting the Brakes: Engineering Design and the Production of Knowledge* (Durham/London: Duke University Press, 2009).

1-2. Christopher Neumaier, *Dieselautos in Deutschland und den USA; Zum Verhältnis von Technologie, Konsum und Politik, 1949–2005* (Stuttgart: Franz Steiner Verlag, 2010).

1-3. Gijs Mom, *The Electric Vehicle: Technology and Expectations in the Automobile Age* (Baltimore: Johns Hopkins University Press, 2004).

1-4. Walter G. Vincenti, *What Engineers Know and How They Know It: Analytical Studies of Aeronautical History* (Baltimore/London: Johns Hopkins University Press, 1990).

1-5. Peter J. Hugill, "Technology Diffusion in the World Automobile Industry, 1885–1985," in Peter J. Hugill and D. Bruce Dickson (eds.), *The Transfer and Transformation of Ideas and Material Culture* (College Station: Texas A&M University Press, 1988), 110–142.

1-6. Richard R. Nelson and Sidney G. Winter, *An Evolutionary Theory of Economic Change* (Cambridge, MA/London: The Belknap Press of Harvard University Press, 1982).

1-7. Andrew Pickering, "Practice and Posthumanism: Social Theory and a History of Agency," in Theodore R. Schatzki, Karin Knorr Cetina, and Eike von Savigny (eds.), *The Practice Turn in Contemporary Theory* (London/New York: Routledge, 2001), 163–174.

1-8. Deborah Lupton, "Monsters in Metal Cocoons: 'Road Rage' and Cyborg Studies," *Body & Society* 5 (1999), No. 1, 57–72.

1-9. Vincent Frigant and Damien Talbot, "Technological Determinism and Modularity: Lessons from a Comparison between Aircraft and Auto Industries in Europe," *Industry and Innovation* 12 No. 3 (September 2005) 337–355.

1-10. Eric Higgs, "Revisiting Determinism," *Research in Philosophy and Technology* 16 (1997) 189–193.

1-11. Edmund Russell, *Evolutionary History; Uniting History and Biology to Understand Life on Earth* (Cambridge/New York: Cambridge University Press, 2011).

1-12. Jonathan Gottschall and David Sloan Wilson (ed.), *The Literary Animal: Evolution and the Nature of Narrative* (Evanston, IL: Northwestern University Press, 2005).

1-13. Gijs Mom and Ruud Filarski, *Van transport naar mobiliteit; De mobiliteitsexplosie (1895–2005)* (Zutphen: Walburg Pers, 2008).

1-14. George Basalla, *The Evolution of Technology* (Cambridge: Cambridge University Press, 1989).

1-15. William J. Abernathy, *The Productivity Dilemma: Roadblock to Innovation in the Automobile Industry* (Baltimore/London: The Johns Hopkins University Press, 1978).

1-16. Jeffrey Robert Yost, "Components of the Past and Vehicles of Change: Parts Manufacturers and Supplier Relations in the U.S. Automobile Industry," (unpublished dissertation, Case Western Reserve University, May 1998).

1-17. Gijs Mom, *The Electric Vehicle; Technology and Expectations in the Automobile Age* (Baltimore: Johns Hopkins University Press, 2004).

1-18. W. Bernard Carlson, "Invention and Evolution: The Case of Edison's Sketches of the Telephone," in John Ziman (ed.), *Technological Innovation as an Evolutionary Process* (Cambridge: Cambridge University Press, 2000), 137–158.

1-19. Harro van Lente, *Promising Technology: The Dynamics of Expectations in Technological Developments* (Delft: Eburon, 1993).

1-20. Nelly Oudshoorn and Trevor Pinch (eds.), *How Users Matter: The Co-Construction of Users and Technologies* (Cambridge, MA/London: The MIT Press, 2003).

1-21. Tim Dant and Peter J. Martin, "By Car: Carrying Modern Society," in Jukka Gronow and Alan Warde (eds.), *Ordinary Consumption* (London/New York: Routledge, 2001) (Studies in Consumption and Markets, eds.: Colin Campbell and Alladi Venkatesh, Vol. 2) 143-157.

1-22. Peter Kroes and Anthonie Meijers, "Introduction: The Dual Nature of Technical Artefacts," *Studies in History and Philosophy of Science* 37 (2006), 1–4.

1-23. Pieter E. Vermaas and Wybo Houkes, "Technical Functions: A Drawbridge between the Intentional and Structural Natures of Technical Artefacts," *Studies in History and Philosophy of Science* 37 (2006), 5–18.

1-24. Carl Mitcham, "Do Artifacts Have Dual Natures? Two Points of Commentary on the Delft Project," *Technè* 6 No. 2 (Winter 2002), 9–12.

1-25. Gijs Mom, "The Dual Nature of Technology: Archeology of Automotive Ignition and the Evolution of the Car," in Christian Huck and Stefan Bauernschmidt (eds.), *Travelling Goods, Travelling Moods; Varieties of Cultural Appropriation (1850-1950)* (Frankfurt am Main: Campus, 2012), 189–207.

1-26. S.J. Liebowitz and Stephen E. Margolis, 'Path Dependence, Lock-In, and History," *Journal of Law, Economics, & Organization* 11 No. 1 (Spring 1995), 205–226.

1-27. Robin Cowan and Staffan Hultén, "Escaping Lock-In: The Case of the Electric Vehicle," *Technological Forecasting and Social Change* 53 (1996) 61–79.

1-28. P. Saviotti, "The Measurement of Changes in Technological Output," in A.F.J. van Raan (ed.), *Handbook of Quantitative Studies of Science and Technology* (Amsterdam/New York/Oxford/Tokyo: North-Holland, 1988), 555–610.

1-29. *Automotive Industries* (17 February 1921).

1-30. Michael Hård, *Machines Are Frozen Spirit: The Scientification of Refrigeration and Brewing in the 19th Century—A Wberian Interpretation* (Frankfurt am Main/Boulder, CO, 1994).

1-31. Edward S. Reed, *James J. Gibson and the Psychology of Perception* (New Haven/London: Yale University Press, 1988).

1-32. Gijs Mom, "'The Future Is a Shifting Panorama': The role of Expectations in the History of Mobility," in Weert Canzler and Gert Schmidt (eds.), *Zukünfte des Automobils; Aussichten und Grenzen der autotechnischen Globalisierung* (Berlin: edition sigma, 2008), 31–58.

1-33. Daniel Sperling and Deborah Gordon, *Two Billion Cars: Driving Toward Sustainability* (Oxford: Oxford University Press, 2009).

1-34. Gijs Mom, *Atlantic Automobilism: The Emergence and Persistence of the Car, 1895–1940* (New York and Oxford: Berghahn Books, forthcoming).

1-35. Octave Uzanne, *La Locomotion à Travers l'Histoire et les Moeurs* (Paris: Sociétés d'Éditions Littéraires et Artistiques/Librairie Paul Ollendorf, 1900).

PART I
STRUCTURE

Chapter 2
The Engine: Mixture Formation

2.1   Introduction: Finding the Car's Basic Layout

The car emerged as a fountain of ideas and solutions in a community where the distinction between producers and users was not very clear-cut: the earliest automotive culture was highly technical. At the very beginning, not even the propulsion type was clear. Some inventors and manufacturers (the Bollée brothers in France are perhaps the best known) decided to miniaturize the English steam buses that had for a short while competed with the train in the 1830s [2-1]. Apart from this steam revival, most engineers believed in the future of electric propulsion as a clean and vibration-free vehicle. But shortly after the start of the new century, the internal combustion engine seemed to win, for reasons we will investigate in detail in Chapter 13. Suffice to say here, that these reasons were a mix of technical, social, and cultural arguments in which the technical did not dominate, to say the least [2-2]. In an earlier study we came to this conclusion by starting with the technical arguments and systematically eliminating them as main factors for the electric's demise, forcing us (and the readers) to acknowledge the power of culture. We will use the same procedure in this book.

Panhard System and Front Wheel Drive

A second source for early evolutionary variation was the question where to place the engine. Here it seemed as if technology itself resisted engineering. One of the technical reasons that may have influenced the choice in favour of the combustion engine car was its limited design freedom: whereas the battery set could be positioned almost anywhere in the car, connected to the motor(s) by simple wiring, the combustion engine not only was bulkier, but it also necessitated a well-designed mechanical connection to the driven wheels ([2-2], p. 19). Indeed, it was the so-called drivetrain configuration that during the following decades formed a hotly debated issue among knowledgeable motorists, especially since around 1905 a consensus had emerged that the Panhard system (after the French make that was one of is first users; see Figure 2.1) was the best: engine in front, propeller shaft to the driven rear wheels.

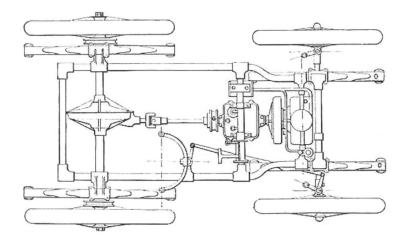

Figure 2.1 Renault 1898 drivetrain layout later known
as the Panhard system ([2-3], p. 63).

Since then, and especially during the second phase of the car's biography (the phase of persistence), sports car enthusiasts marveled at the placement of the engine right at the gravity center of the car, for reasons of dynamic stability, while dissident engineers repeatedly tried to distinguish themselves by proposing alternative configurations, such as the rear engine of the first Volkswagen.

During the second, interwar phase, when the attention of the manufacturers had shifted from technicalities to comfort, the efforts to create as much passenger space as possible made a compact drivetrain configuration attractive, and several manufacturers started to propose front-wheel drive, with the engine in front, as well [2-4]. Before the Second World War, Walter Christie (U.S.), DKW (Germany), and Citroën's Traction Avant (France) were the most famous, but only in the third post–WWII phase of mass motorization (the phase of exuberance) front-wheel drive became really popular, to the disgust of those sporting motorists who would cling to rear wheel propulsion, if only because it allowed a "Chicago spin" in the hands of a skillful driver. Together with the shift from an open to a closed body, no automotive development has ever had so many technical and cultural repercussions, and cost so much, because so many components in the car's structure had to be revised. It was all, the manufacturers claimed, for the benefit of the average, lay motorist, because it was easier to handle [2-5].

The coming two chapters will be dedicated to the core of the car's propulsion system: its engine, in which, from an energetic perspective, potential energy stored in a reservoir is transformed into mechanical energy in a (rotating) form able to propel the driven wheels. Figure 2.2 shows this in the form of an abstract flow diagram, while it also shows the more usual, ideal typical representation in early handbooks of a schematic, quasirealistic but stylized overview of the main components of the propulsion system.

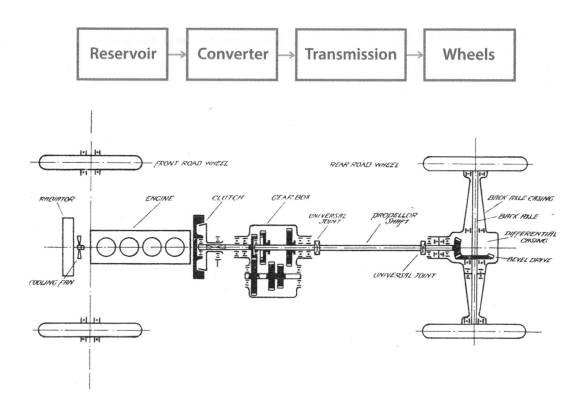

| Reservoir | → | Converter | → | Transmission | → | Wheels |

Figure 2.2 Energy flow diagram and ideal typical
schematic of the driveline ([2-6], p. 223).

Because the history of the internal combustion engine for automobile applications has hardly been touched upon by historians of technology, we will not attempt to cover the entire engine. Instead, as a *pars pro toto*, we will focus on two crucial substructures: mixture formation (Chapter 2) and ignition (Chapter 3). In this chapter, we will first give a taxonomy of engines (Section 2.2) and then deal with their common basic structure (Section 2.3). In Section 2.4, we will look into the changes that governed the development of the carburation systems, and we close this chapter with conclusions (Section 2.5).

2.2 Constructing a Taxonomy of Engines

If we are to believe the technical handbooks, the engine was by far the most important component in the car: most of the handbooks' printed space was dedicated to this subject, suggesting that it was the most complex and deserved detailed treatment. But many things in the life of the car cannot be reasoned away by simply referring to technical rationality: we know from nontechnical histories of early automobilism that braking generated at least as much astonishment as fast driving (see Chapter 8). And yet, early handbooks did not dedicate much space to this crucial substructure. Thus, handbook writing was a selective process: it emphasized those elements that were technically challenging and also seemed to favor those interface technologies that enabled a knowledgeable user to regulate speed through the control of mixture formation and ignition [2-7].

Writing a history of automotive engine technology is a daunting task, indeed. If we follow the lead of Darwin on the Galápagos Islands, we should start with two tables: an (unhistorical) taxonomy of a well-defined cross section in time and geographic space, followed by a genealogy that tries to explain the variation and subsequent selection and retention of a few solutions. The task in the case of the engine is so daunting (not only because of the enormous variety of solutions but also because hardly anything reliable has been published that does justice to the richness of this topic) that we will refrain from such a "static structural diagram" (or structural diagram in short) and a "dynamic structural diagram" (or dynamic diagram), respectively, of the engine. We will do so in the next chapter for a tiny part of the engine: the ignition, of which our knowledge is more extensive.

We cannot refrain from giving at least a basic taxonomy of engines, however. According to Table 2.1, the main distinction among engine types is based upon the inner working process: two-stroke and four-stroke engines complete their working cycles in two and four strokes of the piston, respectively, that is, if those pistons are of the reciprocal type (going up and down in the cylinder instead of rotating). The working cycle of four stroke engines has been designed to take place along two different thermodynamic paths, the most well known being the otto and diesel engines (named after Nicolaus Otto and Rudolf Diesel, whose work we will study in Chapter 12 in more detail).

These basic variants then come in all kinds of mechanical configurations, from single-cylinder to 16-cylinder engines, from V engines to boxers (with cylinders that are placed opposite to each other, their pistons connected to the crankshaft in a special way), and from water cooled to air cooled, to name only a few possibilities. The enormous variety of alternative solutions not only is an expression of the extent of the problems encountered (the more complex the problem, it seems, the more variegated the spectrum of alternatives), it also reflects a sheer lust of variation, as if the signature of the inventor or manufacturer is truly delivered by the engine type he has put in the car. This becomes clear when one realizes that the engine variation was never solved through the lock-in of only one alternative. Although the high-speed four- and six-cylinder in-line engine, working according to the four-stroke otto principle, and adorned with mushroom valves, water cooling, and short-stroke pistons, no doubt became the hegemonic technology, two-stroke engines still dominate the large two-wheeler culture of the globe, whereas hardly any otto engines can be found in heavy trucks, and racing car engines have been successful working with rotating pistons (such as the Wankel engine).

Table 2.1 Taxonomy of the engine	
Dimension	Types
Piston movement	oscillating (reciprocating) vs. rotating
Work process (pV diagram)	otto vs. diesel
Scavenging	two stroke vs. four stroke
Cylinder configuration	line, V, boxer, W, X, star
Number of cylinders	1 . . . 16
Cooling	water (thermosiphon or forced) vs. air
Scavenging control	valves (camshaft in block, ohc, dohc) vs. sleeves
Induction type	atmospheric vs. pressure feed
Stroke/bore ration	short stroke vs. long stroke
Rotational frequency (rpm)	slow vs. high-speed

American versus European Car Culture: Elasticity and Torque Rise

On a more subtle level, for more than half a century an American and European culture could develop based on larger, heavier, and slower engines on the one side and high-revving, smaller, and lighter engines on the other side. Engineering folklore has it that this was due to the different tax regimes on both continents: European automobile tax was levied on the basis of the bore (the diameter) of the cylinder, necessitating manufacturers to develop small-bore, noisy engines. But there were other factors playing here as well, one of them being that American motorists tended to refrain from using the gear box and liked to drive in the highest gear position without gearing down. Why this was so is unclear (before we know, we start stereotyping about "comfort-driven" Americans and more "backward" Europeans, whom we euphemistically call "sporting"), but in order to afford this kind of behavior engineers learned that you need so-called "elastic" engines, delivering smooth work at very low speed as well as high speeds. Such engines tend to be heavier than their less elastic, more "supple" counterparts [2-5].

Elasticity is one of those terms used nowadays by engine designers and knowledgeable users when they want to assess an engine's suitability for a certain application. Both specialists prefer to do so on the basis of graphs enabling them to see in one blink of the eye how an engine behaves and to compare engines' behavior. The torque curve of an elastic engine is as flat as possible, delivering the same energy at a broad range of speeds. The other term is suppleness (or torque rise in American parlance), which characterizes an engine with a torque curve that drops at higher engine speeds, which means that if the car encounters a load (for instance a slope) and the speed of the car decreases, the torque increases automatically, so to speak, enabling a smooth behavior under different load conditions.[1]

---

1. Elasticity is defined as n_{max}/n_{Tmax} (the maximum engine speed divided by the speed at maximum engine torque), whereas suppleness can be defined as T_{max}/T_{nmax} (the maximum torque divided by the torque at maximum speed).

Torque and Power Curve

Three curves giving the engine's output characteristics are normally used for this purpose, but in order to be able to produce them you need sophisticated and expensive measuring equipment as well as knowledge of the internal engine properties (Figure 2.3). In other words, these characteristics are not intrinsic, unhistorical properties of a combustion engine, but they have been constructed by using certain types of measuring instruments. The historical construction of the torque curve, the power curve, and the specific fuel consumption curve has not yet been investigated, but we can readily assume that they became common during the third phase and prepared (as well as theoretically founded) by scientists and manufacturers' engineers during the second phase. There exists a direct, reciprocal relationship between the engine's efficiency (defined as the relationship between the work stored in the fuel as potential energy and the work delivered at the crankshaft's end) and the lowest point in the specific fuel consumption diagram.

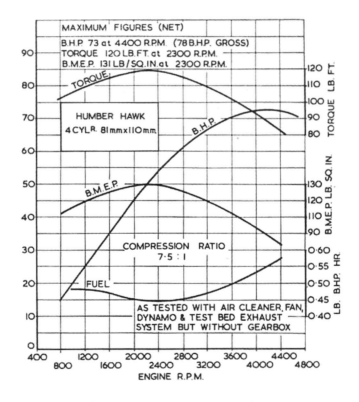

Figure 2.3 Power (B.H.P., brake horse power, or power as measured by an engine brake), torque, and fuel consumption curves (the latter expressed in pounds of fuel per brake horse power hour) of an internal combustion engine. The B.M.E.P. curve (break mean effective pressure, expressed in pounds per square inch) is identical to the torque curve, but enables easier comparison between engines as the torque has been normalized by dividing it by the piston surface, thus making it independent from the specific engine shape. Efforts to standardize technical unities internationally according to the ISO standards have failed so far ([2-8], p. 51).

Figure 2.4 gives an idealized, schematic *power* and *torque curve* of an internal-combustion engine and shows that their relative positions to each other are fixed: it allows one to easily check whether curves provided by manufacturers are reliable.[2] The figure also shows two (normalized) torque curves of electric motors, suggesting that manufacturers still try to mimic the latter's unattainable elasticity. We will encounter this phenomenon in the course of the entire automotive century and have given it a proper name: the Pluto Effect.

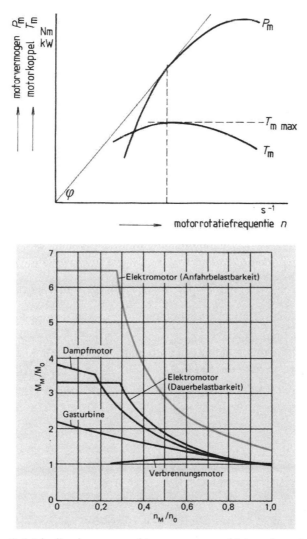

Figure 2.4 Idealized power and torque curves of internal combustion engines (showing the structural relationship between the two) and of other power sources such as gas turbines, steam engines and electric motors ([2-9], p. 143, [2-10], p. 678) *(courtesy of Springer)*.

---

2. This relationship can be derived from the insight that the maximum torque appears when the tangent drawn from the graph's origin is maximum (T = tan φ), as indicated in Figure 2.4. If $P = T.2\pi n/60$, then $T = (60/2\pi).(P/n)$ = c. P/n = tan φ.

Pluto Effect and Sailing Ship Effect

Called after Walt Disney's dog, Pluto stands for the (engineer of the) alternative technology, who is running after a sausage held at the end of a stick in front of his nose by the person on the cart (the engineer of the mainstream, hegemonic technology). Whereas Pluto is the real engine of the unit, he will never reach his goal, but the car driver will. This metaphor, which we will fine-tune in the course of the coming chapters as it is crucial to understand the competition between alternative technologies, explains how innovations, proposed by new contenders, tend to get absorbed in the existing paradigm of automotive technology, a form of *predation* (to use an evolutionary term) that mostly works in favor of the status quo. Yet, the Pluto Effect makes this status quo definitely evolve incrementally, provoked by the newcomers.

Only in very rare cases, Pluto manages to escape from this mechanism and to follow his own route. The sailing ship in the 19th century is a classic example (hence the term Sailing Ship Effect for this exceptional case): it managed to resist the "attack" from the steam and internal-combustion engine boats for more than half a century (for instance, by becoming as fast as its competitor), but in the end had to give in, after a phase in which ships were equipped with both sails and engines. But although it was substituted by another technology, the sailing ship never left totally, as we now observe in the many thousands of pleasure boats. Indeed, coexistence (through a functional shift) instead of full substitution.

Fuel Consumption Graph

The *specific fuel consumption graph* (Figure 2.5) gives lines of constant fuel consumption (according to the ISO standards expressed in grams of fuel per kWh), showing the engine's best point somewhere near the middle of the torque curve.

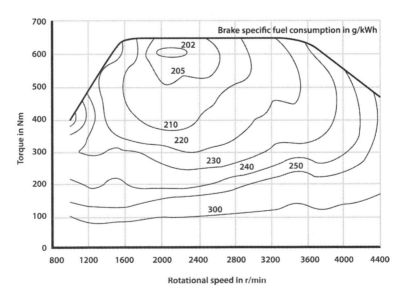

Figure 2.5 Curves of constant specific fuel consumption projected onto a torque curve.

Advertising, as well as the common discourse about cars, often gives the maximum torque and especially the maximum power, while engineers add the engine's best point in terms of fuel consumption.[3] One should be cautious while "reading" these curves: they represent the extreme working conditions of the engine, while the most frequent conditions are usually to be found well below the maximum curves given here.

Figure 2.6 shows how maximum power has, from an early date, become a measure of affluence, an expression of an automotive culture of abundance, especially in the United States, until the 1970s gave way to a more modest, but still significant distinction vis-à-vis Europe. This trend of a *divergence* between a European and an American culture, which we will encounter more often in the course of this book, is followed by a post–WWII (only very partial) *convergence* of the two cultures, in this case even followed again by a slight divergence during the last two decades of the 20th century.

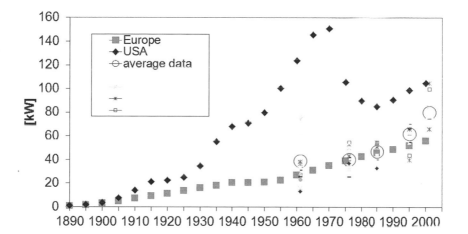

Figure 2.6 Evolution of average engine power along two different trajectories, suggesting the existence of an American and a European car culture; the outlier for the Netherlands in 2001 is probably the result of an error of measurement. (*Source*: See Chapter 1, note 4).

2.3   Making the Engine Work

The third table we should make after Darwin's example is a static diagram of the engine structure itself, allowing us to see which major components constitute a typical engine and which variation exists in these major components. Again, we will not do so for the engine (but see Chapter 5, where we will develop such a diagram for the transmission, of which we know much more), but refer to Table 2.2 where the structure of the engine is laid out in three systemic sublevels: a moving or dynamic core (of pistons, shafts, and

---

3. One can show that the engine's efficiency η is the reciprocal of the specific fuel consumption b_e. For an average hydrocarbon fuel, $\eta = \sim 88/b_e$.

valves realizing the working cycle), a housing structure, and all auxiliary substructures, including the intake and exhaust systems, the cooling system, the ignition, and the mixture formation system.

Table 2.2 Basic structure of a typical engine	
Dynamic core	• Piston and piston rod • Crankshaft • Camshaft • Flywheel • Transmission from crankshaft to camshaft • Valves
Housing	• Cylinder head • Cylinder block and crank case • Oil pan (carter)
Auxiliary and control systems	• Induction system: manifold, air filter • Mixture formation: carburetor, injection system • Exhaust system: exhaust manifold, exhaust pipes, silencer, catalyst • Ignition system • Starter system • Cooling system • Battery charger

Like the car as a whole (see Chapter 1), most of the elements of Table 2.2 have to be present before one can talk of an internal-combustion engine. When the first internal-combustion engines were mounted in a carriage, however, engine technology could already look back on a long tradition of both engineering and scientific analysis (see Chapter 12). Pistons, crank shafts, valves, and sturdy engine housings were known for more than a century as they were applied, and evolved, in engines as well as other machines (such as pumps). Well before the emergence of the combustion engine car, cooling, ignition, and mixture formation had occupied many engineers in their effort to derive internal-combustion engines from steam engines. Whereas in the latter, combustion took place outside the cylinder and pistons were used to transport the energy of the expanding steam and afterwards push the steam out of the cylinder (hence the term external combustion engines for this type of energy transformers), now the piston first had to compress a mixture that only after ignition would expand and do its work on the piston.

This functional shift had design repercussions for the piston's structure and through this for the entire engine setup, the details of which would require a separate book to enumerate. Figure 2.7 gives some archetypical examples of complete engines, while Figure 2.8 focuses on the piston as the core artifact of the engine, for which companies emerged that specialized in its (re)shaping and material constitution, while other companies specialized in the piston rings and the bearings around the gudgeon pin and the connecting rod around the cranks.

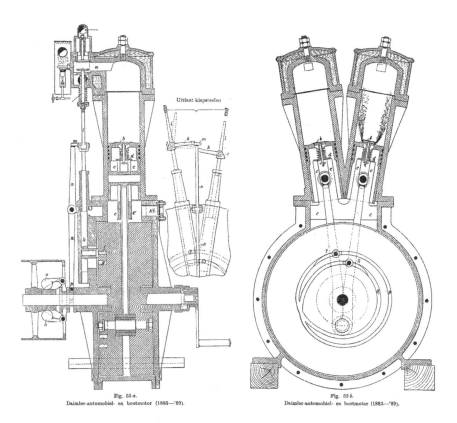

Figure 2.7A ([2-11], p. 136–137)

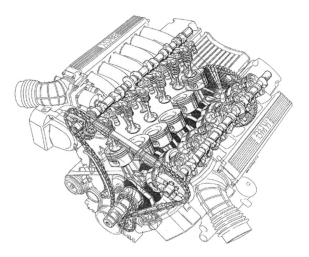

Figure 2.7B ([2-12], p. 117)

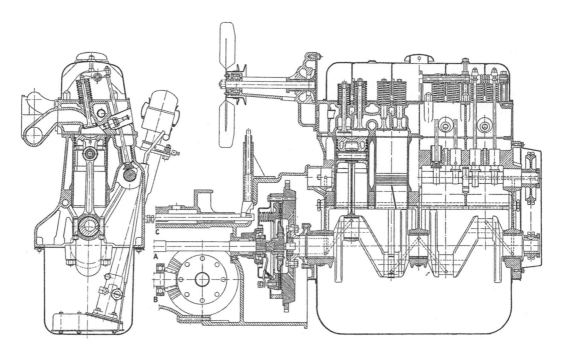

Figure 2.7C ([2-13], p. 107)

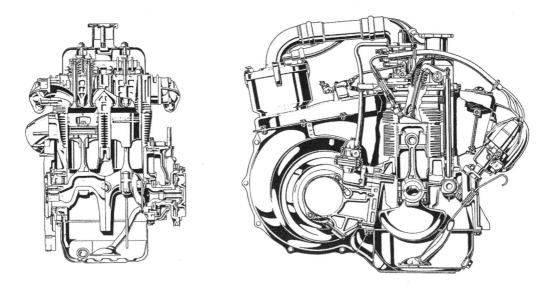

Figure 2.7D [2-14], p. 107

Figure 2.7 Some archetypes of internal-combustion engines.

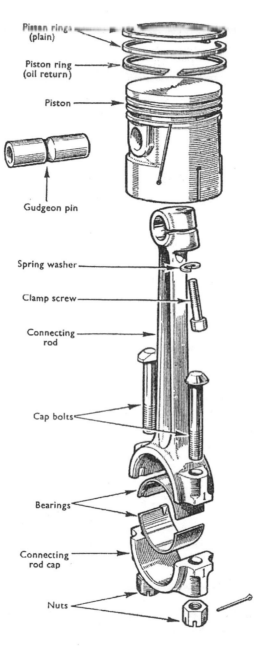

Piston ring (plain)

Piston ring (oil return)

Piston

Gudgeon pin

Spring washer

Clamp screw

Connecting rod

Cap bolts

Bearings

Connecting rod cap

Nuts

Figure 2.8 Piston configuration ([2-15], p. 108).

2.4 Carburation: Constructing the Mixture

What made the automotive engine special, however, was not the mechanics of its moving core. Whereas Karl Benz called the ignition "the problem of problems," mixture formation was at least as difficult to realize. It was a hot item, as these were the two properties that directly influenced the car's functions: it was the accelerator pedal, which enabled the motorist to change the car's speed by varying the amount of combustible mixture drawn into the engine, whereas the manette (hand lever) at the steering wheel allowed the motorist to determine the spark timing and through this the fine-tuning of the speed and load the engine had to cope with. The problem was not so much to get the right mixture in the cylinders at the right time; that was already done in industrial, stationary engines and boat engines through the application of a "governor" with centrifugal weights that controlled the mixture strength in function of a fixed engine speed ([2-16], p. 230, 287–288). The problem to be solved was the fact that what was right changed over time in dynamic situations of changing car speeds under varying conditions of use, such as mountainous terrain, wind resistance, and the number of passengers transported.

Thus, the challenge for early car engineers was to dynamize existing engine technology. It was well known that the fuel had to be metered in the right quantity into the air in order to make a combustible mixture. Depending on the chemistry of the fuel there were rich and lean ignition limits: gaseous hydrogen mixed and ignited much more eagerly, for instance, than liquid petrol from a Sumatra well. The ideal mixture depended on temperature and pressure but was for petrol known to be in the neighborhood of 15 parts (in weight, in kg) of air to 1 part of fuel. Engineers use the term stoichiometric nowadays to characterize this theoretically correct mixture, and they prefer to give it a numerical value: a stoichiometric mixture, then, has the value (lambda, λ) of 1, meaning that the relationship between the real value of the mixture and the theoretical value (also called the equivalence ratio) is identical. Ignition limits of petrol are around λ values of 0.6 and 1.3 for the rich and the lean extremes, respectively, so 40% less than stoichiometric and 30% more, respectively.

But hundreds of specialists in dozens of factories had to learn the hard way that only under very specific conditions (quiet driving, near the best point of the specific fuel consumption graph) stoichiometry was possible (and necessary), and that in all other situations of use the mixture had to be enriched. Only in very specific cases, it could be made leaner than stoichiometric, at the best point of the graph. The very physics of this phenomenon again suggest that engines talk back: they resist efforts to be thrifty in fuel, once one has opted for internal combustion. Starting the engine when cold, for instance, requires a very rich mixture (with a λ value of about 0.1 to 0.2, so ten times as much fuel needed as in the theoretical situation), and also when suddenly accelerating, or when driving up a slope under full load, more fuel has to be metered (in the latter case about 10 to 15% more) than theoretically would be necessary. The more dynamic the car's behavior, the more subtly mixture formation had to be controlled.

Soon, then, the simple surface carburetors (where air was led over the surface of a petrol reservoir, or percolated through it, or flowed past a wick placed in the petrol, as in Figure 2.9) had to be replaced by contraptions that allowed control of both the flow of air and fuel. This became all the more necessary as less-volatile fuels came on the market that evaporated less easily [2-18]. Sucked into the cylinder by the intake stroke of the piston (in fact, pushed by the atmospheric pressure after the piston had created a vacuum in the cylinder), the air was led through a venturi (a narrow passage in the inlet tract), creating a speed increase and a pressure drop that in turn sucked petrol from an orifice placed in the venturi. This spray nozzle or jet type of carburetor, claimed to be invented by Daimler's Wilhelm Maybach by the end of the 19th century, used the well-known atomizing principle of a perfume dispenser to create a mixture of tiny fuel droplets and air before it entered the cylinder.

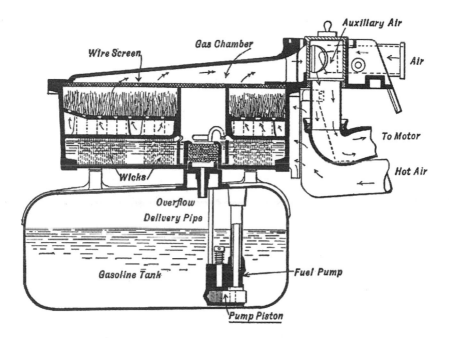

Figure 2.9 Surface (wick) carburetor ([2-17] p. 378).

Carburation: Evaporation versus Atomization

Carburation, then, was the metering of carbon-containing fuel into air and, as so often happened in the history of automotive technology, the improvement came with a price: wick carburetors worked on the basis of evaporation (which eases mixture formation of the two gases over a wide range of speeds), while nozzle carburation worked with atomization (droplets in air), making it much more difficult to create a homogeneous mixture (that is, a mixture of equal strength throughout).

The change was far-reaching. From the very moment they opted for the carburetor, engineers struggled with what they called the mixture problem, the phenomenon that air (as a gas) follows a different nature law than gasoline. Whereas gases follow the law of Bernoulli, where the flow is proportional to the square root of the pressure drop over the venturi, liquids follow the law of Poiseuille (where the flow is directly proportional to this pressure drop). If the flow speed increases, this results in a richer mixture. Also, the difference between the volumetric amounts of flow over the entire range of use of an engine is quite substantial.[4]

A Typical Early Automotive Engineer: Arthur Krebs

The problem was so complex that several "schools" competed fiercely. Arthur Constantin Krebs (1850–1935), for instance, a former commander in the French Army turned general director of the major car manufacturer Panhard & Levassor (with 3,000 employees around 1900), was a proponent of the school that preferred to allow extra air to flow in at increasing speed [2-20]. Another school preferred the restriction of fuel at higher speeds, while a third school tried to regulate both. Krebs (who seems to have been the model for the protagonist in Jules Vernes's novel *Robur The Conqueror*, from 1886) may stand for the typical European automotive engineer of the first phase: educated as a military engineer and fascinated by the new energy form of electricity, his first technical achievements were in building an electrically propelled military airship, followed by an electrically propelled submarine (Figure 2.10).

Figure 2.10 Arthur Krebs on a car of his own design, patented in 1896, with a Panhard & Levassor engine and an electromagnetic transmission, and built in the workshop of the Parisian fire fighting brigade, which he led at the time [2-20] (*Courtesy Philippe Krebs*).

---

4. In more recent engines (with a maximum speed of 5,000 rpm), this difference is in the order of a factor 50 (a factor 5 between the minimum and the maximum engine speed, and a factor 10 between the lowest and the highest load) ([2-19], p. 48).

From 1897 until his retirement in 1916, Krebs led the famous car manufacturer, first as a technical director, from 1899 as general director. He may well have laid the basis for the accelerator pedal as one of the crucial control interfaces of modern car technology, although further research on German inventors needs to confirm this.[5] Up until then, the driver had to overrule the governor by a manette (hand lever) at the steering wheel in order to allow the engine to develop more speed ([2-21], p. 97, 143, 266).

Early handbooks dedicated much space to the problem of regulating both the speed and the mixture strength of the engine. European engines tended to be regulated in speed by a governor, derived from steam engine technology and working on the basis of centrifugal weights. If the engine threatened to run out of control, such governors either kept the inlet valves or the exhaust valves closed (the latter solution was directly derived from stationary engines). American car engines, on the other hand, were often fully controlled by the driver, for instance by pushing a button which pneumatically kept the inlet valves closed, thus enabling the driver to regulate engine speed at constant values of between 100 and 850 revolutions per minute. Packard used a foot-operated inlet-valve controller, which also modified the spark timing. In the course of the last years of the 19th and the first of the 20th century, throttling (first by hand, then by hand and foot, and then by a foot pedal alone) became the norm ([2-22], p. 312–318, [2-23], p. 27–30).

The wide variety of solutions indicates that engineers were struggling with a near-insurmountable problem. In Europe in 1902, Krebs patented an automatic carburetor, while he also developed a high-tension magneto ignition system (see Chapter 3). In the midst of several large-scale projects of electrically propelled taxicab fleets (well covered in the technical press from around the turn of the century), such electrical improvements, also proposed by many others, may well have rescued the still highly unreliable internal combustion engine.[6]

Krebs's carburetor innovation was a nice example of scientifically inspired engineering (which we will deal with extensively in Chapter 12). Whereas others (especially the German pioneers) provided the extra air needed at higher engine speeds through an orifice, of which the opening was determined by a governor (with centrifugal weights, depending on the engine speed), Krebs's solution was governed by the pressure in the inlet system. The solution was clearly part of what we could call a *trend of automation*: previously, knowledgeable motorists had to adjust the mixture strength as well as the amount of mixture by hand. The automation in this case consisted of a scientifically designed profile of an air orifice (of which the contours followed the mathematical derivative of the surface of the orifice), which he subsequently presented at the Paris Academy of Sciences[7] [2-24], [2-25], [2-26] (Figure 2.11).

---

5. I thank Philippe Krebs, Paris, for engaging in an e-mail discussion with this author about the accomplishments of his great-grandfather (September 2012).
6. ([2-20], p. 9). On the unreliability of the petrol car for taxicab application see [2-2].
7. I thank Philippe Krebs for providing me with a copy of these papers (September 10, 2012). For an English description of Krebs's carburetor see for instance [2-27].

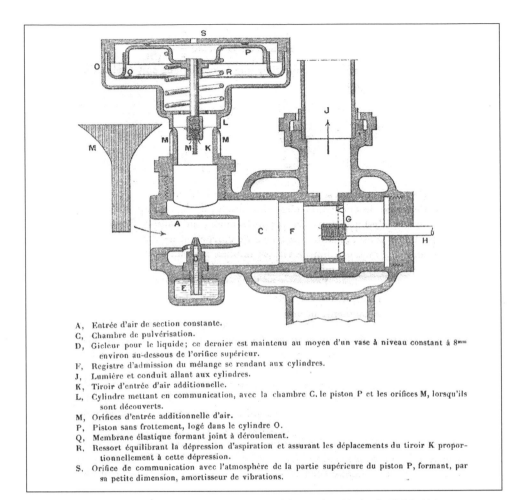

A, Entrée d'air de section constante.
C, Chambre de pulvérisation.
D, Gicleur pour le liquide; ce dernier est maintenu au moyen d'un vase à niveau constant à 8ᵐᵐ environ au-dessous de l'orifice supérieur.
F, Registre d'admission du mélange se rendant aux cylindres.
J, Lumière et conduit allant aux cylindres.
K, Tiroir d'entrée d'air additionnelle.
L, Cylindre mettant en communication, avec la chambre C, le piston P et les orifices M, lorsqu'ils sont découverts.
M, Orifices d'entrée additionnelle d'air.
P, Piston sans frottement, logé dans le cylindre O.
Q, Membrane élastique formant joint à déroulement.
R, Ressort équilibrant la dépression d'aspiration et assurant les déplacements du tiroir K proportionnellement à cette dépression.
S, Orifice de communication avec l'atmosphère de la partie supérieure du piston P, formant, par sa petite dimension, amortisseur de vibrations.

Figure 2.11 The carburetor designed by Arthur Krebs in 1902. M is the shape of the orifices that are opened by lowering the circular slide K ([2-26], p. 177) (*Courtesy of Philippe Krebs*).

Krebs was thoroughly *au courant* within the automotive state of the art: on a trip to Vienna, he took a license on Lohner-Porsche's electric propulsion system on the basis of wheel hub motors exhibited at the Universal Exhibition of 1900 in Paris, and later he built a small, electrically propelled delivery vehicle in 1913 using Thomas Edison's brand-new nickel-iron batteries. But for Panhard & Levassor's luxurious passenger cars he chose the combustion engine, starting a decade of constant, extremely complex improvements with the aim to approach the reliability of electric propulsion. This second example of the Pluto Effect nicely shows how much technological and scientific trouble it caused to opt against the electric vehicle and in favor of internal combustion. Krebs's efforts were crucial in this process leading away from the electric vehicle: his carburetor afforded (in James Gibson's terms; see Chapter 1) an unsurpassed engine suppleness. From now on the accelerator pedal was the only actuator to control the engine's mixture formation ([2-20], p. 47).

Complexifying the Carburetor

Once the basic mixture problem was considered to be solved, it was the subsequent need to control λ in function of the speed and load of the car which led to a plethora of carburetor types, housing a bazaar of jets, pipes, pumps, canals, springs, membranes, and orifices, a Darwinian feast of pneumatic-hydraulic engineering, one even more sophisticated than the other. One study of the history of carburation gives 180 carburetor manufacturers until 1940, not counting the 26 car manufacturers that also ventured into the mixture formation business ([2-28], p. 57-58). The competition between manufacturers was so fierce that, in Krebs's case, court battles between newly founded specialized manufacturers such as Zenith (1908), Solex (1910), and Claudel on the one hand and Panhard & Levassor and other car manufacturers on the other were common. Zenith, for instance, proposed a different solution to the mixture problem without the need of a separate membrane as in Krebs's case. All of the variants proposed, however, had a throttle valve, which allowed the engine to speed up when opened, enabling the air to be sucked in and the fuel automatically (because of the air's pressure drop) metered to the air flow. The sophistication and elegance of the solutions was stunning as most added mechanisms functioned automatically, triggered by pressure drops, the clever placing of springs, and so on. It shows clearly that the trend of automation started well before the post–WWII electronic "revolution" (see Chapter 9). Indeed, automation started along a mechanical, pneumatic, and hydraulic trajectory.

Once the very basics of fuel and air metering were realized, small pumps were added to meter extra fuel when *accelerating* (actuated by the motorist's foot on the accelerator pedal), while other devices enriched the mixture when the engine was *idling*. In the latter situation the flow speed of air was so low that other orifices had to be added to get the fuel into the passing air when the venturi system was hardly working.

Indeed, the emancipation from its industrial heritage as a stationary engine demanded much creativity: the higher the differences between low and high speed, the more the "mixture problem" caused flow speeds in the venturi to drop so far that at low speeds the pressure drop was not high enough to suck petrol out of its orifices. Making the venturi narrower was no solution, as the engine would then experience a lack of air at high speeds and loads, so elegant double venturis (patented by Zenith in 1906[8]) were introduced, where a smaller venturi was used to allow petrol to be mixed with air at very low speeds. For a *cold start* an extra valve (the choke) restricted the flow of air such that extra fuel could be drawn through additional openings in the central carburetor barrel.

The handbooks of the time deal with this ever more complex carburation technology depending on their educational level: some included combustion theory (see Chapter 12), others functioned like a catalogue, giving carburetor after carburetor, every make and type differing slightly from the other. In the United States, Holley, Schebler, Marvel,

---

8. I thank Philippe Krebs (October 2, 2012) for this information.

and Stromberg, next to several car manufacturers, dominated the market, while in continental Europe Solex, Zenith (in France) and Pallas and Pierburg (in Germany) became well-known and powerful players on the supplier market (Figure 2.12).

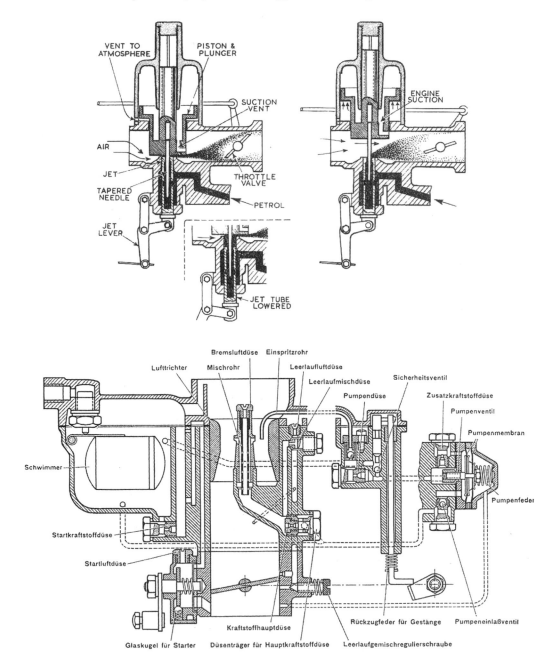

Figure 2.12 Examples of carburetors, including a British S.U. carburetor working according to a different principle of constant vacuum, and a Solex AIP ([2-29], p. 60, [2-30], p. 144).

Constant Venturi versus Constant Vacuum Carburetors

A special role in this game of evolutionary variation played the British S.U. carburetor (Skinners Union, referring to the brothers Georg, John, and Carl Skinner), as it fundamentally differed from the constant venturi systems dealt with so far. With a monopoly on this distinct *species* of constant vacuum carburetors, the British company claimed that its system better served high-performance cars, as it did not have a venturi (which restricted air flow at high speeds) but instead varied the amount of fuel directly metered into the passing air stream ([2-28], p. 43-44).

What happened then was a nice example of the Pluto Effect: threatened by the newcomer, mainstream carburetor companies had to devise other solutions (within the vocabulary of their own technology) to allow high-speed mixture flows through their venturi systems, for instance by doubling the amount of barrels and using both pipes to channel the high-speed flows. Very large engines, in the third post–WWII phase of carburation history, even had four barrels (two double-barrel carburetors), causing a new problem of fine-tuning, at the same time representing a new way for tuning specialists to distinguish themselves from the crowd (Figure 2.13).

Carburetors versus Injection Systems

The carburetor represents a nice example of a technology that became ever more complex to satisfy increasingly sophisticated motorists, exposing its initial weaknesses until it seemed to hurt at a technological limit, an example of the Sailing Ship Effect. One of these weaknesses was the central position of the carburetor on top of the inlet manifold, causing distribution problems of the right mixture among the cylinders, especially those farthest away, where bends in the tract caused the liquid droplets to become separated from the air flow. Whether this has triggered the search for a radically different solution is not known, but the (successful) efforts to substitute carburetors by injection systems make abundantly clear that more precise fuel metering under increasingly subtle speed and load changes seemed to be realized more easily through injection (artificially generated pressure, easily controllable) rather than atmospheric pressure. However, it is often highly unlikely that technical restraints were the real cause of such substitutions. More detailed studies are necessary to confirm our assumption that such disadvantages of the existing, mainstream technology are only "constructed" after an alternative has been proposed (the classic example here is the "limited action radius" of the electric vehicle). The popular shortlists of advantages and disadvantages in technical handbooks (also used in engineering education) are to be mistrusted: they are a-historical, post-hoc constructions, and while they suggest objectivity and neutrality, they are in fact highly ideological constructs formulated with one or two possible solutions in mind (and others excluded).

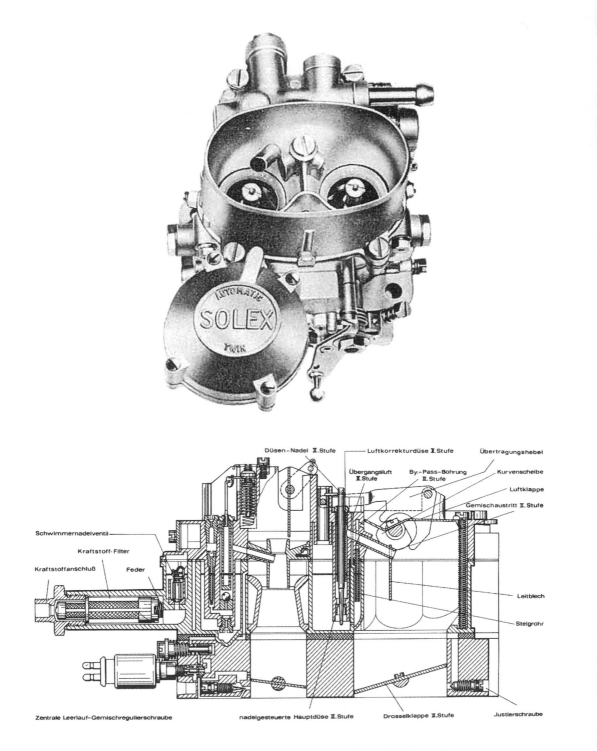

Figure 2.13 Double-barrel carburetor (Solex PAIA) from the third, post–WWII phase ([2-28], p. 107, [2-31], p. 290).

Precise metering is not the same as electronic control: petrol engine injection systems were already used on aviation engines well before the electronic era, and the first automotive injection systems were mechanical-hydraulic, such as the Bosch K Jetronic system (only its successor, the KE Jetronic, was electronically controlled), as we will see in Chapter 9.[9] However this may be, the carburetor lost its central position during efforts to cope with the increasingly severe antipollution standards of the 1970s (see Chapter 11). It lost the battle to the multijet injection systems, which were eventually integrated into the current fully electronically controlled motor management system. Until today, we lack a thorough comparison between these systems and the very sophisticated last generation of electronically controlled carburetors. Only then can we establish whether technical factors or more contingent factors (such as the struggle between two companies, as told in Chapter 11) caused the demise of the carburetor. Until then, it seems safe to assume, as in so many other cases, that causes in the history of (automotive) technology are multiple, a mix of technical, social, economic, and cultural influences.

2.5 Conclusions

In this chapter, we started our focus on the combustion engine as the core of the car's propulsion system. We gave a taxonomy of engine types and explained the basic structure of such engines. Although car engineers and users alike tend to see the car as an engine on wheels (and the engine as an encased piston on the move), from an evolutionary perspective, the engine was not so much new in its basic configuration but rather in its regulation, in the need to adjust its behavior to the dynamic conditions on the road.

Crucial for this regulation was the mixture formation substructure, which for automotive purposes had to be equipped with fine-tuning structures enabling the enrichment of the fuel air mixture under conditions of cold start, acceleration, and climbing slopes. In the course of these nonlinear trends of increasing *complexification* and *automation*, the accelerator pedal emerged, which made the foot, instead of the hand, become the main interface between the driver and the car.

The foot adjusted the car's speed and load through regulating the amount of mixture, while all other adjustments had meanwhile been automated in a compact "mixture factory", except for the choke, for which a separate hand-controlled interface remained, to be used in case of a cold start.

A similar adjustment was necessary for the ignition system, the topic of the next chapter.

---

9. On the K and KE Jetronic see for instance ([2-32], p. 99–107).

References

2-1.   K. Kühner, *Geschichtliches zum Fahrzeugantrieb* (Friedrichshafen, 1965).

2-2.   Gijs Mom, *The Electric Vehicle: Technology and Expectations in the Automobile Age* (Baltimore: Johns Hopkins University Press, 2004).

2-3.   Gijs Mom, "De aandrijflijn in historisch perspectief," in G. Mom and H. Scheffers, *De aandrijflijn* (Deventer, 1992) 19-80 (Part 3A of *De nieuwe Steinbuch; de automobiel; handboek voor autobezitters, monteurs en technici onder redactie en coördinatie van drs. ing. G.P.A. Mom*).

2-4.   Peter J. Hugill, "Technology Diffusion in the World Automobile Industry, 1885–1985," in Peter J. Hugill and D. Bruce Dickson (eds.), *The Transfer and Transformation of Ideas and Material Culture* (College Station: Texas A&M University Press, 1988), 110–142.

2-5.   Gijs Mom, "Orchestrating Car Technology: Noise, Comfort, and the Construction of the American Closed Automobile, 1917–1940" *Technology and Culture* 55, no. 2 (April 2014) 299–325.

2-6.   A.W. Judge, *Modern Motor Cars Vol. I, Their Construction, Maintenance, Management, Care, Driving, and Running Repairs: With Special Sections on Cycle Cars, Commercial Cars, and Motor Cycles* (London: Caxton Publishing Company, 1922).

2-7.   Gijs Mom, "The Dual Nature of Technology: Archeology of Automotive Ignition and the Evolution of the Car," in Christian Huck and Stefan Bauernschmidt (eds.), *Travelling Goods, Travelling Moods: Varieties of Cultural Appropriation (1850–1950)* (Frankfurt am Main: Campus, 2012), 189–207.

2-8.   K. Newton and W. Steeds, *The Motor Vehicle: A Text-book for Students, Draughtsmen and Owner-drivers* (London: Iliffe Books, 1966).

2-9.   G. Mom and H. Scheffers, *De aandrijflijn* (Deventer, 1992) (Part 3A of *De nieuwe Steinbuch; de automobiel; handboek voor autobezitters, monteurs en technici* onder redactie en coördinatie van drs.ing. G.P.A. Mom).

2-10. Olaf von Fersen, *Ein Jahrhundert Automobiltechnik; Personenwagen* (Dusseldorf: VDI Verlag, 1986).

2-11. B. Stephan and S. Snuyff, *Verbrandingsmotoren; Beschrijvend handboek voor het onderwijs en voor zelfstudie* (Leiden: A.W. Sijthoff's Uitgeversmaatschappij, n.y., [1927]).

2-12. *Motorklassik* (1988) nr. 12.

2-13. G. F. Steinbuch, *De motor; de Automobiel; handboek voor autobestuurders, monteurs, reparateurs en technici.* Bewerkt door Ir G.J. Tonkes en J.S. Buyze (Deventer – Djakarta: E. Kluwer, 1954).

2-14. FIAT 126 (Fiat Press Documentation, collection of the author).

2-15. A.W. Judge, *Motor Manuals: A Series for All Motor Owners and Users, Car Maintenance and Repair Volume IV* (London: Chapman & Hall, 1955).

2-16. Arnold Heller, *Motorwagen und Fahrzeugmaschinen für flüssigen Brennstoff; Ein Lehrbuch für den Selbstunterricht und für den Unterricht an technischen Lehranstalten aus dem Jahre 1912* (Moers: Steiger Verlag, 1985).

2-17. Victor W. Pagé, *The Modern Gasoline Automobile: Its Design, Construction, Operation and Maintenance; Invaluable to Motorists, Students, Mechanics, Repair Men, Automobile Draughtsmen, Designers and Engineers, Every Phase of the Subject Being Treated in a Practical, Non-Technical Manner* (London/Toronto/New York: Hodder and Stoughton, 1918).

2-18. "Zur Geschichte des Vergasers," *Kraftfahrzeugtechnik* 37 (1987) No. 1, 7–9.

2-19. Erich Werminghoff, "Die Einrichtungen zur Vergasung von Otto-Kraftstoffen," *Automobil Industrie* (1959), 48–60.

2-20. Philippe Krebs, "Arthur Constantin Krebs (1850–1935): Autorité et stratégie à la direction de Panhard & Levassor (1913–1916)" (Master's thesis, Conservatoire Nationale des Arts et Métiers, 2009).

2-21. Louis Baudry de Saunier, *Das Automobil in Theorie und Praxis; Band 2: Automobilwagen mit Benzin-Motoren; Mit einem Vorwort von Peter Kirchberg* (Leipzig: Reprintverlag, 1991) (first ed.: 1901).

2-22. James E. Homans, *Self-Propelled Vehicles: A Practical Treatise of the Theory, Construction, Operation, Care and Management of All Forms of Automobiles* (New York: Theo. Audel & Company, 1905–1906) (2nd ed., revised).

2-23. W. Poynter Adams, *Motor-car Mechanism and Management: Part I: The Petrol Car* (London: Charles Griffin, 1907).

2-24. M.A. Krebs, "Sur un carburateur automatique pour moteurs à explosions" (Paris: Gauthier-Villars, 1902) (paper presented at the Académie des Sciences, Paris, November 24, 1902).

2-25. Baudry de Saunier, "Le carburateur à réglage automatique Panhard & Levassor" (Paris: Ch. Dunod, n.y.) (reprint from *La Locomotion*, December 27, 1902).

2-26. L. Baudry de Saunier, "Le carburateur Krebs," *La Nature* 31, No. 1552 (February 21, 1903), 177–179.

2-27. W. Worby Beaumont, *Motor Vehicles and Motors: Their Design, Construction and Working by Steam, Oil and Electricity* (Westminster/Philadelphia: Archibald Constable & Company/J.B. Lippincott Company, 1906), Vol. II.

2-28. Günter Böcker, *Auf die Mischung kommt es an; Technik für die Mobilität: Erfinden–Entwicklen–Verwirklichen* (Meerbusch/Neuss: Lippert-Druck & Verlag/Pierburg, 1990).

2-29.   J.R. Singham, *The Autocar Handbook: Complete Guide to the Modern Car* (London: Iliffe & Sons Ltd., 1960), 22nd ed.

2-30.   E. Blaich e.a., *Internationales Automobil-Handbuch; Umfassendes Lehr- und Nachschlagewerk für alle Gebiete der Kraftfahrt* (Lugano: J. Kramer, 1954).

2-31.   K. Newton and W. Steeds, *The Motor Vehicle: A Descriptive Text-book for Students, Draughtsmen and Owner-drivers* (London/Birmingham/Coventry/Manchester/Glasgow: Iliffe & Sons, 19504) (1st ed.: 1929).

2-32.   J. Trommelmans, *Het moderne auto ABC: Constructie en werking* (Deventer: Kluwer Technische Boeken, 1993).

Chapter 3
The Engine: Ignition

3.1 Introduction: Regulating the Engine from Its Industrial Application

Borrowing the internal combustion engine from its industrial, stationary application and converting it into a propulsion source for a highly dynamic and mobile machine did not change so much its very basics, such as the pistons and the crank shaft, or the valves and the engine's housing. Rather, it was the auxiliary systems, such as the cooling, and especially the mixture formation and the ignition systems that had to be adjusted to conditions of highly variable and rapidly increasing loads and speeds. And because of the close relationship between early buyers and manufacturers, both technically well informed, this adjustment led to a higher appeal on the driver's skills. Indeed, driver skills were constructed, so to speak, together with the new technologies. For mixture formation this skill consisted of the accurate tuning of the mixture strength (the lambda value, λ; see Chapter 2). For ignition it was the timing of the moment that the combustion had to start in order to generate so much heat and subsequent pressure in the cylinders at the right moment that the pistons could perform their work of propelling the crank shaft and, through this, the car.

Flame versus Incandescent Ignition

This chapter is dedicated to the "problem of problems," as Karl Benz called it ([3-1], p. 269), the question of how to ignite the combustible mixture in the cylinder, and the related question of how much of the control power of the machine to leave to the user and how much to build into the hardware as automation. The combustion engine's roots in industrial, stationary applications with easy access to gas outlets had made the gas flame ignition a popular device.[1] Viewed from an evolutionary point of view, it was the steam engine that had formed the inspiration: designers of light, industrial internal combustion engines had simply taken the flame from the continuous external

---

1. The following paragraphs are taken from [3-2].

combustion of the steam engine, where its heat was used to convert water into steam, which does the work within the cylinder after entering through valves.

When it came to igniting the fuel-air mixture in an internal combustion engine, however, this concept of continuous heat provision had to be adjusted because the flame could not be in constant contact with the mixture in the cylinder. Ingenious valve systems driven by the engine itself through rods and springs were developed to catch the flame and bring it into the cylinder to ignite the mixture. Sophisticated systems worked with two flames, the one intermittently in contact with the mixture in the cylinder reignited by the second, outer flame every time it had ignited the mixture and was subsequently blown out by the cylinder pressure (Figure 3.1).

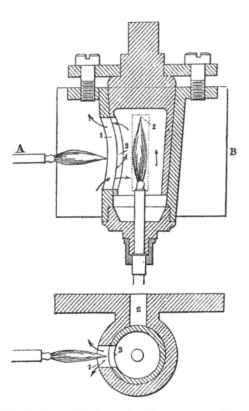

Figure 3.1 Early double-flame ignition system with a rotating valve mechanism to intermittently bring the heat source in contact with the combustible mixture within the cylinder; burner A reignites the inner flame whenever it is blown out by the combustion pressure ([3-3], p. 10).

Evolutionary variation made slides compete with rotating devices, but neither of these could keep up with the user's apparent wish to have the engine's rotational speed increase, let alone cope with the requirement that an engine should be able to change speeds continuously in the context of an automobile trip. Also, the tendency to increase the cylinder pressure (to generate more work out of a given quantity of mixture) did not help either. The extra flame in Figure 3.1 may be interpreted as the Pluto Effect: the effort to allow higher cylinder pressures so as to lower the chance of customers and manufacturers to opt for alternative ignition systems. When, in the end, this did not help to keep the threatening alternative technology at bay, the Pluto Effect turned into a Sailing Ship Effect.

One such alternative option was incandescent ignition, although this also worked on the basis of a flame that kept a metal tube heated from outside the cylinder. Glow-tube ignition, in all sorts of variations, was derived from an English patent for use in automobiles by Gottlieb Daimler, who cherished a deep distrust against electricity.

Because German automotive engine technology was leading in the fledgling car industry, with French and British car manufacturers taking licenses, glow-tube ignition became the mainstream ignition during the first decade of automotive technology, at least until 1893. As late as 1896, no single British-produced car was equipped with an electric ignition. The reason was the unreliability of the lead-acid accumulators, while their alternative, primary dry batteries, discharged too quickly and had to be replaced too often; at least that is the explanation we find in the handbooks. Generally, as we have seen in the previous chapter, such mono-causal technical explanations have to be distrusted, but in this case it seems indeed that once such batteries became available (around the mid-1900s), electrically heated glow-plugs were introduced. By then, however, sophisticated regulating systems were necessary and the glow-plug (which glowed or glowed not, without much regulation) disappeared from the automotive scene, only to reappear during the inter-war years as a starting device for diesel engines. This *migration of a device* from one application field to another belongs to the characteristic phenomena of technological evolution.

It seems that higher car speeds, and the air rushing past, prevented the flame and flame-incandescent ignition systems from becoming universal. In a way, the same applies to the nonflame incandescent ignition, as neither the pure flame ignition nor the incandescent ignition systems in all their variants were able to cope with the apparent wish to control the moment the combustible mixture in the cylinder should ignite. This, engineers knew as soon as they started thinking about how to speed up the car, could only be accomplished through a spark, a discontinuous, adjustable (in terms of timing) heat source.

This chapter covers the efforts to regulate and automate the ignition timing such that new car drivers and their "chauffeurs" could be deskilled into motorists. To this end, we will first describe the successful activities of Robert Bosch to develop an electric ignition by circumventing the awkward battery (Section 3.2). We will then see a remarkable revival of the spark ignition system in the United States, led by Charles Kettering and his Delco system (Section 3.3). The next section describes the struggle of the systems between the two competing options (Section 3.4), and we will close this chapter with an epilogue into the post–WWII era of electronization, followed by the conclusions (Section 3.5).

3.2   Robert Bosch and the Magneto Ignition

Early automotive handbooks give taxonomies of ignition systems. They often distinguished, at one level, between nonelectric systems (gunpowder ignition; catalytic glow ignition; flame ignition) and electric systems (Figure 3.2). The latter were mostly spark ignition systems, in which the fuel-air mixture in the cylinders was ignited by an electric spark that jumped between electrodes. Electric spark ignition systems were said to break down into two major subspecies, magneto ignition and battery ignition; interestingly, these were associated with the respective automotive cultures of Europe and the United States. Magneto and battery ignition systems each split into a low tension subgroup, in which the spark was created by pulling a mobile electrode away from a fixed one (thus drawing a spark), and a high tension subgroup, in which enough high-tension energy was created to get the spark to jump between fixed electrodes.

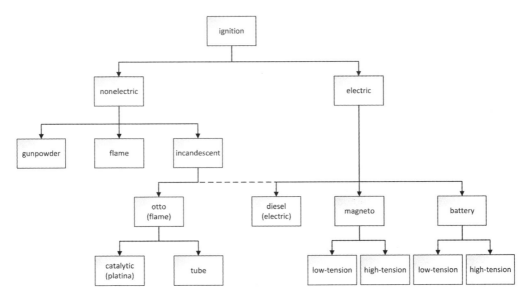

Figure 3.2 Hierarchical classification of the population of ignition systems until 1940 as taken from automotive handbooks. In terms of evolutionary history, this taxonomy is a-historical, and hence we call this a static structure diagram.

Once motorists started to define their hobby as something characterized by speed, in evolutionary terms the flame and incandescent ignition subspecies died out. At first sight, the shift to electricity thus seems to have been determined by technically informed arguments (such as about the vulnerability of flame ignition to wind pressure). But whether engineers opted for magneto ignition or not depended at least as much on culture, as we will see. Whereas the lead-acid battery was developed to a comparable degree of sophistication on both sides of the Atlantic, European car producers highlighted reliability (as a proxy for high quality) and opted for the magneto that did not need a battery, although the original incentive to go for this option might have come from engineering considerations.

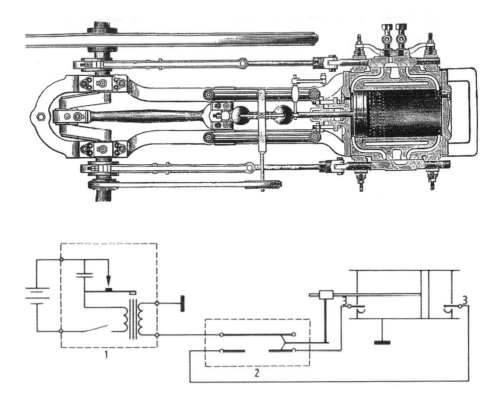

Figure 3.3 Lenoir's engine (1860) with battery ignition and trembler coil
(1 in the schematic below) causing a high-tension spark to jump over the fixed
electrodes of what perhaps can be considered as the first spark plug (3).
The engine was double-acting (reminiscent of steam engine practice):
on every side of the piston a mixture could be ignited, their sparks
distributed by the slide 2 ([3-4], p. 23, [3-5], p. 95).

Like most automotive technologies, both battery and magneto ignitions had their pre-
automotive ancestors. Spark ignition on itself was popular in some early engines such as
the Lenoir engine from 1860, developed by the French Nicolaus Otto, Étienne Lenoir
(1822–1900) (Figure 3.3). Karl Benz's first car engines a quarter-century later had a
similar ignition principle based on a trembler-type coil (Figure 3.4). The principle of
ignition of a combustible mixture by a spark goes back to toylike laboratory contraptions
such as Volta's gun from around 1800, in which a combustible mixture was ignited by a
spark, causing a cork to be forced from the container's mouth. The word *gun* is symp-
tomatic for early engine technology, as the roots of the combustion engine itself may
well be related to military thinking: engine technology improved substantially by the
cylindrical boring techniques developed for canons. After all, from an evolutionary
perspective, the engine cylinder is not much more than a canon that pulls its grenade
back once it is launched. Not coincidentally, several very early stationary engines were
ignited by gunpowder ([3-6], p. 7, 33).

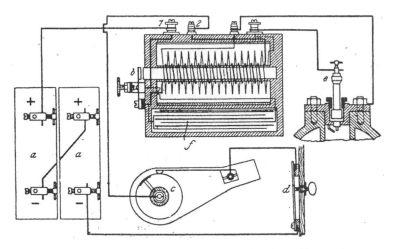

Figure 3.4 Electric battery ignition system developed by Karl Benz with trembler coil ([3-7], p. 96) (*courtesy Robert Bosch GmbH*).

The trembler coil used in these early types of ignition systems was in fact a direct-current transformer, based on the principle that the interruption of a current in a coil with a low number of windings generates a high-tension current in an adjacent coil with a high number of windings, a principle realized for laboratory purposes around the mid-century. It had the same disadvantage as electrically fed incandescent systems: the need for electrical energy was continuous and emptied the dry batteries quickly. Dry batteries lasted about 10 km before they had to be replaced: it seems that early French motorists greeted each other by shouting "Bon allumage!" (have a good ignition!). ([3-8], p. 48). Here the Pluto Effect may well have played a role in an odd way: the constantly sparking plug was an electric mimicry of the constantly burning flame. In other words: the Pluto Effect works irrespective of any purpose; it is always active.

A Second Typical Car Engineer: Robert Bosch

Indeed, it was this problem that Robert Bosch (1861–1942) tried to circumvent when he started, as early as 1887, working on his low-tension magneto. Like the flame ignition, magneto ignition (which did not need a battery, because it generated its own electricity by rotation in a magnetic field) also had its nonautomotive predecessors, in this case systems used to detonate mines. Here, electricity for the spark was generated by a manual cranking mechanism. Bosch's first system was built after an example shown to him by a client, but what he subsequently did can be called typical for the pioneering time in the German South with its fine-mechanical tradition [3-9].

As was usual for young entrepreneurs in those days, Bosch first undertook a trip of apprenticeship to the United States, where he not only met with Thomas Edison, but also became a member of the secret proto-labour union Knights of Labour. This would inspire him to introduce the eight-hour day as early as 1906 as one of the first in his country, and granting his workers a day off on the 1st of May. "Red Bosch" (who visited his clients on a bicycle) later would become a member of the reform movement, wandering in nature and consuming healthy food.

In close collaboration with the nearby engine developers (including those of stationary engines: the magneto was first applied on these engines, not on cars), Bosch constructed an apparatus of unsurpassed reliability, propelling his company into a giant of high-quality precision technology for automotive applications. His reputation was so solid that in the middle of the First World War, a British trade journal wrote: "Whatever, from a racial point of view, can be said against the Germans and thus against mister Bosch, he knows how to cut screw thread. I have used many magneto ignition systems and never a nut came loose, although I have driven with fully unbalanced engines at the most abominable speeds. Even if you take them apart a couple of times, one can tighten those tiny screws exactly as if they are brand new." ([3-10], p. 27). Bosch's reputation was so good that even Rudolf Diesel, who in 1894 briefly lost faith in his ability to create a new type of engine based on self-ignition induced by compression within the cylinder, approached him for a trial of electric ignition, as we will see in Chapter 12 ([3-1], p. 270, [3-11], p. 24).

The Bosch low-tension magneto ignition forced all other alternatives out of the European market, including the low- and high-tension battery ignition systems employed by Karl Benz in his first car in 1886 (improved in 1893) or by De Dion-Bouton in 1895 (Figure 3.5). When the Simms-Bosch magneto ignition (1898) came on the market (developed together with his British-German partner Richard Simms, whom he later would force out of his company), the battery version seemed doomed to extinction.

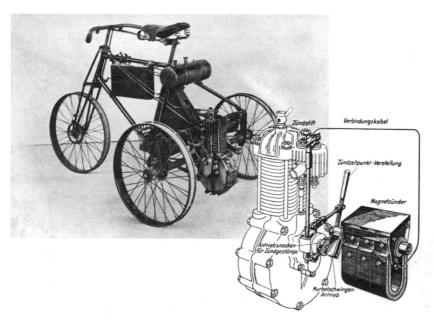

Figure 3.5 Probably the first application of a low-tension magneto on an automobile: a De Dion-Bouton three-wheeler was changed, in 1898, from battery to magneto ignition (*courtesy Robert Bosch GmbH*).

Ignition Timing

The low-tension version of the magneto ignition system, known as the make and break system in English-speaking cultures (as opposed to the jump spark high-tension variants), was considered very beneficial to car engine designs because of the "fat" spark it produced (Figure 3.6). ([3-12], p. 89, 92). Bosch and its competitors, such as Eisemann (Germany), Lucas (UK), Magneti Marelli (Italy), SEV Marchal and Ducellier (France), devised various mechanisms to enable the moment at which the spark was produced to vary relative to the piston position in the cylinder. This ignition timing soon became the core focus of research and of descriptions of the various systems in the handbooks, because it was through this mechanism, called spark advance, that the user could influence the behavior of the engine. It was a principle also applied to the filling of the cylinders of steam engines and later also of internal combustion engines: the dynamic character of the engine as opposed to the rather constant time it took to ignite, or to fill the cylinder, necessitated that the process (of starting and of filling) had to be initiated well in advance.[2]

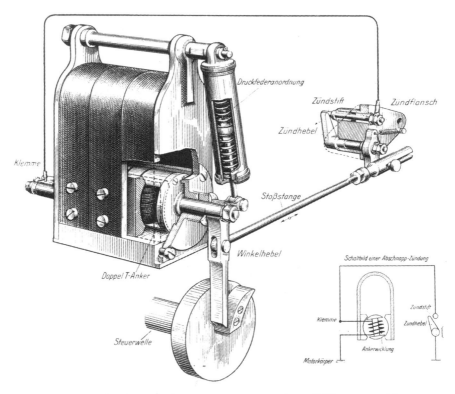

Figure 3.6 First low-tension magneto system from 1887 produced by the Robert Bosch company: the spark is drawn by moving the mobile electrode (*Zündhebel*) away from the fixed electrode (*Zündstift*), which both protrude into the cylinder (*courtesy Robert Bosch GmbH*).

---

2. I thank Philippe Krebs for this suggestion (October 2, 2012).

In terms of James Gibson, introduced in the first chapter, this meant that however different their technical properties, the proposed ignition systems all had to *afford* the user the ability to change the ignition timing depending on the situation (starting; climbing a hill; high-speed cruising). Starting the car, for instance, had to begin with cranking the engine by hand at a late spark setting, achieved by turning a manette on or near the steering wheel; but as soon as the engine caught, the motorist had to hurry back to the manette to advance the spark and thereby get the engine to run regularly. Increasing the engine (and thus the driving) speed initially did not even involve an accelerator pedal but could be done partially by advancing the spark. This ensured that, even at higher engine revolutions, the fuel-air mixture still had enough time to burn and generate pressure upon the pistons, which, through the clutch and transmission, turned the wheels. When climbing a hill, the skilled motorist had to advance the ignition timing to compensate for the fact that a richer mixture takes longer to burn ([3-13], p. 289).

3.3 Charles Kettering and the Systemic Approach of Technical Problems

Whether it was the near-monopoly by Bosch or the complexity (and thus the high costs) of the device, or even simple indifference to or ignorance of what went on in Europe that drove American inventors to chose another route is not known. What is known is that once the battery ignition was revived by Charles Franklin Kettering (1876–1958) (Figure 3.7) by the end of the first decade of the last century, followed by a transition period just before the First World War, a fierce competition emerged, resulting in a fairly quick conquering of the American market by Kettering's and related systems ([3-8], p 96).

Kettering, who represents the classical American rags-to-riches type, studied electrical engineering but did not much consider himself an inventor. Hired by the National Cash Register Company (NCR), he "didn't hang around much with other inventors or the executive fellows," as he remembered when he was already busy constructing his later self-image. "I lived with the sales gang. They had some real notion of what people wanted" ([3-14], p. 55).

Figure 3.7 Charles Kettering (1876–1958) behind the steering wheel (*photo courtesy of Kettering University Archives, Flint, MI USA*).

A Typical American Car Engineer

Kettering's invention is remarkable because he approached the Benz's problem of problems as a systemic challenge, not as an apparatus, a carefully designed fine-mechanical artifact, as Bosch did. Not without some stereotyping, automotive history describes both engineers as typical representatives of their respective automotive cultures. Whereas Bosch was the precision maniac, Kettering was the practical engineer who answered to a remark about the simplicity of his inventions: "We simply didn't have time to make it complicated" ([3-10], p. 30). That remark was made when he had developed, at NCR, a cash register drawer pushed open by an electric motor instead of a spring, as well as an electrically controlled counting machine. This embodied exactly the same principle he later would apply to the car, when it came to devising an apparatus to start the combustion engine when cold, a nice example of *functional drift* (the phenomenon that a function is transferred from one application to another).

Claiming that Kettering "invented the self-starter" is incorrect, however: electric starter motors were proposed earlier, but during a competition in 1905 at the Paris automobile show (in which also electric versions played a role) it was a pneumatic device that had won the main prize. Apparently, the apparatus did not cause much enthusiasm among engineers or buyers, probably because many of them had a chauffeur, well trained in handling the hand-crank (Figure 3.8) ([3-4], p. 35–36, [3-14], p. 52). This shows again that *automation* is not a simple, linear trend: as long as the owner-driver could delegate this skill to his mechanic, motorists apparently were not very interested in this type of technical "progress." It shows, too, that technological development is not a linear process to ever more "perfect" configurations: as long as the chauffeur's muscle power was enough to turn the "moving core" of the engine, there was apparently no reason for change.

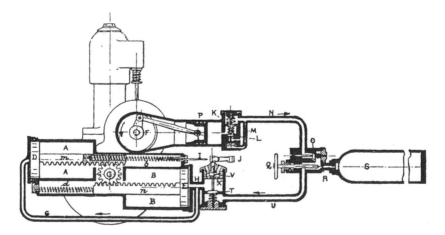

Figure 3.8 Pneumatic starter system used by Renault (1906): bottle S is filled with air by the running engine at 12 bar, enough for 20 cold starts. At the start, a foot pedal opens, through lever J, the valve T, and the pistons E and D move a rack-and-pinion gearing system connected to the engine's crankshaft via the gear C. The springs b and d push the pistons back ([3-15], p. 158).

Kettering's starter-generator was the endpoint of a development that has been described in all its romantic heroism as a three-year team effort. It started in 1909 when a colleague at NCR, who tinkered on a car, asked him to help design a better ignition than the battery-destroying trembler-type he used. Several NCR colleagues joined Kettering, who thus started a type of team research for which he later would become famous and which started in the barn of the tinkering colleague. At that moment, the lead-acid battery was just far enough developed to function as a reliable power source for electric vehicles. The relatively light grid-plate version could also be used to feed the ignition, but now the battery energized an ignition coil that only produced one single spark at the right moment, increasing the longevity of the battery by a factor 10 (Figure 3.9). The high tension was generated by using an interrupter driven by the engine, whereas the capacitor mounted over the interrupter had to decrease the tension precipitously, thus enhancing the transformer effect of the coil [3-16].

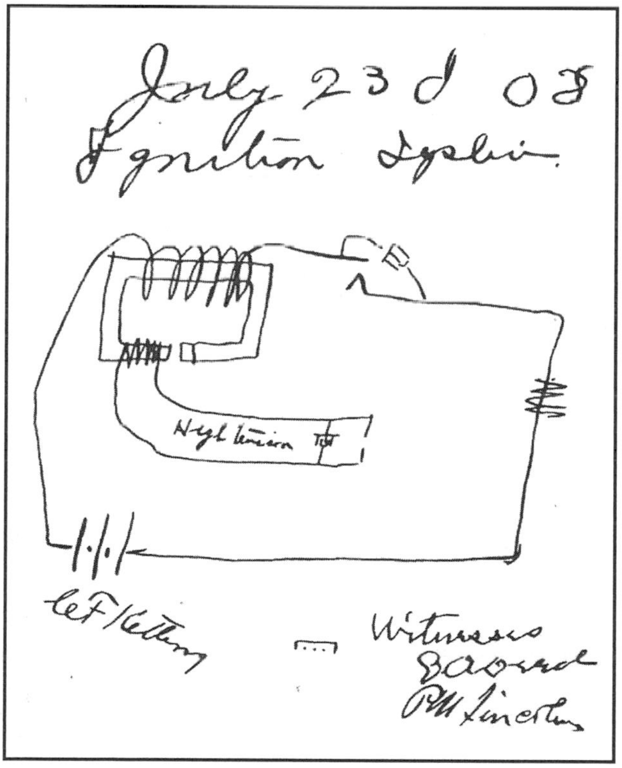

Figure 3.9 Charles Kettering's sketch of his high-tension battery ignition system (*photo courtesy of Kettering University Archives, Flint, MI USA*).

Kettering's Starter-Generator as System

When Cadillac ordered 8,000 units, the "barn gang" founded the Dayton Engineering Laboratories Company (DELCO). A year later, Kettering and his team started research on the starter problem upon Cadillac's request. The story goes that others were skeptical because they calculated that an electric motor propelling a cold engine had to have a power of at least 5 hp (that is nearly four times as much as current starter motors), which would mean that its size would be comparable to that of the combustion engine. But Kettering saw that the high power necessary to crank a cold engine was only requested during a very brief period of time, so he deliberately designed a rather small, high-speed motor, just like in the cash register case, a nice example of cross-industrial transfer (or *migration*) of technical concepts. As an example of his practical thinking, he later justified his conviction by explaining that he had never realized to have generated five horsepower when cranking an engine by hand. "Never mind about the experts," he said later, when he was busy creating his own image. The starter motor, he later commented, had to be "90 % car and 10% electric machine' ([3-17], p. 39). It was an insight many engineers (especially those working at suppliers to the car industry) had to acquire the hard way (as we will see in the case of the ZF *Einheitsgetriebe* transmission, in Chapter 5).

What is most remarkable, however, is how Kettering was forced, by the technology itself, as it were, to think systemically. When starting a cold engine, not only was the oil thick and sticky, and the spark had to jump in a cylinder with cold walls, but magneto ignition (if applied) produced a weak spark at low revolutions. This was drastically improved when using battery ignition. The core of this system was a powerful electric starter motor, which drove the engine's crank shaft through the same friction clutch as Kettering had devised for his cash register, thus allowing the first ignitions to take place, which subsequently took over the energy supply to the crankshaft through the combustion in the cylinders. No need, then, to design a complex magneto. Instead, the same battery could feed a high-tension battery ignition, as well as the head and rear lights, and other lighting if necessary [3-18].

After testing, in August 1911 Cadillac ordered 12,000 systems to be mounted in its 1912 models. Delco's SLI (starting, lighting, ignition) system was so successful that the company was integrated into the General Motor concern by 1916, where Kettering became a celebrated, although authoritarian leader of research teams (see Chapter 12).

The accepted history of the automobile relates how the Delco system and its derivatives (produced by American supplier companies such as Splitdorf and Remy) conquered the market after it had first been mounted in Cadillacs. But the irony of this history is that it took nearly a decade for the system to spread through the car population, as Ford's Model T, on the market since 1909, did not offer a starter motor as a standard component until 1919. Instead it was equipped with an idiosyncratic low-tension magneto ignition system built in the flywheel of the engine (Figure 3.10). This was only changed to battery ignition as late as 1921 ([3-10], p. 31). By then, the well-known Bendix spring (allowing to disengage the pinion at the end of the starter motor axle from the flywheel gear after the ignition had "caught") was already part of the state of the art: it appeared on the Chevrolet Baby Grand of 1914 ([3-18], p. 591).

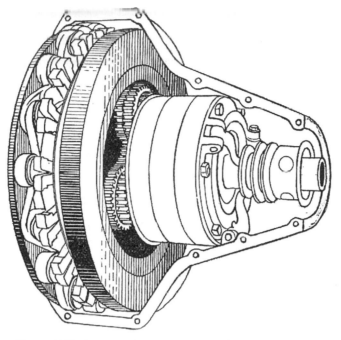

Figure 3.10 Low-tension magneto ignition in the flywheel of the Ford Model T ([3-21], p. 273).

3.4 Struggle of the Systems: Constructing Both a Winner and His Story

It was consumers' desire to drive faster (or at least, this desire as perceived by the producers) while maintaining control of the car that subsequently drove innovation. It set in motion an interesting evolutionary competition between alternatives and a transfer of functions between them.

Increasing car speeds meant higher compression pressures in the cylinders, which led to sealing problems involving the mobile shafts of the make and break low-tension magneto protruding into the cylinders (allowing maximum engine speeds in the order of only 200–300 revolutions per minute).[3] High-tension battery ignition did not have this problem because it worked with two fixed spark plug electrodes, so proponents of the magneto solved the sealing problem by introducing a high-tension version of this ignition type. Thus the functional requirement reliability, which could be achieved with nonleaking spark plugs, was adopted by (European) mainstream technology from an alternative that had threatened to take the lead in the market, a nice example of the Pluto Effect.

---

3. ([3-8], p. 52). Stationary engines had a constant speed in the order of 120 revolutions per minute. Benz's first engine had 300–350, and Daimler's 650 ([3-11], p. 22).

Spark Plugs

Production figures from Bosch indicated that the low-tension magneto had its best year in 1906, but it stayed in production until 1925, while the high-tension magneto, developed in 1902, took over the market quickly after 1906, helped by a Grand Prix victory in that year ([3-8], p. 65). At the moment that Kettering started working on his Delco system, most American cars were equipped with magneto ignition (half of them provided by Bosch). From that moment onwards, a separate and fascinating history of the spark plug sets in, characterized by several specialized companies that tested hundreds of porcelain, mica, and ceramic mixtures and dozens of electrode configurations in a spectacular technological dance around each other, one solution even more sophisticated than the next. American spark plug manufacturer Champion alone tested more than 25,000 insulation materials and 2,000 electrode shapes and materials during its existence in the 20th century ([3-4], p. 28). Here, indeed, the Darwinian mechanism of prolific variation can be witnessed, followed by a repeated selection of a few alternatives, over cycles of generations (Figure 3.11).

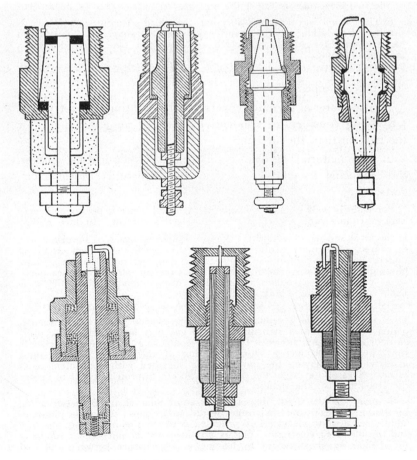

Figure 3.11 Variations of spark plugs ([3-19], p. 293)

Magneto versus Battery Ignition

In 1912, Bosch sold its millionth magneto and three years later its second millionth, including local production in the United States by the Bosch Magneto Company from 1910 ([3-10], p. 27). Whereas 87% of the models at the 1920 Paris car show and 80% of the models at the 1926 London show were equipped with a high-tension magneto ignition, by then the majority of American cars used battery ignition. Because no one could deny that the magneto was more reliable, dual ignition systems were in use on luxury and racing cars (the battery system for starting, the magneto for normal running). From the end of the 1920s onward, however, even European carmakers started to introduce the less-expensive battery ignition, although this resulted in the ignition becoming the most common defect in daily automobile use (Figure 3.12).

Figure 3.12 At the Grand Prix of France of 1908, German racing ace Otto Lautenschlager drove a 120 hp Mercedes with double spark ignition developed by Bosch.[4] Two years before, Renault had won the Grand Prix equipped with Bosch's high-tension magneto ignition (*courtesy Robert Bosch GmbH*).

The struggle between the systems has been recorded quantitatively by journalists who counted the cars equipped with the different ignition types in several countries (Figure 3.13; also see Figure 1.7).

---

4. Ten tire changes were necessary during that race ([3-11], p. 64).

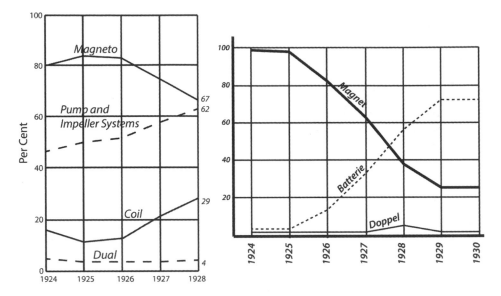

Figure 3.13 Construction of average types of ignition systems by calculating the frequency of the subspecies represented at automobile exhibitions (from the United States on the left, Germany on the right); A: ([3-20], p. 303) B: ([3-21], p. 15).

There were also technical reasons why producers shifted toward the battery, such as better spark quality at low engine speeds compared to magneto alternatives. The handbooks deal with these extensively. In European handbooks the arguments against battery ignition were piled upon each other the more the high-tension Bosch magneto became the norm. But in the end, production costs and the relative simplicity of the design led to the extinction of the magneto in cars, although we will see that geopolitical motives may have played a role in the magneto's demise for automotive purposes. The emergence of aviation, however (and, again, the influence of military engineering) enabled Bosch to compensate for its market losses until well into the inter-war years ([3-10], p. 31). In motorcycles and other light road vehicles, as well as in airplanes with piston engines, magneto ignition systems also were in use until long after WWII.

Nonetheless, the statistics of the cars on display on motor shows document the demise of the battery ignition unmistakably. In 1921, an American overview of European car technology stated that on "small cars without self-starter, the magneto is almost universal and seems correct, since in this way the car can always get home, battery or no battery" ([3-22], p. 551). As late as 1926, according to the handbook writer P. M. Heldt, "practically all of the cars [in Europe] carry ignition magnetos, those with simple battery ignition being the exception" ([3-23], p. 254).

Germany versus USA and the First World War

This struggle of the systems took place against a background of an industrial struggle started during the war. In the UK, authorities realized that "the magneto is the vital spark of the engine of national security," so they declared the English Bosch patents invalid. In the United States, the Bosch Magneto Company was put under local custody. The new owner sold the company to an American, who simply founded the American Bosch Magneto Corporation and subsequently started to advertise its products with the slogan "I am an American!" In Italy, too, Ercole Marelli started his imperium by copying the Bosch magneto. After the war, Bosch managed by clever maneuvering to grow into the automotive industrial giant it still is today, controlling the European market of electrical automotive equipment to a large degree, but it could not stem the tide of the American battery ignition ([3-8], p. 92, [3-10], p. 26).

Indeed, the remainder of the pre–WWII ignition history is also about control, but technical in this case: the spread of the car to less knowledgeable users had increased the desire to automate, that is, to delegate the skill of spark timing to the machine.[5] Automatic spark advance was first accomplished by introducing centrifugal weights that advanced the spark as a function of the rotation speed of the engine. By the end of the 1930s, this had been fine-tuned with the addition of a vacuum control, which corrected the rather straightforward rotation-speed control algorithm to one of so-called engine load, measured by the amount of vacuum in an engine's inlet system (Figure 3.14).

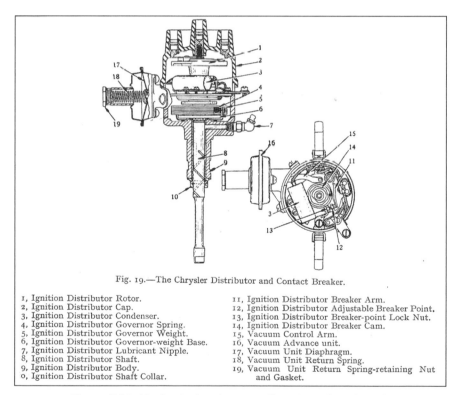

Fig. 19.—The Chrysler Distributor and Contact Breaker.

1, Ignition Distributor Rotor.
2, Ignition Distributor Cap.
3, Ignition Distributor Condenser.
4, Ignition Distributor Governor Spring.
5, Ignition Distributor Governor Weight.
6, Ignition Distributor Governor-weight Base.
7, Ignition Distributor Lubricant Nipple.
8, Ignition Distributor Shaft.
9, Ignition Distributor Body.
0, Ignition Distributor Shaft Collar.

11, Ignition Distributor Breaker Arm.
12, Ignition Distributor Adjustable Breaker Point.
13, Ignition Distributor Breaker-point Lock Nut.
14, Ignition Distributor Breaker Cam.
15, Vacuum Control Arm.
16, Vacuum Advance unit.
17, Vacuum Unit Diaphragm.
18, Vacuum Unit Return Spring.
19, Vacuum Unit Return Spring-retaining Nut and Gasket.

Figure 3.14 Mechanical and pneumatic automation of spark timing in a Chrysler distributor ([3-24], p. 17, Fig. 19).

5. For efforts by the American car industry to push knowledgeable users out of the realm of automotive technology, see [3-25].

Technical Competition and the Pluto Effect

But this story of control is also about the question of who controls our memories. During the 1920s the American car industry became more self-confident with respect to the European example and came to see its own solution, the battery ignition, as the universal solution. They were right, but more in the sense of a self-fulfilling prophecy and not in the technical sense: Bosch only very reluctantly gave up its superior technology in face of an "invasion" of cheap, American technology that it could not resist. In Gibsonian terms, the two systems afforded largely the same practice: replacing, or automating, skillful spark advance. Though the battery system was less reliable, this was easily compensated in the United States by a more extensive service system. Figure 3.15, a dynamic structure diagram of the early automotive ignition (and as such the dynamic counterpart of the static diagram of Figure 3.2), illustrates this sequence of events: the arrows between the alternatives represent a transfer of "functions" and suggest that by borrowing from encroaching technologies, some other technologies managed to survive and eventually achieve dominance.

Figure 3.15 illustrates the Pluto Effect, which we already have seen functioning on several occasions. It is the transfer of functions between competing technologies that forms an important driving mechanism of technical evolution. By borrowing functions from the competition through the re-introduction into its own technological vocabulary, mainstream technology mostly manages to stay in power.

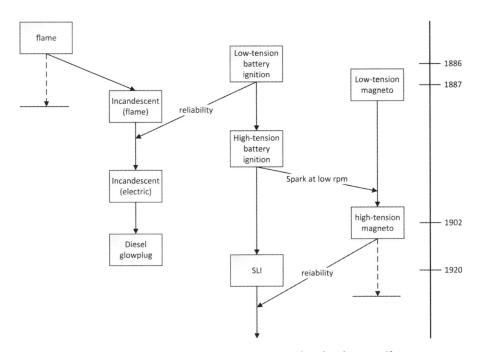

Figure 3.15 Dynamic structure diagram showing inter-artifact competition (the Pluto Effect) between three subspecies of ignition systems; the transfer of functions is indicated through arrows.

Constructing the Ignition "Story"

It is remarkable that in the case of the ignition this process is accompanied, in the handbooks, by a gradual rewriting of the system's history. If we take the Dutch handbook *De automobiel en haar behandeling* (*The Automobile and Its Handling*) by former Spyker employee J. W. Brand as an example, we can trace this modification of the system's history as a repeated rolling-back of the new winner into the past, because multiple editions of this handbook appeared over nearly half a century until at least 1950.[6]

In the first available edition, thought to date from 1911, Brand presented the battery as an auxiliary component only used for starting.[7] In the 1921 edition the battery ignition was "not anymore applied as separate ignition, unless on very old engines." Only one year later, in the "improved" 13th edition of 1922, Brand presented the battery ignition as a system that "witnesses a comeback during the last years. Modern cars are all equipped with a complete electrical installation." ([3-26], year 1921, 104, year 1922, 103–104). In 1932, Brand formulated the definitive, consolidated story by presenting the magneto ignition as an aberration on the battery ignition's road to triumph: "Although the magneto ignition had supplanted battery ignition almost completely, the battery has reconquered its place again. […] Because […] a good and solid electricity source is […] present, the magneto ignition is very often entirely left out" ([3-26], year 1932, 151). In the post–WWII 25th edition of 1950, at last, battery and magneto ignitions were both presented as alternatives, without any comments on the frequency of their use. Knowledgeable readers would have known that the battery was a car system, while the magneto was to be found here and there on other vehicles, such as mopeds ([3-26], year 1950, 151).

Remarkably as well, Brand's history is compatible with a story constructed on the other side of the Atlantic as early as 1917. As a result, the American story became the universal story, and the magneto's European past was more or less forgotten. In 1917, when sales manager H. Rice, of ignition manufacturer Atwater Kent, in the journal of the Society of Automotive Engineers wrote about "Problems in Ignition Development," he presented battery ignition as the "first ignition used for automobile engines," altogether forgetting or ignoring all flame and glow ignition variants. He emphasized that "today out of a total of 109 manufacturers of cars, 87 use battery ignition exclusively," while 22 used either low- or high-tension magnetos ([3-27], p. 154–155). Ironically, one of the latter 22 was the Model T, which at that time comprised about half of the world's total car production.

Three years later, most American engineers agreed that there was no difference in quality and performance between the competing systems. After 1920, both systems managed to eliminate whatever differences had existed between them in quality and performance, and ignition became a matter of national identity rather than of

---

6. Brand had already published the 4th edition of his handbook in 1911; the last traceable edition (the 25th) dates from 1950. See ([3-26], year 1911). Other editions consulted for this analysis are those of 1921 (12th ed.), 1922 (13th ed.), 1926 (15th ed.), 1932 (19th ed.), and 1950 (25th ed.).
7. ([3-26], p. 74). If an engine was still warm, the handbook recommends restarting it on the ignition alone (without using the hand-crank) ([3-26], p. 133).

technology, supported by deep-rooted nationalistic sentiments on both sides of the ocean. It disappeared as such from the pages of the major American engineering journal and became a part of the users' discourse on comfort: the system's unreliability was a constant pain in the neck of many a motorist.

3.5 Conclusions

After the Second World War, the battery ignition with spark plugs, distributor, centrifugal and pneumatic spark advance, and battery witnessed its heydays before it disappeared, as it were, in larger, electronically controlled, motor management systems, together with the mixture formation system of the previous chapter. In the process, these systems got electronized as well in an integrated process of automation, which we will describe in more detail in Chapter 10.

During its heydays, the ignition still belonged to the largest causes of defects in cars, so a thorough campaign was set up within Robert Bosch GmbH in Germany to provide engineering schools with high-quality wall drawings, booklets, and auto-technical handbooks that taught young engineers and mechanics the intricacies of German car technology, especially its ignition and mixture formation systems. Again, a co-construction of technology and its related discourse took place: we learn car technology while at the same time learning to speak about it, and we speak an adjusted history, in which the eventual winner gets all the credit. Unjustly so, as we have seen.

The intriguing ignition history on the previous pages also teaches us that we learn to speak about car technology in a highly nationalistic way, seeing cars as representative of a nation's spirit, despite the continuous converging tendencies on both sides of the ocean. Thus, *converging* and *diverging* forces, as well as *homogenizing* and *diversifying* forces, exert their influence on the evolution of automotive technology, a bundle of nonlinear trends that not necessarily lead to ever more perfect artifacts. A part of these converging forces is the Pluto Effect, which in this chapter we have learned to appreciate as a mechanism privileging mainstream technology, and helping it become ever more hegemonic.

References

3-1.   Eugen Diesel, "Rudolf Diesel," in Eugen Diesel, Gustav Goldback and Friedrich Schildberger, *Vom Motor zum Auto; Fünf Männer und ihr Werk* (Stuttgart: Deutsche Verlags-Anstalt, 1958) 205–255.

3-2.   Gijs Mom, "The Dual Nature of Technology: Archeology of Automotive Ignition and the Evolution of the Car," in Christian Huck and Stefan Bauernschmidt (eds.), *Travelling Goods, Travelling Moods: Varieties of Cultural Appropriation (1850-1950)* (Frankfurt am Main: Campus, 2012), 189–207.

3-3.   J.C.B. MacKeand, *Sparks and Flames: Ignition in Engines: An Historical Approach* (Montchanin, DE: Tyndar Press, 1997).

3-4   Gijs Mom, "De elektrische installatie in historisch perspectief," in: H. de Boer, Th. Dobbelaar, and G. Mom, *De elektrische installatie* (Deventer, 1986) (Part 8 of *De nieuwe Steinbuch; de automobiel; handboek voor autobezitters, monteurs en technici onder redactie en coördinate van drs.ing. G.P.A. Mom)*, 17–51.

3-5.   B. Stephan and S. Snuyff, *Verbrandingsmotoren; Beschrijvend handboek voor het onderwijs en voor zelfstudie* (Leiden: A.W. Sijthoff's Uitgeversmaatschappij, n.y. [1927?]).

3-6.   Horst O. Hardenberg, *The Middle Ages of the Internal-Combustion Engine 1794–1886* (Warrendale, PA: Society of Automotive Engineers, 1999).

3-7.   Arnold Heller, *Motorwagen und Fahrzeugmaschinen für flüssigen Brennstoff; Ein Lehrbuch für den Selbstunterricht und für den Unterricht an technischen Lehranstalten aus dem Jahre 1912* (Moers: Steiger Verlag, 1985) (reprint of 1922 edition, 1st ed.: 1912).

3-8.   *Bosch und die Zündung* (n.p., n.y. [Stuttgart: Robert Bosch GmbH, 1952]) (Bosch-Schriftenreihe, Folge 5).

3-9.   Theodor Heuss, *Robert Bosch; Leben und Leistung* (München: Wilhelm Heine Verlag, 1981²) (1st ed.: 1946).

3-10.   Gijs Mom, "De vitale vonk (Techniekhistorie: de elektrische installatie-18)," *Auto en Motor Klassiek* 3 No. 11/12 (November/December 1987), 25–33.

3 11.   *75 Jahre Bosch; 1886–1961, ein geschichtliche Rückblick* (Stuttgart: Robert Bosch GmbH, 1961) (Bosch-Schriftenreihe Folge 9).

3-12.   Roger B. Whitman, *Motor-Car Principles: The Gasoline Automobile* (New York: D. Appleton and Company, 1907).

3-13.   A. D. Libby, "Advantages of Magneto Ignition," *The Journal of the Society of Automotive Engineers* 7 No. 3 (September 1920), 277–290.

3-14.   Oliver E. Allen, "Kettering," *Invention & Technology* (Fall 1996), 52–63.

3-15.   *Der Motorwagen* (1907).

3-16.   Gijs Mom, *The Electric Vehicle: Technology and Expectations in the Automobile Age* (Baltimore: Johns Hopkins University Press, 2004).

3-17.   Stuart W. Leslie, *Boss Kettering* (New York: Columbia University Press, 1983).

3-18.   T. A. Boyd, "The Self-starter," *Technology and Culture* 9 (1968), 585–591.

3-19.   James E. Homans, *Self-Propelled Vehicles: A Practical Treatise of the Theory, Construction, Operation, Care and Management of All Forms of Automobiles* (New York: Theo. Audel & Company, 1910).

3-20.   *Journal of the Society of Automotive Engineers* 23, No. 3 (September 1928).

3-21.   *ADAC-Motorwelt*, 27 Nr. 33 (August 15, 1930).

3-22.   Maurice Olley, "Why American, British and Continental Car Designs Differ," *Automotive Industries* 44, No. 10 (March 10, 1921), 547–553.

3-23. P.M. Heldt, "What Is a European Light Car?" *Automotive Industries* 55 No. 7 (August 12, 1926), 252–254.

3-24. Arthur W. Judge, *The Modern Motor Engineer: A Practical Work on the Maintenance, Running, Adjustment, and Repair of Automobiles of All Types, and on the Management of Garages, with Special Contributions by Experts* (London: The Caxton Publishing Company, n.y. [1937]) (new and revised edition), Vol. IV.

3-25. Kathleen Franz, *Tinkering; Consumers Reinvent the Early Automobile* (Philadelphia: University of Pennsylvania Press, 2005).

3-26. J. W. Brand, *De automobiel en haar behandeling* (Rotterdam: Nijgh & Van Ditmar).

3-27. H.E. Rice, "Problems in Ignition Development," *The Journal of the Society of Automotive Engineers* 1, No. 2 (August 1917), 154–158.

Chapter 4
The Drivetrain: How to Get the Energy from the Engine to the Wheels

4.1 Introduction: Shaping the Drivetrain Configuration

Even more so than the subsystems dealt with in previous chapters, the transmission of the energy from the engine to the driven elements (in the case of the car: the wheels) had been developed in its principal traits well before the emergence of the car. If we define drivetrain as the ensemble of clutch, gearbox, shafts, final drive, and wheels (see Figure 2.2 for an overview of the layout),[1] then it had been the bicycle and the industrial tooling machine which had given the most: the former its chain drive and wheels, the latter the rest. These pre-car sources on their turn could draw from an extensive tradition of gearwheel cutting and calculation, the design of all kinds of couplings and clutches between moving parts, as well as shafts, belts, and chains. The only exception is the so-called constant-velocity (CV) joint developed expressly for automotive use from the 1920s onwards in order to enable the greatest conversion ever undertaken in the history of the car: the shift from rear- to front-wheel drive. It is to this component that we will turn our attention in this chapter (Section 4.5), after we have given an introduction to the basic layout (Section 4.1) and main component assemblies (Section 4.2) of the drivetrain, and have looked into the nonlinearities of the developments of the clutch (Section 4.4)

---

1. The wheel has multiple purposes: apart from being the last in the drivetrain, it also plays a role in the suspension (as a "carrier" of the car, in static and dynamic situations) as well as in braking and cornering. The wheel will be dealt with in Chapter 7.

and the final drive (Section 4.4). We will bypass the transmission, however, as this will be the subject of the next chapter.

Pre-automotive developments

It is customary in popular histories of the car to highlight predecessors that go back deep into the pre-modern period. In this case the harvest is very prolific, as gearwheels have been used as jewelry (perhaps representing the sun) since times immemorial, while the differential, the device that allows two different speeds while turning, seems to have been used in China as early as the 12th century to enable a human figure on a so-called south-pointing chariot (*Zhinanche*) to always keep it pointed toward the south, functioning as a compass [4-1].

Many of these so-called predecessors are just isolated inventions and re-inventions that have no bearing on the history of the automobile, if only because they often were not known at the time the knowledge incorporated in them would have been needed. They were rediscovered, as it happens so often, well after the re-invention had already taken place and a new history was rolled back into the past. Nonetheless, although their influence on later developments cannot be confirmed because of a lack of research, the early modern phase of mobility history (the late 18th and the 19th century) has been very prolific, especially the two periods in the first half of the 19th century, when British pioneers like John Blenkinsop, William Henry James, and David Gordon were trying not only to get the steam locomotive running but also to derive a steam bus, a road vehicle, from this development (in vain, as we will see in Chapter 7) [4-2].

In 1828, for instance, the French steam bus builder Onésiphore Pecqueur proposed a differential to be used in his steering system, to allow the wheels to be turned independently of each other. Pecqueur claimed in his patent that the British steam locomotive builders had to design their rails in straight lines because they were not familiar with the differential ([4-3], p. 31–32).

One of the best researched ancestors is the universal joint proposed by Italian mathematician, philosopher, and physician Girolamo Cardano (1501–1576) and hence also known as cardan joint (Figure 4.1). Another name is Hooke's joint, called after the British physicist Robert Hooke (1635–1703) who discovered the nonuniformity between the movements of the first and the second shaft and proposed a solution, namely to add a second joint that would do the same and so would result in the third shaft turning synchronous with the first. Such a constant velocity transmission was later called homokinetic. It was French military officer and mathematician Jean-Victor Poncelet who in 1824 proved this nonuniformity of the joint by using spherical trigonometry, triangular geometry applied to the surface of a sphere ([4-4], p. 3–9).

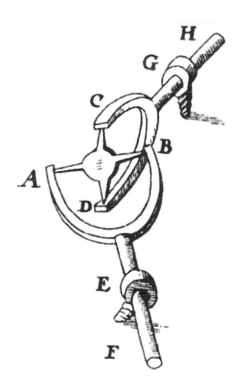

Figure 4.1 Universal joint as proposed by an unknown author called Amicus (friend) and described by Kaspar Schott (1664) ([4-3], p. 26).

Translation and Transformation of Functions

It would, however, be an intolerable simplification of history if we would assume that the drivetrain was ready to be taken into the automobile at the moment of the latter's emergence. It is exactly the reverse: with the emergence of the car, the drivetrain became (re-)assembled and its parts' functions were redefined, their properties redesigned. Several components had never before been used for the functions they came to enable in the car, and it was the *translation and reformulation* of these functions, the subsequent adjustment of the properties to afford them and their ultimate assemblage into a layout that fitted the car which formed the real innovation in these early years of automotive drivetrain history.

In the following section, we will investigate the main components in the drivetrain (except the transmission, as we have said) for the adjustments that were undertaken to integrate the different strands of mechanical heritage into car technology worthy to be taken up in the handbooks. As car technology is a historical concept, the adjustments were an ongoing process.

4.2   From Belts and Chains to Prop Shafts (and Belts and Chains)

The problem of getting the engine's energy to the wheels was especially acute in cars with internal-combustion engines. In principle, both electric and steam vehicles did not need complex transmission systems, as they had a high torque at low speeds (enabling them to drive away from standstill without any intervention by a de-multiplication of their speed and a concurrent multiplication of their torque). A steam engine could start by letting steam into its cylinder, so a clutch was not necessary either. Only some shafts and some gears were used, for instance to get the energy change direction from a longitudinal shaft to the shafts that directly drove the wheels, as is the case of the so-called final drive.

Once manufacturers and users started to opt in favor of the internal-combustion engine, however, a clutch was added to interrupt the energy flow at standstill or when gears had to be changed. Indeed, because of the limited speed range of such an engine also a gearbox was called for, enabling the application of this speed range at ever higher levels of speed at the vehicle side, the driven wheels. In other words, the price for the choice in favor of the combustion engine again appeared to be very high in terms of technological effort: it illustrates the attraction this noisy, smelly, vibrating machine had for the fledgling macho culture around the car. It makes this choice all the more enigmatic from an engineering point of view that believes in efficiency and simplicity.

Belts

Look how they struggled to get it done: the earliest combustion engine cars had a mixture of belts and chains to solve this problem.[2] A belt, mounted on a pulley driven by the engine and transmitting energy to another pulley on an intermediary shaft, could be shifted to a free-wheeling pulley to enable a clutch function. The intermediary shaft allowed the torque to become multiplied through a gear or another type of reduction, so the further energy flow at a higher level of energy transfer had to be undertaken by a more sturdy element such as a chain. Belts slipped and had to be shortened during use now and then, but by leading them over pulleys with a different diameter, a speed reduction (and torque multiplication) could be added at the same time (Figure 4.2).

---

2. See for the following pages especially: ([4-3], p. 35–38).

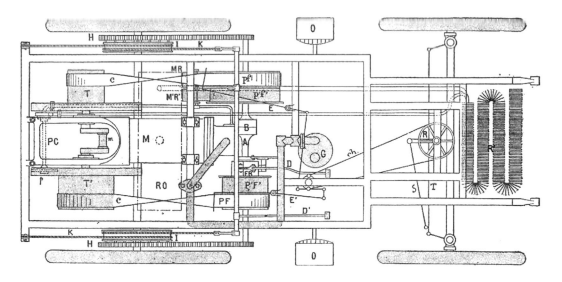

Figure 4.2 An early combustion engine car with belts as transmission to the driven wheels ([4-5], p. 268).

Chains versus Prop Shafts

Unprotected chains, however, wear rapidly. To avoid the extra noise this caused, manufacturers could have chosen to enclose them in casings (as did happen in the bicycle). Instead, they often mounted them directly in the chassis frame, as Chapter 5 will show (see Figure 5.1). Some manufacturers (especially those who made lighter cars) opted instead for a rigid propeller shaft. Heavier cars remained loyal to the chain drive much longer, such as the Mercedes of 1901 (which won the Gordon Bennett race in Ireland in 1903), and nearly pushed the luxury car producer Adler into bankruptcy, as nobody wanted to buy its innovative prop shaft cars anymore. This shows how misleading it can be to interpret historical events as inevitable "facts": when Mercedes in 1908, as one of the last big German brands, decided to change to the shaft drive, users and engineers alike may have realized that the race victory had not been won because of but despite the chain ([4-6], 19).

The historicity of this story must be apparent: encased chains are still widely used in a multitude of automotive applications (such as in four-wheel drive cars, or in engines between the crank shaft and the cam shaft, until recently) and are still near-universal in motorcycles (and in the racing cars of Honda, originally a motorcycle manufacturer, between 1962 and 1969), so it would be a historical mistake to call them *intrinsically* worse than shafts. This case is comparable to the controversy between carburetors and injection systems (see Chapters 2 and 9). Likewise, toothed belts are still used as alternatives to the distribution chain in the engine. In other words, substituted components may be worse in certain aspects (such as noise and mechanical efficiency) but often compensate that in other aspects. In this case, changing from chain to shaft made a final reduction necessary, a function previously accomplished by the reduction ratio of the chain sprocket wheels. The gain of higher efficiency thus was compensated by a more complex layout, less efficient overall, and governed by *functional split*.

Advantages and disadvantages are socially constructed, and intrinsic technical properties mostly play a subordinate role in this construction. The new system, which we have introduced in Chapter 2 as the Panhard system, was also known in France as the chainless system (*système acatène*), usually a negative way to characterize an innovation by what it does not have. Apparently, that is one of the few certainties contemporaries have when encountering technological newness, such as in the case of the horseless carriage.

Torque Tube versus Hotchkiss Drive

The shift from belt and chain to shaft for the purpose of engine torque transmission was an international phenomenon. There were two schools: one tried to avoid the use of the delicate and costly universal joints by encasing the shaft in a torque tube that took up a part of the torsion forces occurring between the engine and the driven wheels and so enabled the shaft to be made simpler and lighter. But as we will see in Chapter 7, engineers did not like to increase the unsprung weight (the weight of the wheels and other parts that follow the road pavement profile separated by and connected to the rest of the car through the springs). Furthermore, the hollow tube worked as a sound amplifier in a time (the 1920s) when the noise of the car became a hot topic [4-7]. It was the open driveline type, also called the Hotchkiss drive (after the American-French weapon and car manufacturer located in Paris), that became universal ([4-8], p. 4).

As always in the history of automotive technology, when a problem appears to be difficult, a plethora of alternatives for the universal joint were proposed, and many of them were adopted, if not in the prop shaft then in other subsystems of the car. One challenge appeared to be the combination of the angular changes and the changes in length, while at the same time maintaining the homokinetic properties of the transmission. Olive couplings, nut couplings, ring joints, and what came to be known in the UK as pot type joints were applied, but the most popular (and lasting) solution was the elastic or flexible joint, initially designed as a leather disk, later universally accepted as a combination of rubber and textile. It was the Hardy disk (made fully from textile and called after the British manufacturer), which was widely accepted in all sorts of applications (Figure 4.3).

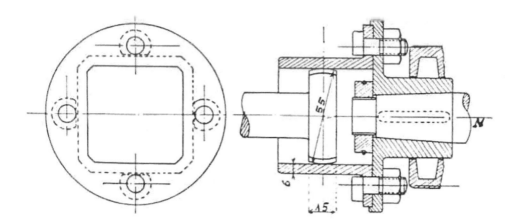

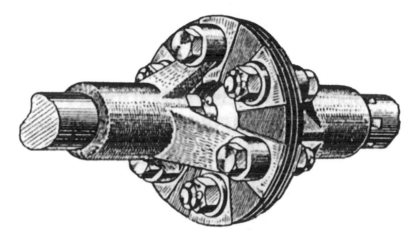

Figure 4.3 Pot type joint and Hardy disk as replacement for a universal joint ([4-3], p. 65).

4.3 The Clutch: Substitution and Coexistence

Once the necessity of a gearbox was acknowledged as the means to reach higher driving speeds, a separate clutch appeared in the drivetrain, borrowed from industrial applications in the form of a conical clutch, equipped with friction lining made of animal hair or hide. Jamming, burning, and problems of gear shifting because of jerking at the friction surfaces sent a host of inventors and engineers to their desks and sheds to come up with alternatives (Figure 4.4).

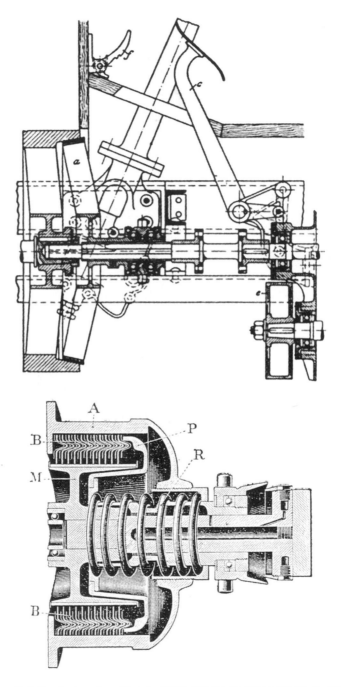

Figure 4.4 Conical and multiplate (Hele-Shaw) clutches ([4-3], p. 34, 35).

Cylindrical clutches were tried, including a spring band version used on the Mercedes of 1901. Multiplate clutches were also popular, the most well-known being the Hele-Shaw clutch, a wet clutch developed by English professor Henry Selby

Hele-Shaw, a pioneer of early research on rolling resistance and an inventor of a continuously variable transmission (see Chapter 5). When the wet multiplate clutch was superseded by the single dry plate clutch, the former did not leave the automotive field, as it reappeared in gearboxes and final drives, and was used as the main clutch in mopeds until this very day, a nice example of *migration* (or *crossover*) and of nonlinear evolution. The conical clutch did not die out either: it is still found inside the manual gearbox as part of the synchronizing units (see Chapter 5). In all of these cases (if one zooms out to the proper birds-eye view) *coexistence rather than substitution* was the case, although the properties of the components had to be adjusted to the new functionalities.

But even when the single-plate clutch drew more adherents (most likely because of its promise of simplicity), its trajectory of development was not linear, caused first by the problematic friction behavior of the plate lining material. De Dion-Bouton, for instance, as early as 1904, applied a plate with graphite lubrication, while Morris in the 1920s adorned the plate with cork pads (Figure 4.5).

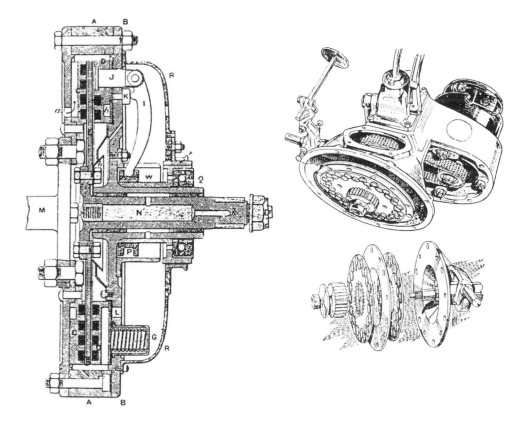

Figure 4.5 Single-plate clutches proposed by De Dion-Bouton in 1904 (left) with graphite lubrication and by Morris in the 1920s with cork pads ([4-3], p. 36).

Emergence of the Dry Single-Disc Clutch

It was the asbestos lining that eventually enabled the decisive step (as we can observe only from hindsight). Invented in the UK (by Herbert Frood, 1921), the version developed in the second half of the 1920s, especially for mass production by the American company Borg & Beck, made this clutch type into the hegemonic version all over the automotive world. George Borg of Borg & Beck was also one of the founders of the Borg-Warner conglomerate, which in the second half of the century acquired nearly a monopoly on this clutch type, supplying three-quarters of the American market (see Chapter 5).

It was Borg-Warner in 1936 that added the last major innovation to the single-plate clutch as standard equipment, substituting the helical spring that pressed the plate against the flywheel by a so-called diaphragm spring. The helical spring had first been designed as a central spring, but was later replaced by a circle of smaller springs (in the US, hidden under the clutch cover, in Europe placed in metal cylindrical cases sticking out of this cover). The diaphragm spring was a direct result of General Motors' laboratory work aimed at making the undulated disc spring, known as the Ingersoll spring, fit for application in a clutch. This spring was named after Roy C. Ingersoll, who had devised a special plate roller technique for the production of plough discs and whose company had been merged into the Borg-Warner conglomerate when the latter was founded in 1928.

In Europe, the diaphragm spring clutch became known during and after the Second World War through the military trucks of General Motors, especially when the French occupying forces in 1945 ordered the German company LuK (*Lamellen- und Kupplungsbau*, Stuttgart) to manufacture spare clutches for American trucks.

The success of this type of clutch was later explained by pointing out its properties: the force required to declutch was not linear (as in the coil spring version) but decreased after having reached a maximum (Figure 4.6). A second argument found in the handbooks in favor of this type of clutch was its symmetrical design, which caused less vibration at high engine speeds and thus was beneficial when during the 1950s and 1960s, smaller, higher-revving engines were developed. While such arguments sound like a-historical reconstructions from a hindsight position, perhaps unjustifiably cementing advantages, gradually discovered, as the motives for their introduction, we do not have better arguments at the ready because of a lack of research in this domain. However this may be, the first application in large quantities in Europe was the Opel Rekord B from 1965 [4-9], [4-10], [4-11].

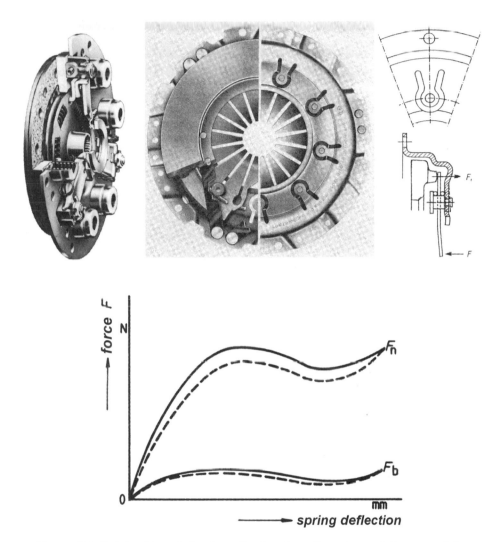

Figure 4.6 Clutch with a circle of small coil springs (in casings sticking out of the cover) and diaphragm clutch with actuation characteristic ([4-12], p. 186, 180, 182).

A Trend of Automation

The remainder of the single-plate clutch's history is (apart from the efforts to get the asbestos out of the lining material) a story of *automation*. In the American automatic transmission (see Chapter 5) it was replaced by a new type of clutch (a fluid coupling, later integrated in a torque converter, also wet) that better fit the characteristics of the planetary gearing used in that type of gearbox. In Europe, hesitating to follow the American example, car manufacturers after the Second World War introduced semiautomatic transmissions in which all sorts of automated clutches were proposed, but most of them stayed within the lineage of the dry single-plate design, such as Opel's Olymat, a combination of a conventional plate clutch and a centrifugally operated clutch. The

Citroën 2CV, however, was equipped with a centrifugal segment clutch, with segments directly expanding to an outer friction ring (Figure 4.7). It was only quite late in the 20th century that efforts to smooth the interruption of the torque flow were proposed, such as the double clutch (Figure 4.8).

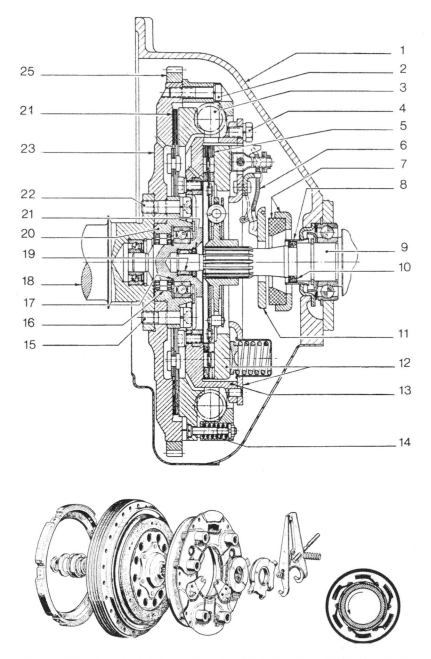

Figure 4.7 Automated clutches from Opel (Olymat; 3: centrifugal weight) and Citroën (direct segment clutch on the 2CV) ([4-3], p. 42, 40).

Figure 4.8 Double clutch by LuK, Germany [4-13] (*courtesy of LuK*).

4.4 The Final Drive: A Feast of Gearwheel Designs

A comparable collective search for the "right" solution took place for the final drive, where the torque and speed of the crank shaft are adjusted to what the wheels can master when propelling the car. Here the fact that the end reduction was not adjustable but fixed led to a divergence of final drive families, as heavier automobiles such as trucks became equipped with double drives through the addition of an extra gear reduction in the hub of the driven wheels.

For the final drive mounted in-between the driven wheels, whether for trucks or for passenger cars, the first phase (the phase of emergence) witnessed a frenzy of gear and gear teeth geometry research. The main development criteria were gear noise, longevity, and costs, criteria that were not always compatible.

The smoothest gearwheel configuration was the worm wheel drive, preferably equipped with a small "worm" of a globoid shape (also called hour glass worm, as applied by Frederick Lanchester in his cars as early as 1896, who had to design his own cutting machine for its production). In this type of gearing the tooth surfaces glide rather than roll upon each other, leading to serious wear and tear. That was programmed in, so to speak, as the worm was made from hard material (such as chrome-nickel or vanadium steel), while the worm wheel was made from softer phosphorous bronze. The underslung version (where the propeller shaft could be placed low, beneath the passenger cabin) was very popular, although quite expensive. Some brands remained loyal to this configuration until well after WWII, such as Peugeot, until the 1960s.

A second way to get the noise down was to design a perpendicular conical wheel drive with helical teeth. The French steam car producer Bollée used this in 1896, whereas Citroën developed a conical wheel with arrow or fishbone-shaped teeth (which the French manufacturer then used as a logo: two arrow teeth, one above the other). Palloid, circular and other mathematically developed shapes were proposed, sometimes leading to fierce patent battles between companies because the end results were often much more similar than the calculations would suggest.

Emergence of the Hypoid Pinion Drive

One of the major manufacturers was the American Gleason Works in Rochester, Michigan, founded as far back as 1865 and since the 1870s the manufacturer of tooth cutting and fraising machines as well. It was Gleason that made the decisive step in this development (as we can observe from hindsight) by applying the Pluto Effect: in order to allow the passenger cabin to be constructed as low as possible (and thus attacking one of the declared advantages of the underslung worm version), the small pinion wheel was lowered. The resulting hypoid tooth shape (kinematically speaking a merger of the conical and worm gear shapes, resulting in the decisive advantage that the pinion could be made larger than the gear reduction would predict) was patented in 1927 [4-14]. Packard was the first in the United States in that year, followed by Bentley in the UK two years later.

Now, an interesting divergence in automotive culture occurred: American manufacturers opted for a larger displacement from the final drive's centerline (borrowing more from the worm wheel option and allowing more gliding friction and thus less noise, but more wear), but as car speeds increased further, the growing risk of tooth damage triggered a reversed *convergence* of both versions toward near-universality of a limited off-set of the pinion wheel after WWII.

But unconventional solutions always pop up in the history of car technology, such as the Pontiac Tempest's (1961) flexible prop shaft, which can be interpreted as an ultimate attempt to mobilize the Pluto Effect against the onslaught of the front-wheel drive in Europe: by proposing a solution in its own "grammar," so to speak, rear-wheel drive tried to offer an alternative that promised to avoid the costly conversion to front-wheel drive, to no avail, as we now know, turning the Pluto Effect in this case into a Sailing Ship Effect (Figure 4.9). After all, front-wheel drive promised to afford more space for the passengers, because it got rid of the prop shaft tunnel in the base of the cabin. It was this front-wheel drive, combined with a transverse engine, that did not need a perpendicular transmission of the energy flow from the engine to the driven wheels and thus enabled a new subspecies to develop. The final reduction in this type of drivetrain configuration revived the original cylindrical gears that automotive gearing started with more than a century ago (Figure 4.10).

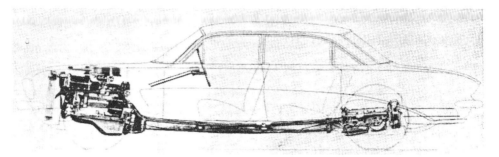

Figure 4.9 Pontiac Tempest (1961) with flexible prop shaft: An effort to postpone the shift towards front-wheel drive by mobilizing the Pluto Effect ([4-3], p. 64).

Conical wheels were also applied in the earliest differentials. Also called balance gear or jack-in-the-box, they competed with cylindrical gear wheels from an early date. When, in the 1920s, cars for the first time started to encounter traction problems because of ever increasing propulsion torques at the driven wheels that threatened to transgress the border of maximum friction between the tire and the road (thus causing instability while accelerating), engineers devised ways to increase the internal friction of the differential. They did so, for instance, by designing gearwheels with helical teeth, such as in the American Powerlok differential proposed in the 1920s. Multiplate clutches were added and even full-stop mechanisms in case the speed difference between the two outgoing wheels became too big. The development here was accelerated when, after WWII, four-wheel drive cars became popular and differentials had to be added between the two driven axles. Such systems became part of a general *automation* trend and were eventually included in electronic control systems that covered the entire drivetrain (Figure 4.10).

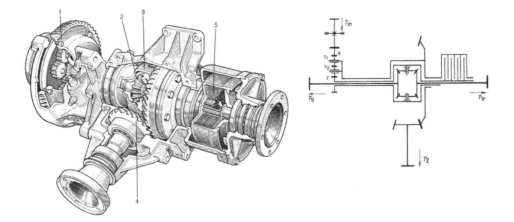

Figure 4.10 Differential of a four-wheel driven Lancia, with an epicyclic final reduction (1) and a perpendicular reduction to the rear wheels (4). The schematic on the right has been drawn according to internationally standardized symbols ([4-12], p. 426).

4.5 Universal Joints, Front-Wheel Drive, and the Reconfiguration of the Drivetrain

In the midst of this erratic evolution toward less noise, less weight, and lower costs, the transition from rear-wheel to front-wheel drive added extra complexity. The need to combine the propulsion and steering functions in one compact drivetrain configuration made the development of a true homokinetic solution into what technology historian Thomas Hughes has coined a *reverse salient*, a military term indicating a position in the front (at war, or in the struggle of engineers and managers to "advance" technology) that lags behind. The existence of a reverse salient implies that the lack of a solution is felt throughout the *state of the art*. As a consequence, everybody's attention is drawn to this point in an effort to solve the problem and enable further "progress" [4-15].

The first reflex, of course, was to miniaturize the double universal joint into a homokinetic arrangement (the solution proposed in pre-automobile times as explained in Section 4.1), but soon many independent inventors focused their creativity upon finding more compact solutions.

Jean Albert Grégoire

One of these inventors, French mechanical engineer Jean Albert Grégoire (1899–1992), may stand for a fourth type of automotive engineer (after Krebs, Bosch, and Kettering presented in the previous chapters). Backed by a rich friend, Pierre Fenaille, this champion on the 100 m and lover of what he himself called mechanical sports bought a garage and converted it into a research laboratory. There, the two friends developed a racing car that won the 24-hour races at Le Mans. Both driven by what Grégoire himself called "the passion of the automobile," they were attracted by the heroism and nationalism of the racing culture. But while Fenaille fostered extravagance ("what he wants is surprising the crowds"), Grégoire initially opposed his friend's proposal to go for front-wheel drive (a "crazy adventure," he called it later) in a culture that was fully geared to the classic drivetrain configuration. Not surprisingly, Fenaille won. During the subsequent decades, Grégoire not only patented a CV joint based on two simple gliding pieces that could easily be made on tooling machines (Fenaille's idea, as Grégoire later admitted), but he also became known as a pioneer of the light, aluminum body, a gas turbine car and an electric vehicle ([4-16], p. 9, 33, 107, [4-17], [4-18], p. 10).

Their Tracta joint (Figure 4.11) dominated front-wheel drivetrain technology for the next 40 years, but it may well have precipitated Citroën's bankruptcy in the early 1930s because the joint heated up under large angles and loads. In his memoirs, Grégoire acknowledges his error, which led to a "terrible catastrophe": at the last minute it was replaced by a double CV joint from Glaenzer. Nonetheless, Citroën adopted the Tracta joint on its all-terrain war vehicles since 1938. During WWII it also was applied to four-wheel drive military vehicles (in the in-board position, where smaller angles were needed), through the British daughter of the American Bendix Aviation Company ([4-4], p. 12, [4-16], p. 323, 329, [4-19], [4-20]).

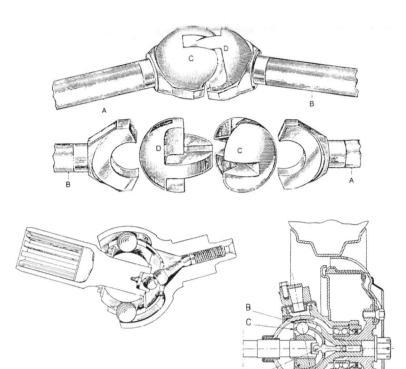

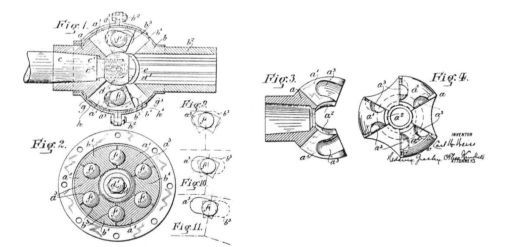

Figure 4.11 Tracta, Rzeppa, and Bendix-Weiss constant velocity (homokinetic) joints ([4-3], p. 69, 71, 73).

Bendix-Weiss, Rzeppa, and Delta Joints

As usual in the history of car technology, seen from an evolutionary point of view, when a problem is experienced as complicated, a *variation of solutions* develops. In this case, the inspiration often came from insights in the kinematic properties of force transmission under a certain angle. One solution, for instance, was to take the bevel gear reduction as a conceptual starting point (which was intrinsically homokinetic, of course) and replace the teeth by a set of balls in the same plane. Here, another French inventor (who immigrated to the United States), Carl W. Weiss, developed in 1923 an earlier quasi-homokinetic proposal from William A. Whitney (patented in 1908) into a true homokinetic solution. He did so by adding a self-centering element in the form of nonconcentric grooves fraised in the wall of the ball housing. An exclusive license to the Bendix Aviation Corporation enabled this first truly homokinetic ball-type joint to be produced in massive numbers (4 million) during the war. It was applied in jeeps, trucks, and armored cars. In Europe, the Bendix Weiss CV joint (see Figure 4.11) became popular after the war because it also allowed a small variation in length.

The third type of CV joint, the Rzeppa joint (called after a Ford engineer with a Czech background, Alfred H. Rzeppa, whose name is pronounced *sjeppa*), was a true sliding joint that also allowed an extremely large angle of 45 degrees (see Figure 4.11 and Figure 4.12). Patented in 1927, Rzeppa was accused of having copied the ball concept of the Bendix Weiss layout. After the war, however, it was acknowledged as a separate design of "a self-supported constant velocity universal joint which consists of an outer and inner race drivably connected through balls positioned in the constant velocity plane by axially offset meridionally curved grooves, and maintained in this plane by a cage located between the two races," as an automotive design handbook formulated it in the 1970s ([4-21], p. 145). Its production, however, was even more complicated than that of the Bendix Weiss version, which was the reason the American government at war time opted in favor of the latter. Thus, whereas Tracta and Bendix Weiss dominated war mobility, Rzeppa found its application after the war, starting with the Austin Morris Mini of 1959.

The fourth type is fully post-war, but consists of a revival of the pot joint encountered in Section 4.2: the tripot joint (also called Delta joint by its British manufacturer Birfield) was first applied in 1956 and then 1958 in the East German cars Wartburg and Trabant, in the Citroën DS since 1959, in 1963 in the Peugeot 204 (after which both Peugeot and Renault started their own production under Glaenzer Spicer license), while Chrysler and AMC in 1977 were the first American makes that converted to this simpler, cheaper solution, and that same year were accompanied by Nissan, Toyota, and NTN in Japan (Figure 4.12) ([4-4], p. 238, [4-22], [4-23], p. 7).

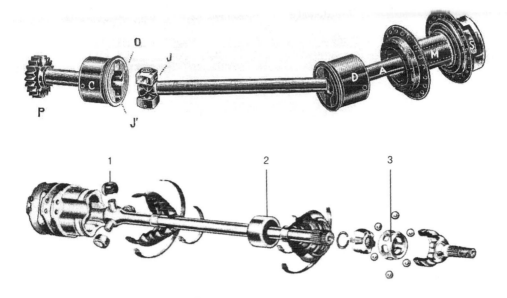

Figure 4.12 Early bipot joint from De Dion Bouton and later tripot joint (1) as applied by Citroën; 3 is a Rzeppa joint ([4-5], p. 108, [4-12], p. 340).

All in all, about 800 patents have been filed on the homokinetic joint; of these, 60 to 70 have been actually applied.[3] Efforts to make the drivetrain lighter have always, as a continuous background requirement, governed its evolution, but this became a real challenge with the increasing popularity of four-wheel drive systems in the 1980s and 1990s, when more differentials had to be added, because speed differences between front and rear axles had to be balanced. One of the solutions was the option to change to a different shaft material, such as mixtures of glass and carbon fibers embedded in a matrix of epoxy resin, as was done with the rally car Lancia Delta S4. Another solution was a combination of an inner tube of aluminum covered with a carbon fiber mantle, as has been tried by the American Dana Corporation ([4-12], p. 338–340). Their high costs, and perhaps also some fear about their long-term reliability, prevented them from being implemented in "normal" cars.

4.6   Conclusions

If we would analyze the history of automotive technology from the perspective and at the level of the component and the component group only (clutch, shaft, final drive), we would indeed have to conclude that the only real automotive innovation of the first half century was the CV joint. But such an approach would neglect a crucial level: that of the subsystem, the next-higher level of the ensemble of components and component groups that constitute the drivetrain (see Figure 1.1). It would also neglect the myriad

---

3. Personal notes of the course [4-24].

incremental innovations in gear shape design and materials that made the final drive of the end of the last century quite a different object than its pre-war counterpart. In the case of the drivetrain, the *translation* of pre-car technology to automotive applications seems to have been predominant.

Indeed, the nature of most of the innovations dealt with here is based on at least a century (if not more) of scientific, especially mathematical, practice at a structural level of the basic components such as the teeth of a gearwheel, the balls of a CV joint, and the bearings or case of a universal joint. Perhaps because of this generic level, where the automotive application is far away, "amateur" inventors (such as Carl Weiss) could play a decisive role, although they had to be well-versed in basic mechanical engineering, as Jean Albert Grégoire's career testifies. Technological evolution at such a deep level may not be very visible to even the skilled user: as we said in Chapter 1, the deeper one descends into the bowels of technology, the more difficult it is to relate changes at that level to societal changes, let alone explain them.

In the case of the CV joint as the major innovation within the drivetrain, however, this connection is easily made: most of the minor and major changes in this subsystem from, say, the early 1920s, and all changes in the joint itself were the consequences of the shift from rear-wheel to front-wheel drive, and this shift was triggered by what the manufacturers observed as the users' wishes. As the handbooks of the period testify, front-wheel drive was attractive for less-skilled drivers, while the skilled drivers also benefited from its introduction because of the extra space it enabled in the passenger cabin. Would this mean that the American car culture, where front-wheel drive encountered much more difficulty to get acknowledged as the configuration of the future, was less driven by consumer wishes? Did American engineers have more to say on the shape of the car than their European colleagues? And could they do so, because American motorists were less well versed in technology than their European counterparts? We do not know.

What we *do* know, however, is that also in this highly mechanical domain of the car (in which *weight reduction* seems to be the primary incentive for change), the same trends occurred that also drove previously described evolutions: *variation*, *divergence*, and *convergence* of separate (American and European) car cultures, as well as *automation*.

References

4-1.   A. Wegener Sleeswijk, "Reconstruction of the South-Pointing Chariots of the Northern Sung Dynasty; Escapement and Differential Gearing in 11th Century China," *Chinese Science* 2 (January 1977), 4–36.

4-2.   K. Kühner, *Geschichtliches zum Fahrzeugantrieb* (Friedrichshafen, 1965).

4-3.   Gijs Mom, "De aandrijflijn in historisch perspectief," in G. Mom and H. Scheffers, *De aandrijflijn* (Deventer, 1992), 19–80 (Part 3A of *De nieuwe Steinbuch; de automobiel; handboek voor autobezitters, monteurs en technici onder redactie en coördinatie van drs. ing. G.P.A. Mom*).

4-4. F. Schmelz, H.-Ch. Graf von Seherr-Toss, and E. Aucktor, *Gelenke und Gelenkwellen: Berechnung, Gestaltung, Anwendungen* (Berlin/Heidelberg/New York/ Paris/Tokyo: Springer Verlag, 1988).

4-5. Louis Baudry de Saunier, *Das Automobil in Theorie und Praxis; Band 2: Automobilwagen mit Benzin-Motoren; Mit einem Vorwort von Peter Kirchberg* (Leipzig: Reprintverlag, 1991) (1st ed.: 1901).

4-6. Erik Eckermann, *Nathan S. Stern, Ingenieur aus der Frühzeit des Automobils* (Düsseldorf: VDI Verlag, 1985).

4-7. Gijs Mom, "Orchestrating Car Technology: Noise, Comfort, and the Construction of the American Closed Automobile, 191 –1940," *Technology and Culture* 55, no. 2 (April 2014) 299–325.

4-8. E.R. Wagner, "Driveline and Driveshaft Arrangements and Constructions," in *Universal Joint and Driveshaft Design Manual* (Warrendale, PA: SAE International, 1979) (Advances in Engineering Series No. 7), 3–10.

4-9. H.L. Gleist, *Borg-Warner: The First 50 years* (Chicago: Borg-Warner Corporation, 1978).

4-10. R. Simons, *Geschichte der Automobile mit Frontantrieb* (München: Deutsches Museum, 1981).

4-11. Olaf von Fersen (ed.), *Ein Jahrhundert Automobiltechnik; Personenwagen* (Düsseldorf: VDI Verlag, 1986).

4-12. G. Mom and H. Scheffers, *De aandrijflijn* (Deventer, 1992) (Part 3A of *De nieuwe Steinbuch; de automobiel; handboek voor autobezitters, monteurs en technici* onder redactie en coördinatie van drs.ing. G.P.A. Mom).

4-13. http://www.luk.de (consulted on November 24, 2012).

4-14. A. Maier, *Zur Geschichte des Getriebebaues* (n.p., n.y. [1962]).

4-15. Thomas P. Hughes, *Networks of Power: Electrification in Western Society, 1880–1930* (Baltimore/London, 1983).

4-16. J. A. Grégoire, *50 ans d'automobile; la traction avant* (Paris: Flammarion, 1974).

4-17. J. A. Grégoire, *50 ans d'automobile; 2 la voiture électrique* (Paris: Flammarion, 1981).

4-18. Jean-Albert Grégoire, *Toutes mes automobiles; Texte présenté et annoté par Daniel Tard et Marc-Antoine Colin* (Paris: Ch. Massin, 1993).

4-19. E. R. Wagner, "Tracta Universal Joint," in *Universal Joint and Driveshaft Design Manual* (Warrendale, PA: SAE International, 1979) (Advances in Engineering Series No. 7), 127–140.

4-20. M. R. Clements, *King of Stop and Go—The story of Bendix: A history, 1919–1963 in South Bend, Indiana* (n.p., n.y. [South Bend: Bendix Aviation Corporation, 1963]).

4-21. F. F. Miller, D. W. Holzinger, and E. R. Wagner, "Rzeppa Universal Joint," in *Universal Joint and Driveshaft Design Manual* (Warrendale, PA: SAE International, 1979) (Advances in Engineering Series No. 7), 145–150.

4-22. J. H. Dodge, "Tripot Universal Joint (End Motion Type)," in *Universal Joint and Driveshaft Design Manual* (Warrendale, PA: SAE International, 1979) (Advances in Engineering Series No. 7), 131–140.

4-23. *Glaenzer Spicer 1838–1988* (brochure Glaenzer Spicer, Poissy, n.y. [1988]).

4-24. Gijs Mom, personal notes from the course *Automotive Technology* (Voertuigtechniek) at the HTS-Autotechniek, Apeldoorn, The Netherlands, by ing. G. Révèsz, 1978–1979.

Chapter 5
The Drivetrain: Multiplying Energy, De-Multiplying Speed

5.1   Introduction: Costs and Cheapness

Whereas the driveshaft, and especially its joints, became the realm of the creative inventor and the supplier, the transmission, like the engine, was seen as something to deal with by and within the car manufacturer. The reason was that the transmission and engine have a special interface toward the user, that is, special compared to the other component assemblies also adorned with an interface, such as the braking and steering systems. While users did not normally notice the developments taking place in the driveshaft and did not seem to be very much interested in the technical intricacies of steering and braking, they did notice the efforts of the manufacturers to try to get the small speed range of the engine spread over the much larger speed range of the car. Perhaps because of this, car manufacturers were reluctant to source out the development work on the transmission, out of fear that the car would lose its "character."

A car's "character" is a set of *functions*, realized by the motorist and the passengers but constrained by the car's technical *properties* as designed by the manufacturer.[1] Users became aware of this character through pedals (accelerator pedal, clutch pedal) and levers (ignition, mixture strength, transmission gear shifting) in an active way: they had to recreate the character of the car every time they drove in it, thus assembling the driver-car cyborg repeatedly, until the car became part of their motor nerves, so to speak (see Chapter 9 for more details on the distribution of skills over the human and the mechanical part of this cyborg).

It is no surprise, therefore, that it took a while before renowned independent transmission suppliers gained a foothold in the automotive world. This chapter focuses on their

---

1. For the distinction between properties and functions see Chapter 1.

story of production and costs. Although suppliers, like car manufacturers, are run by people, this chapter focuses more on the impersonal business side of automotive technology.

Before we deal with this, however, we first have to go into the evolution of the transmission itself, a seeming labyrinth of options, proposals, and solutions. Remarkably, it was here, because the internal-combustion engine fitted so badly in the energy requirements of the car, that the electric vehicle, left behind shortly after the beginning of the century and popping its threatening head up again for a brief phase shortly before and during the First World War, seemed always to be at the dark edges of the engineers' minds: many transmission proposals seemed to aim at mimicking the steam engine's and the electric's ideal of stepless driving energy deployment.[2]

From an evolutionary point of view, the mimicking goes even further back, to the pre-car times when steplessness was a matter of course in mobility, for horse, pedestrian, and bicyclist alike: they could in a continuous manner adjust their energy production to the required incremental changes in speed and load.

In Section 5.2 we will first explain the spectrum of options available to early car manufacturers. We will then look into the efforts to automate the motorist's speed change practice, a wish prevalent especially in the United States, giving rise to a *divergence* of automotive cultures in the industrializing, Western world (Section 5.3). The next section (Section 5.4) will be dedicated to the relationship between suppliers and manufacturers, with the German *Zahnradfabrik Friedrichshafen* (ZF) and the American Borg-Warner (BW) as examples. We will close this chapter with conclusions (Section 5.5).

5.2 The Transmission: How to Circumvent Gear Shifting

We will use the word transmission rather than gearbox to refer to the sets of properties enabling a torque multiplication (accompanied by a similar speed de-multiplication), because of two reasons. First, the evolution of car technology in this realm was much more than a question of gear wheels: other technologies (hydraulics, pneumatics, nongear mechanics such as friction discs) were also proposed and used for this purpose. The inclusion of a hydraulic coupling or torque converter in the automated version of the gearbox (see Section 5.3) is also a reason why this version has been called automatic transmission.

Second, the earliest gear transmissions did not come in a box: they were often directly suspended in the car's frame, a clear sign that early manufacturers considered the transmission as an inseparable part of the car structure (Figure 5.1). In other words, the use

---

2. In the case of the electric, this was a rather theoretical advantage, as the real electric vehicles were equipped with controllers that enabled a multistep (often up to eight steps) adjustment of the motor's torque to the car. See [5-1].

of the term gearbox is a typical, linguistic self-delusion from hindsight hiding the original, much wider spectrum of transmission versions.[3]

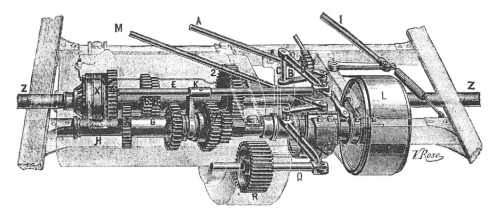

Figure 5.1 Gear transmission mounted directly in the frame ([5-2], p. 346, Fig. 201).

Adding a gear transmission in the car resulted in a *split of functions* of early belt and chain drives: these earlier transmission systems combined the functions of the bridging of distance (between the engine and the driven wheels) and multiplication of torque (through the use of different sizes of pulleys and sprocket wheels). In the case of gear transmission, distances between shafts had to be as small as possible, if only to gain weight and, indeed, to allow the gear sets to be housed in a box. This not only protected them better from street dust and allowed a more secure form of lubrication (by having the teeth splash through a pool of oil at the bottom of the box), but it also dampened at least to a certain extent the "horrendous noise" produced by the spur wheels with their straight, nonhelical teeth ([5-3], p. 45). The price for this "improvement" was the inclusion of a new contraption to bridge the distance between the output shaft of the transmission and the final drive (or between the engine and the transmission, if the latter was placed near the final drive), as we will see.

The Problem of Noise Production

It was the problem of irregular cutting of the teeth (causing noise) and the kinds of steel (causing tooth failures) of gearwheels that made engineers stick to the belt and the chain for a surprisingly long time. And when, slowly, production methods were adjusted to the "high-tech" requirements of the car, the differences in gearwheel quality between manufacturers and countries were enormous: the *module*, for instance (enabling more than one tooth per wheel being in contact with the other wheel, thus lowering the load on the tooth surface), was between 3 and 4 in Europe, whereas in the United States much coarser teeth were used (with a module in the order of 8 to 12) ([5-4], p. 47). As a matter

---

3. Some handbooks use the term transmission to refer to the entire transfer mechanism of the energy, including what we have called the drivetrain (see Chapter 4). We will not follow this habit.

of fact, it was the fledgling car industry that revolutionized general gearwheel production, and because of the fact that this took a while, it was only after 1918 when helical teeth were introduced that the gear wheel transmission started to become the *dominant design* (see Chapter 1 for this concept).

Another solution to attack the noise problem was the option, proposed as early as 1899 by Louis Renault in a patent of a three-step gearbox, to place the in- and outgoing shafts not next to but behind each other, and connecting both through a dog clutch to enable what he called a *prise directe* (a direct drive) for the highest step. In a design from about 1917, Renault also used constant-mesh gearing to combat another source of noise: the clashing of gearwheels of the sliding-mesh type of transmission, despite efforts to make the gear change easier by tapering off the sharp edges of the wheels (Figure 5.2).

Constant-mesh, the phenomenon that gear wheels were engaged all the time and had to be made to work by connecting them to their shafts, seems to have been used already in the steam car period, but around the *sliding-mesh* principle (where the wheels had to be shifted over their shaft in order to be engaged) an entire macho car culture emerged, audible through the noisy clashing of the gearwheels when their respective speeds were not yet fully equalized, and the skilful double-declutching necessary when choosing a lower gear step. Car driving during the first phase was defined as sport, and sport had to be noisy and violent.

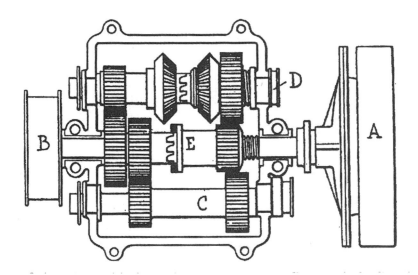

Figure 5.2 *Prise directe* (1899) and gearbox with constant mesh gearing by Louis Renault. In his patent, a third meshing option was applied, which could be called swivelling mesh: the gears were engaged through the swivelling of the eccentric shafts, but these were insufficiently supported in their casings, causing a lot of noise ([5-3], p. 10).

Gear shifting had to be done by a lever, either through a sequential pattern or through an H-shaped pattern. The latter was introduced as early as 1895 by Wilhelm Maybach (the first designer of Mercedes, 1846–1929) in a four-step sliding mesh gearbox (Figure 5.3). Mostly, the gearbox was mounted directly to the engine, the clutch encased between the two; only in sport cars was the weight of the gearbox shifted to the rear axle for a better balance, thus creating what has been called a transaxle.

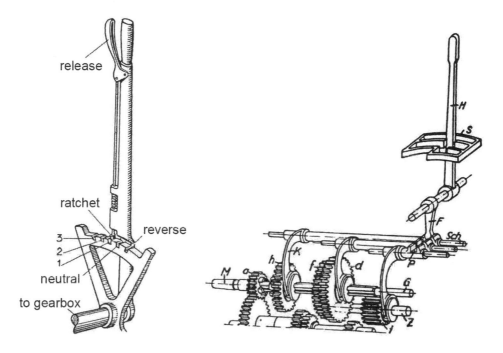

Figure 5.3 Sequential and H-shaped gear shifting patterns, the latter introduced by Wilhelm Maybach as early as 1895 in a four-step sliding-mesh gearbox ([5-4], p. 50).

In a way, we see here the *domestication of the car* at work at a moment (the 1910s) that members of the middle class, with their modest income compared to the aristocrats and high bourgeoisie of the first phase and their ideology of civilized citizenship, became the new target of the car industry. The very wealthy motoring pioneers had approached the car as a machine of pure and fast sport. Gear shifting was experienced as a masculine, noisy part of that sport, as we have seen. The new customers, often less versatile in automotive technology, started to formulate different requirements in their affordable family cars (see also Table 1.1 for an overview). Quiet, efficient running was one of these requirements [5-5].

The noise and longevity problems apparently seemed so insurmountable that radical alternatives were investigated, such as epicyclic (or planetary) gearing, which not only spread the load over more teeth (allowing a very dense energy flow), but also because their constant mesh produced less noise. This was an American specialism, starting with the Duryea brothers around the turn of the century, but also Cadillac applied a double epicyclic gear set, adorned with two conical clutches for gear shifting. It was

Henry Ford who, with his "compound epicyclic gear train," brought this type of gearwheel sets into the mainstream by offering two fixed gear steps in the Model T's planetary gearbox with its huge production figures (from 1908 to 1927). The Model T thus distorted the average type based on a comparison of individual models, which clearly was in favor of spur gears. The compound solution (combining two sets of planetary gear wheels) had been developed before Ford by the British car manufacturer Frederick W. Lanchester (1868–1946) ([5-4], p. 47–53).

Early CVTs

Slowly, gear shifting practices on both sides of the ocean started to reveal a *divergence* between a fledgling American car culture and a bit older European car culture. This can clearly be observed in the development of the number of gear steps on both continents. From an evolutionary point of view, the increase of steps from two or three to four around 1910 can be interpreted as an effort of the population (in the Darwinian sense) to mimic the continuously variable adjustment of speed and torque of pre-car animal and human traction, as well as the broad speed ranges of the electric and steam alternatives. American manufacturers, however, did not follow this strategy of increasing the number of steps; instead, they opted for ever more "elastic" engines (as we saw in Chapter 2). They even returned to the three- and even two-step boxes, thus enabling their customers to perpetuate their bad habit (as American engineers saw it) to "stay always on top," or refrain from gear shifting and accelerate from 5 km/h to the car's top speed in the highest, direct gearing. It was the ever larger American engine that allowed the gearbox to decrease in size ([5-3], p. 45–47). This example shows again that technical development does not (always) follow a linear path but in this case oscillates between more and less complexity.

The co-construction of early automotive technology (meaning: constructed by producers *and* users, which, from an evolutionary point of view, coincides with a co-evolution of properties and functions) was nowhere so pronounced as in the realm of the transmission. In fact, it also is expressed in the frantic search for radical alternatives to the stepped transmission in order to take the burden of gear step choice from the motorist's shoulders. The amount of options investigated was stunning, especially in Europe (Figure 5.4). Hydraulic, pneumatic, mechanical, electric: all technological strategies to enable a continuously variable transmission (CVT) were attempted.

Early examples were the hydrostatic drive, such as proposed in the German *Hydromobile* (Pittler, 1898), or Louis Renault's design with two hydraulic plunger machines to be controlled by hand. An electric CVT was developed by Ray Owen with his Car of a Thousand Speeds (United States, 1914). Mechanical CVTs came in all kinds of options as well: flat belt on single conical pulleys (Mahout, France, 1904–1909) or on double-conical pulleys (Buchet, France, 1899). A CVT version that enjoyed some attention from transmission specialists in Europe, especially for lighter types of cars, was the friction disc, equipped either with one or two sets of (two) discs (Figure 5.5) ([5-6], p. 307–309).

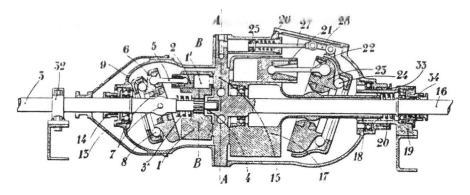

Figure 5.4A

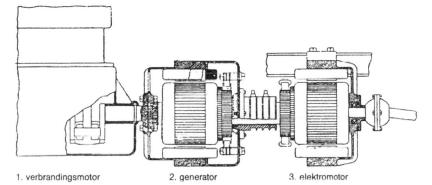

1. verbrandingsmotor   2. generator   3. elektromotor

Figure 5.4B

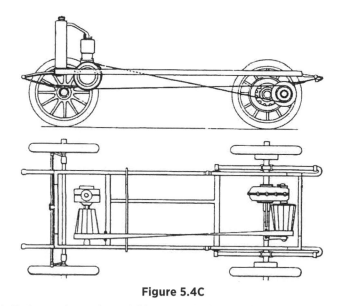

Figure 5.4C

Figure 5.4 Early continuously variable transmissions. A: hydrostatic (Renault, 1905);
B: electric (Owen, 1914); C: flat-belt conical pulley (Mahout, 1904); ([5-7], p. 38–41).

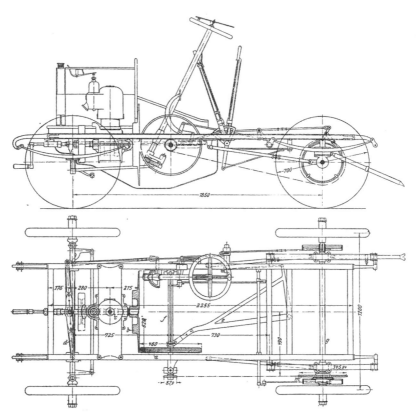

Figure 5.5 Friction disc CVT on a German car of the Nürnberger Motorfahrzeugfabrik Union ([5-6], p. 307).

In the United States, especially during the 1920s, the attention seemed to go to more sophisticated mechanical CVTs for heavier cars, such as the Hayes transmission, a toroidal traction drive, called so after the shape of the discs in which special massive rollers were moving and whose varying positions dictated the continuously variable changes in torque multiplication. American Frank Hayes offered this design to General Motors, which worked on it between 1928 and 1936 (and tested it in 30 Buicks), but in 1932 he left for England, where Austin applied it between 1933 and 1935 in its models 16 and 18 under the name of Self Selector. The fact that "Boss" Kettering himself (see Chapter 3), in a specially created General Motors Research Corporation (founded 1920), was asked to develop this option, shows how desperately car manufacturers were looking for a contraption that would allow American motorists to maintain their "bad habit" of "staying on top." ([5-4], p. 45, [5-3], p. 67–68, 117–122).

The *Journal of the Society of Automotive Engineers* (SAE) abounds with proposals to circumvent gear shifting, including very complex mechanical "variable freewheel transmissions" and "inertia transmissions" developed by individual inventors, such as the Brazilian Robert Dimitri Sensaud de Lavaud (who worked at the factory of the French Gabriel Voisin since 1921) or the British M. Constantinesco (Figure 5.6).

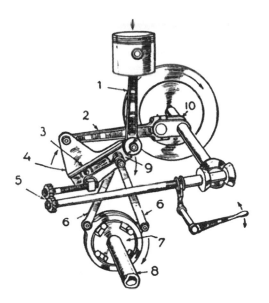

Figure 5.6 Inertia transmission (continuously variable freewheel transmission) from M. Constantinesco ([5-4], p. 47).

Although none of these proposals made it into the mainstream (where the spur wheel with helical teeth still reigned supreme, certainly after production of the Model T was discontinued in 1927), the freewheeling idea suddenly got hold of the fantasy of car engineers: by 1931 nearly half of all American cars were equipped with freewheeling transmissions in a desperate effort to lower the necessary skills of handling the transmission ([5-3], p. 82). Now that car speeds started to increase because of the highway building programs in the 1920s and motorists seemed to be stuck with the stepped gearbox, they might as well shift easier because of the freewheel in one of each set of two gears to be engaged. In this case, however, motorists voted with their feet: they did not like to lose the possibility to brake on the engine, which was the case when freewheels interrupted the energy flow in one direction. As a result, freewheeling disappeared from the market by 1936, a nice example of an early hype among automotive engineers rejected by car users, and again an example of nonlinear development of automotive technology ([5-8], p. 26).

5.3   The Automatic Transmission: Diverging Car Cultures

Just when it seemed that car manufacturers had to give in to the technology's stubbornness that seemed to prevent them from developing a shiftless and stepless transmission, a single inventor, Earl Avery Thompson, proposed to several manufacturers to try out his idea of an automated synchronization of gearwheels while shifting. Sometimes, history develops through the intervention of one person, who finds just the right circumstances to gain disproportional influence on the course of developments. It was the Cadillac division of General Motors that took up the challenge in 1924, because of Thompson's promise that his solution would halve the shifting time compared to the double de-clutch procedure. By 1928, when Cadillac and La Salle introduced Thompson's Synchromesh (equipped with a conical clutch, which decelerated the gearwheel when clutching; see Figure 5.7), Thompson had meanwhile been made assistant chief engineer at Cadillac.

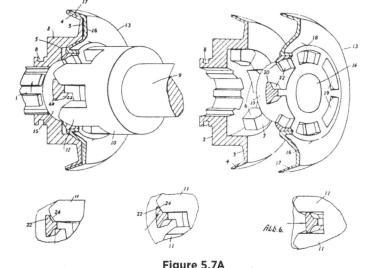

Figure 5.7A

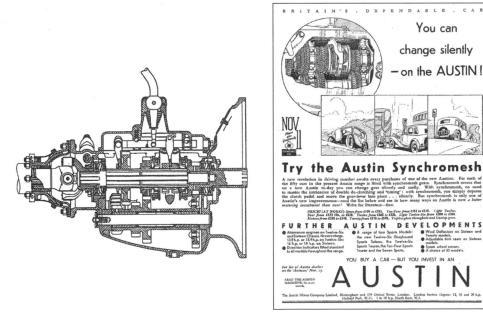

Figure 5.7B   **Figure 5.7C**

Figure 5.7 Thompson's synchromesh patent and application in a "synchronized" gearbox in the 1928 models of Cadillac and La Salle; C: the synchromesh was presented as a remedy against car noise. *Sources:* A: ([5-4], p. 57); B: ([5-3], p. 89); C: ([5-9], p. 20, picture used courtesy of *Autocar*).

Thompson became instrumental in the next major step toward automation: the introduction of the hydraulically controlled shifting of sets of epicyclic gears. Such gear sets consisted of a ring wheel with inner teeth, enclosing a number of satellite wheels turning around a central sun wheel. Depending on which of the wheels was fixed, different reductions could be realized, thus enabling a dense and compact solution. Even Ford had abandoned the more costly planetary principle in 1927, but now that production methods of gear teeth had considerably improved, even for the difficult inner teeth of the ring wheel, it was revived as another road toward the ideal of "crashless shifting" (or no shifting at all). The first American make to reintroduce them was Auburn in 1931 ([5-3], p. 88, 111).

The dilemma facing the manufacturers was to adjust the *properties* of the transmission to enable a new *function* the American motoring public of the 1910s and 1920s was discovering: long-range, high-speed tourism. High-speed tourism suddenly seemed to be enabled by a frenzy of road building activities starting immediately after the First World War.[4] One of the ways to reach that goal was through the construction, by leading American engineers, of an ideal type (see Chapter 1) of the new car. As early as 1925, Herbert Chase, one of the senior SAE members, gave a presentation at one of the society's many conferences on "How passenger cars can be improved," presenting a picture of an idealized car in which an "Infinitely Variable Transmission" was integrated, adorned with the comment: "No Gear-Shift Required."

The "Transmissionless" Ideal

How can the construction of such an ideal be explained? The answer to this question lies hidden in the specific American car culture that had meanwhile evolved during the 1910s. Supported by their very powerful engines, American motorists had adopted the custom of driving in the city in one gear, which during the 1920s led to proposals by engineers to get rid of the gearbox altogether and develop a "transmissionless car." As we have seen, many engineers who saw this solution (theoretically found in cars with steam and electric traction) as a step too far openly flirted with the second-best option of a CVT ([5-11], [5-12], [5-13], p. 16). This formulation of an ideal type in the form of a transmissionless or CVT car was not a vague fantasy; it had very real repercussions during the design process. For instance, a test drive all across America with the gearbox fixed in the highest position showed that a transmissionless car was possible, at least in the hands of a skilled driver ([5-14], p. 10). According to some engineers the CVT approached the transmissionless ideal because it "eliminates 90% of gear shifting," if combined with a very elastic engine ([5-15], p. 26).

---

4. See for the following paragraphs especially: [5-10].

The enormous effort invested to create a kind of electrified gasoline car (in the sense of mimicking the electric's properties within the vocabulary of the combustion engine and transmission, a typical example of the Pluto Effect) on both sides of the Atlantic Ocean (and especially in General Motors' central laboratories) ended in a victory of the stepped transmission, but in a radically altered (epicyclic) form and added with a CVT-like contraption called a hydraulic coupling or a torque converter.

It may not have been a coincidence that several engineers working on novel CVT versions were electro-technical engineers. The hydraulic CVT was originally developed for ship propulsion before WWI by German engineers. One of its inventors, Hermann Föttinger of the Vulkan shipyards in Stettin (the later Gdansk in Poland), called it a hydraulic transformator. Hermann Rieseler, a cooperator of Föttinger, was the first who connected it to an epicyclic gearbox. And although the system was first tested in 1928 in a Mercedes, its low efficiency and high costs prevented it from becoming a mainstream application in Europe, with its small cars and high-revving engines.

It became a mainstream application, however, on the other side of the ocean, where Earl Thompson and his team at General Motors developed it further under highly secretive conditions (the project's code name was Military Transmission). It was first developed into the Automatic Safety Transmission for Buick (which mounted it in 30,000, or 7%, of its cars between 1937 and 1939), then as Hydra-Matic, applied in Oldsmobile (1940) and Cadillac (1941). Chrysler offered its Fluid Drive (a fluid coupling for starting, with a conventional gearbox for the higher speeds) in its top 1939 models. In other words, the CVT function was only created at low speeds by a torque converter. Remarkably, Thompson took the control philosophy for the rest of the transmission from the toroidal CVT, developed earlier under Kettering, a nice example of *crossover* (or *migration*) or *functional shift*. During the war, the new system was built into tanks, and by 1952 3 million units had been sold for passenger car application, also by GM's competitors ([5-3], p. 122–130, 136–140, 154, 170–172, [5-8], p. 30–33).

Thus, slowly, a "typical" American technology evolved, typical in comparison to the European example, so much so that it hindered American export to Europe. Next to the automatic transmission, engine size seemed an expression of this *diverging* subculture. But while American engineers were convinced that large engines were "what the public wants," because the priority of the American users was "comfort," and learning to shift gears was certainly no part of this, the British magazine *Autocar* claimed that "the industry, more than the user, is responsible for the top gear craze in America." Driving always in the highest transmission gear had been advertised so frequently and massively, they claimed, that "it is now almost a cardinal sin to change down" ([5-16], p. 181).

However this may be, on both sides of the ocean manufacturers had to enable the new type of car use on highways and freeways (in the terminology of our evolutionary approach: they had to design new or adjust existing *properties* affording new *functions*). They did so by introducing overdrive systems, first integrated in the final drive assemblies, later added to the stepped gearboxes as a fifth step. It started with Rolls Royce as early as 1907, which had a direct connection (*prise directe*) in its third step, and which added a real (speed) multiplying step as a fourth step. A separate overdrive box was added to the famous Maybach Zeppelin of 1926, which could cruise on the *autobahn* with an engine speed reduced to one-third of its maximum speed. The principle was taken over by Mercedes from 1930 ([5-4], p. 54–55). In the United States, a boom in overdrive system occurred from 1933.

5.4   Transmission Manufacturers and Automotive Production

It is understandable that in an uncertain atmosphere of trying to match *properties* with the perceived and expected changes in *functions*, car manufacturers wanted to keep the development of their transmissions in their own hands.[5] This was especially important as traditional gearwheel manufacturers had developed a different quality culture. Cars, American manufacturers realized once middle class motorists got interested in this type of mobility, had to be produced in large series, where quality and costs always were in a tense relationship. Therefore, it took quite a while until independent gearbox suppliers gained a foothold in the industry.

Zahnradfabrik Friedrichshafen and Borg-Warner

The German *Zahnradfabrik Friedrichshafen* (ZF), for instance, founded by Count Ferdinand von Zeppelin (1838–1917) of airship fame, offered a fully developed gearbox with constant-mesh, dog clutches, and synchronization to the car industry and found the latter not particularly interested. Car manufacturers judged the ZF designs, based on a precision technology for the production of so-called involute teeth by Swiss engineer Max Maag, too good and too expensive. The systemic character of the car system is illustrated very well in this case, as the car industry appeared incapable of mounting the gearwheels in a solid enough way to prevent the gearbox from producing noise. Instead, ZF managed to sell its boxes to the just emerging aviation industry, where a different quality philosophy reigned, more akin to ZF's own point of view.

Only from 1924, with a deliberately simplified construction realized after a careful study of American Fordist production practices, ZF managed to sell its *Einheitsgetriebe* in unsurpassed numbers (300,000), a nice example of a *nonlinear* development not

---

5. The following paragraphs are based upon ([5-4], p. 58–62).

primarily aimed at increasing "perfection." It will not have been a conscious decision, but the name of the box resembled communist and socialist parlance (*Einheitsfront*, united front) as if to suggest that the new product was anti-elite. Soon thereafter, ZF's European fame was definitively established when it developed, in 1929, its *Aphon* gearbox, a four-stepped gear system with helical teeth (*a-phon* being Greek for without sound), followed in 1932 by a fully synchronized gearbox inspired by American practice.

In the United States it was the Borg-Warner conglomerate that developed into a pivotal position, but as in Europe, not without a long struggle of adjustment to the requirements of automotive technology, albeit along a different trajectory. At the moment that car manufacturers started to change from assemblage to in-house production four suppliers started to merge: Borg & Beck, Mechanics, Marvel, and Warner Gear. Borg & Beck produced single dry-plate clutches, Mechanics made universal joints, Marvel was a carburetor factory, and Warner Gear had produced its first gearbox in 1909 and like ZF had to change to a standardized production regime in 1921.

Within a year of its consolidation, five other companies joined, among them Ingersoll, producer of diaphragm springs (see Chapter 4) and Morse Chain, soon to become world famous because of its "silent chain." The banks and other investors proved right: while the Depression of 1929 and following years made American car production collapse to one half, BW survived through the selling of Norge refrigerators, on another market. The competition between the companies was legendary, but according to the official Borg-Warner history this resulted in some radical innovations such as the synchromesh, the overdrive, and the universal joint turning in needle bearings. But BW nearly missed the automatic transmission, as its subsidiary Warner Gear claimed as late as 1938 that it was impractical and inefficient. After WWII, its transmission became a big success. In 1977 Bosch took a 10% share in BW.

European Refusal to Automate

After the war, the European motorists' reluctance to follow the Japanese example of embracing the American automatic transmission was a constant source of frustration to European car manufacturers. They tried to lure their customers to automation, as it were, by introducing semi- or fully automated clutches in the late 1950s and early 1960s. These came in all sorts: mechanically, hydraulically, and electrically operated single dry-plate clutches as well as centrifugal segment clutches (Figure 5.8; also see Chapter 4, Figure 4.7). When the German car manufacturer Borgward started to offer an automatic transmission in 1952, and Daimler-Benz, Renault, and others followed in the early 1960s (more aimed, it seemed, at the American market than at the European), American manufacturers began building them in factories in Europe to reduce costs [5-17], [5-18].

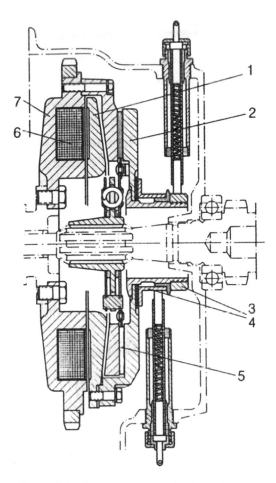

Figure 5.8 Electromagnetic clutch Ferlec from Ferodo-France (Renault, 1955) ([5-4], p. 44).

The perceived "desire of the masses for an automated mechanism" may also have influenced the Dutch manufacturer DAF to bring out its first passenger car (1958) exclusively equipped with an automatic CVT, based upon the V-belt principle with expanding pulleys. After an initial success (which brought the percentage of automatic transmissions on the Dutch market for a short period to heights unparalleled in Europe), the Variomatic was ridiculed by a part of the fast motorizing Dutch as a *"truttenschudder met jarretel-aandrijving"* (a maid shaker with garter propulsion), the latter addition being a reference to the rubber-belt transmission. The joke is revealing: it was ushered at a moment that women were just beginning to motorize on a massive scale [5-19].

Analysis after analysis, comment after comment, the major Dutch trade journal *Auto- en Motortechniek* was dedicated to the motorist's refusal to automate. Three arguments against automatic transmissions existed, one journalist summed up: technical problems of adjusting the complicated machines to the small European engines, causing high power losses (and high fuel consumption) and cars that behaved "lazy"; problems of production costs, causing an average price increase of the car in the order of 15 to 22%; and psychological problems, caused by a fear of losing control over the handling of the car. By the mid-1970s Dutch journalists pointed at the development of new, more efficient transmissions designed for use on European cars to counter the first argument; they pointed at a decrease in extra costs to less than 10% of the average car price; and they kept treating the psychological resistance as being based on "unfounded prejudices" ([5-20], p. 53, [5-21], p. 535). This case shows how automotive journalists may be considered to be *intermediaries between production and consumption*, but they often (and certainly the more technically educated kind as found in the Dutch mainstream technical journal) consider themselves representatives of "the technology" (the manufacturers) rather than the users. The latter should be "educated" to rational behavior.

And yet, perhaps reminded of his (and increasingly: her) "culture" by the first energy crisis, the Dutch motorist kept refusing to switch over to the new transmission, just like most other European users. Despite the reflex among contemporary journalists with an engineering background to explain this by pointing at the higher costs, it seems that cultural resistance, especially fear among masculine users to lose control, has played a role, too. Why male American users did not seem to be haunted by such fears remains to be investigated. Whatever the reason, in 1965, only a half percent of European cars were equipped with an automatic transmission, and even today its diffusion is still remarkably modest in comparison to the United States and Japan. Although European production figures doubled between 1982 and 2000 (from 8.2 to 16%) and optimistic (as they usually are) market watchers overestimated the future growth, claiming that in 2010 half of all cars would have automatic transmission, the production figures only doubled again to 31.2% in 2010, half of this meant for export. The European average in registrations is less than 10%, with Switzerland the most "American" country, with nearly a quarter of its national car fleet equipped with automatic transmissions.[6]

Manual versus Automatic

European motorists stuck to their manual transmissions, which, since the beginning of the 1950s, became expanded with a fifth gear set, and nowadays even can be had with six, seven, or more steps: whether stepped or stepless, transmissions are still mimicking their CVT ideal, thus realizing an ongoing Pluto Effect.

This example of the Pluto Effect is illustrated in Figure 5.9, where the function of steplessness is shown to be taken over by the mainstream technology (the stepped manual gearbox) in an effort to diminish the differences with the competing technology, the

---

6. [5-22], [5-23], [5-24] All data are from Automotive Industry Data; the 2006 registration data are from Polk Automotive Intelligence.

CVT. This was and is realized by introducing more steps (in other words: in the "grammer" of the stepped gearbox), but lately also by electronically masking the tiny shock previously informing the driver that the transmission was shifting.

Conversely, in a culture where manual shifting is highly appreciated (such as in Europe), CVTs are electronically reshaped into stepped transmissions, even if, from an engineering point of view, this seems absurd. The core of the Pluto Effect is that this interartifactual transfer takes place through the functions, not (necessarily) through the properties, as the stepped pattern is recreated through electronic control algorithms while the stepless internal properties of the CVT remain in place.

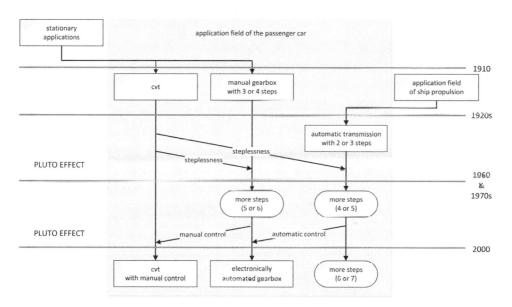

Figure 5.9 Dynamic diagram showing the Pluto Effect: manual and automatic transmissions mimicking the CVT by increasing the number of their gear steps and the reverse effect of simulating manual control.

5.5   Conclusions: Expectations and the Distant User

The case of the transmission helped unearth two important tendencies in automotive technology history: the *divergence* of automotive subcultures and the insight that what we tend to call improvements should be analyzed in a much more careful, and *non-linear* way.

Whether the *bifurcation* (to use a Darwinian term) into national and even continental car cultures is an expression of a kind of innate exceptionalism of the cultures under consideration, we leave unresolved for the moment (but see Chapter 14). We cannot deny, however, that properties and functions co-evolved into clearly distinguishable cultures of automotive technology, aimed at comfort on the one side and sportiness on the other. Such differences not only are visible under the hood, for the engineers and knowledgeable users, but they become elements of the automotive experience through the interfaces between driver and technology, most particularly the steering wheel, the accelerator pedal, the brake pedal, and the clutch pedal (or its absence, in the case of automatic transmission), and, most of all, the transmission actuation lever (or its absence). The increasing sensitivity for comfort has many aspects: it is expressed in the efforts to diminish noise, but also in the tendency to *driver deskilling*, such as in the case of gear shifting.

Automation can be seen as the technical counterpart of deskilling. It can be interpreted in several ways. One way is to see it as part of a struggle between engineers and a certain, knowledgeable type of users. Engineers actively (and successfully) pushed these users away by integrating an increasing amount of user activities into the properties. This process of automation, which is still going on, began when the starter motor was introduced and the delicate process of starting by hand crank was replaced by an electric motor controlled by a push button. It was further refined, to give only one of many examples, when the levers at the steering wheel, meant to control the fuel-air mixture and the spark ignition advance, were replaced by a fully automatic centrifugal and vacuum-operated system. And in the United States it reached its provisional zenith in this period with the taking over of the gear shift procedure by the automatic transmission. We will come back to the automation phenomenon extensively in Chapter 9.

Whether such changes, incremental or radical, are *improvements* remains to be seen and largely depends on the position one takes in these struggles, in other words: they are socially constructed. Was ZF's decision to simplify the gearbox into an *Einheitsgetriebe* an improvement? Not from the perspective of "increasing perfection" as an engineers' Utopia. Was the decision to opt for the rack-and-pinion steering over the much more sophisticated indirect steering boxes (as we will see in the next chapter) an improvement? No doubt it was, especially if one takes the workings of the Pluto Effect into account, which made sure that the simplification underwent all subsequent phases of material and constructive sophistication that characterized the much more complex competing solutions. Whether a technology is an improvement, in other words, depends on the goal one formulates: automotive technology is a compromise of engineering, marketing, and cultural considerations in an always changing mix. This mix takes shape primarily at the level of the system, the arrangement of component assemblies, and its properties are tuned to dynamic behavior within multiple contexts, of low speed (city traffic) and high speed (highway driving). Automotive technology is the hardware side to *afford* these types of dynamic functions. Technological improvements, thus, follow a meandering, bifurcating, nonlinear trajectory. Perfection is Utopia, part of the world of *expectations*: they drive our behavior, but they are not the same as the technology proper, and should be clearly distinguished as such.

References

5-1.   Gijs Mom, *The Electric Vehicle; Technology and Expectations in the Automobile Age* (Baltimore: Johns Hopkins University Press, 2004).

5-2.   Louis Baudry de Saunier, *Das Automobil in Theorie und Praxis; Band 2: Automobilwagen mit Benzin-Motoren; Mit einem Vorwort von Peter Kirchberg* (Leipzig: Reprintverlag, 1991) (1st ed.: 1901).

5-3.   Philip G. Gott, *Changing Gears: The Development of the Automotive Transmission* (Warrendale, PA: Society of Automotive Engineers, 1991).

5-4.   Gijs Mom, "De aandrijflijn in historisch perspectief," in G. Mom and H. Scheffers, *De aandrijflijn* (Deventer, 1992), 19–80 (Part 3A of *De nieuwe Steinbuch; de automobiel; handboek voor autobezitters, monteurs en technici onder redactie en coördinatie van drs. ing. G.P.A. Mom*).

5-5.   Gijs Mom, *Atlantic Automobilism: The Emergence and Persistence of the Car, 1895–1940* (New York and Oxford: Berghahn Books, forthcoming).

5-6.   Arnold Heller, *Motorwagen und Fahrzeugmaschinen für flüssigen Brennstoff; Ein Lehrbuch für den Selbstunterricht und für den Unterricht an technischen Lehranstalten aus dem Jahre 1912* (Moers: Steiger Verlag, 1985) (reprint of 1922 edition, 1st ed.: 1912).

5-7.   Gijs Mom, "De complexe aandrijflijn in historisch perspectief," in G. Mom and H. Scheffers, *De complexe aandrijflijn* (Deventer, 1992), 19–80 (Part 3B of *De nieuwe Steinbuch; de automobiel; handboek voor autobezitters, monteurs en technici onder redactie en coördinatie van drs.ing. G.P.A. Mom*), 17–62.

5-8.   Byron Olsen, "The Shift from Shift to Shiftless: Transmission Advances in U.S. Cars (1929–55)," *Automotive History Review* nr. 46 (Fall 2006), 25–40.

5-9.   *Autocar* (November 3, 1933).

5-10. Gijs Mom, "'The Future Is a Shifting Panorama': The Role of Expectations in the History of Mobility," in Weert Canzler and Gert Schmidt (eds.), *Zukünfte des Automobils; Aussichten und Grenzen der autotechnischen Globalisierung* (Berlin: edition sigma, 2008), 31–58.

5-11. Edward B. Sturges, "A Mechanical Continuous-Torque Variable-Speed Transmission," *Journal of the Society of Automotive Engineers* (July 1924), 86–92.

5-12. P. M. Heldt, "Some Recent Work on Unconventional Transmissions," *Journal of the Society of Automotive Engineers* (July 1925), 127–141.

5-13. "What We Really Need Is a Prime Mover with the Torque Advantages of Steam." *Journal of the Society of Automotive Engineers* (January 1940).

5-14. *Journal of the Society of Automotive Engineers* (July 1924).

5-15. *Journal of the Society of Automotive Engineers* (April 1939).

5-16. W. F. Bradley, "Automobilism in America: How European Influence in Engine Design Is Affecting American Designers," *Autocar* (July 31, 1920), 180–182.

5-17. A.Y., "De massa motorisering in Europa en de automatische transmissie (II)," *Auto- en Motortechniek* (1963), 150–154.

5-18. W. Beel, "De automatische gangwissel historisch gezien," *Auto- en Motortechniek* (1970), 514–526.

5-19. K. Revers, "Kapers op de kust van D.A.F.," *Auto- en Motortechniek* (1962), 517.

5-20. H. Bouvy, "De toekomstige onvermijdelijkheid van automatische transmissies," *Auto- en Motortechniek* (1969), 49–53.

5-21. H. Bouvy, "Over automatische transmissies en hun actualiteit," *Auto- en Motortechniek* (1974), 534–543.

5-22. Wim Oude Weernink, "Europa blijft 'sportief' schakelen," *NRC Handelsblad* (February 3, 2006), 10.

5-23. "Helft Auto's in 2010 Voorzien Van Automaat," *rai voorrang* 5 No. 8 (November 14, 2001), 1.

5-24. "Automatische transmissie schakelt op," *Go! Mobility* 2 No. 2 (April 2012).

Chapter 6
The Chassis: Getting Around the Corner

6.1 Introduction: Pre-Car Developments

From an engineering point of view, the properties of the chassis are quite simple. It has to carry the systems described in the previous chapters plus some extra systems needed to move: suspension, braking, steering. And oh yes, the passengers also need some space.

From an evolutionary point of view, the situation for contemporary engineers also looked quite simple: as they needed to invest so much energy in constructing around, so to speak, the electric propulsion, they must have been glad to borrow most of the chassis basics from two earlier road vehicles with animal traction: the bicycle and especially the horse-drawn carriage. From an evolutionary perspective, the automobile emerged by getting rid of human and animal traction (Figure 6.1).

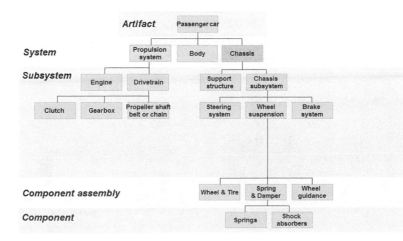

Figure 6.1 Structural diagram of the car, focusing on the chassis.

Wheels and Brakes

And yet, it would inflate the evolutionary metaphor a bit too much if we would argue that the wheeled vehicle evolved from the sled, as has sometimes been done in grand histories of mobility [6-1]. For such a claim, pre-modern technological practices were too dispersed. But the sled and cart, as archetypes of mobility, must often have been very present in ancient mechanics' mind. How else to interpret, for instance, the idea of a horse-carriage brake shaped as two curved wooden poles sticking underneath the carriage and the rear axle and used while descending a mountain slope, which lifted the rear axle and turned the backside of the carriage into a sled? Nowadays, engineers would say that such designers played with the difference in energy loss between gliding and rolling friction, and would describe the development of the wheel as a continuous process of friction decrease, by changing from wheels fixed to a rotating axle to wheels rotating on a fixed axle, or by changing lubrication from water to animal fat, or by lowering the weight of the wheels by changing them from wooden discs into spoke wheels.

Pre-car developments in this realm included the artillery wheel, in which the ends of the wooden spokes are tapered such that they can be mounted close to each other, a solution probably applied for the first time by steam bus pioneer Walter Hancock in 1839.[1] When applied to horse-drawn carriages and carts, wheel diameters were often chosen as a function of the load: bigger wheels would be mounted on the rear axles; the smaller front wheels also gave the axle more space to be turned around the central pivot.

Braking on or near the wheels, too, was developed well before the automobile appeared on the mobility scene, as soon as it became clear that the adhesion between the tire (whether iron or rubber) could be used for decelerating as well: carriages were braked with brake blocks, and very early bicycles were braked with spoon brakes (after the shape the blocks took when used) and band brakes around the wheel axle. The internal drum brake and its mechanical and hydraulic actuator system are automotive developments (see Chapter 8), but the pneumatic brake system was already used on trains, as proposed by George Westinghouse (1846–1914). Although contemporaries had to get used to the idea that braking a heavy vehicle could be accomplished on the basis of air, heavy trucks nowadays are all equipped with this system.

Suspension

The other part of the automotive chassis, the suspension system, cannot be understood without approaching it as part of a cooperative system of vehicle and road. Earliest suspension systems were literally aimed at keeping the shocks of the road away from the goods or passengers carried. This was done through the use of leather straps or large, cast-iron S or C springs.

---

1. See also for the following paragraphs: ([6-2], p. 27–32).

In the history of pre-car carriage technology, the introduction of the leaf spring in the early 19th century (reputed to be proposed by the British Obediah Elliott, 1805) stands out as a turning point. Although known to be very detrimental for the horses (because of their weak longitudinal stability), the springs allowed the cabins to be designed much lighter. Sometimes even the metal beams between the axles could be left out, resulting in a kind of *monocoque* construction, as later car engineers would have called it.

Another early iconic predecessor is Scottish reverend and railway engineer Robert William Thomson who, as early as 1845, patented a pneumatic tire (Figure 6.2) and who, when this idea did not catch on, developed a massive rubber tire for carriage applications. Thomson's patent was forgotten, and the pneumatic tire was reinvented by Irish veterinarian John Boyd Dunlop in 1887 for bicycles, as we will see in the next chapter.

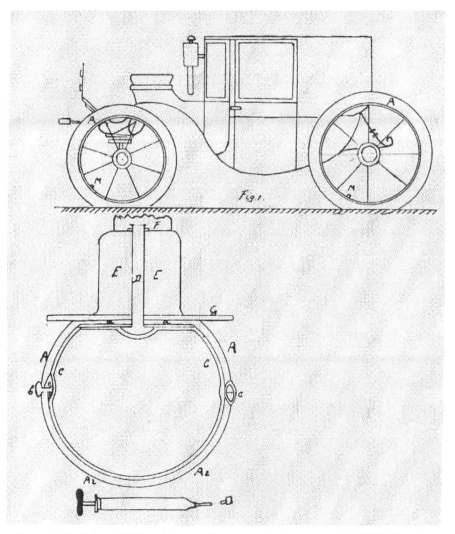

Figure 6.2 Robert William Thomson's pneumatic tire patent, 1845 ([6-2], p. 31).

Steering

Another example of multiple (re)invention is the steering principle, indicating that the construction of a technological lineage can be quite misleading, and often only would strengthen the false belief that technological development is a linear accession to ever higher efficiency. This steering principle was proposed by Georg Lankensperger, a carriage builder in Munich, in an effort to get rid of the central pivot steering. First, he reinvented the stub axle or kingpin steering, the idea that the wheels should be turning around a nearby pivot, the steering knuckle. Thus, the front axle did not have to be turned in its entirety and the frame and cabin could be lowered, and thus made more stable. Second, he added a system of rods that forced the outer wheel in a curve of the road to turn less than the inner wheel. He reasoned that a slip-free turning of both wheels could only be achieved when the geometry of the rods obeyed to what later would be called the Ackermann principle, after German book seller and publisher Rudolph Ackermann, living in London, who patented Lankensperger's idea in 1818 (Figure 6.3) ([6-3], p. 252–261).

Whether the subsequent reinventors of the slip-free principle were not familiar with Lankensperger's idea, or whether they tried to benefit from the weak protection of patent rights is not known, but in the same year a French carriage builder (Arnold Haucisz) patented a similar principle, followed by an Austrian a year later. When Amédée Bollée applied stub axle steering on his steam vehicle in 1873, he borrowed from Haucisz, but when Charles Jeantaud five years later introduced the Ackermann principle to his electric vehicles, French handbooks started to use the term *épure de Jeantaud* (Jeantaud plan) to suggest that he was the real initiator. That was probably because he re-calculated the principle thoroughly and even got a patent in 1889. Despite (or perhaps because of) the chauvinism involved in this dance around the intellectual source of the steering principle, this history also shows that car technology and the *discourse* among engineers and technicians on its basic principles was a transnational affair from the very onset.

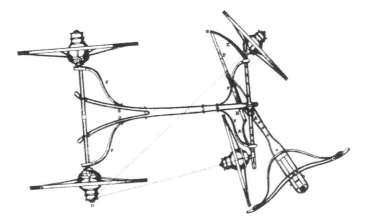

Figure 6.3 The Ackermann principle as shown in the patent of Lankensperger: The lines through the front wheel centers should meet somewhere on the extension of the line through the rear axle, not necessarily (as indicated in this drawing) in the rear wheel hub ([6-2], p. 34).

6.2   Constructing the Automotive Chassis

Although most principles and ideas to construct an automotive chassis were part of the state of the art in the fields of carriage and bicycle technology, and were eagerly borrowed by car pioneers for the new application, their integration into a systemic layout in a vehicle with its own propulsion took another decade or two. Initially, car designers were falling behind the comfort of carriages in the heydays of their deployment (in the last quarter of the 19th century) as they focused their attention upon the propulsion system, another example of the *non-linearity* of automotive technology development. By the end of the first phase, around 1920, car design was still criticized for its failure to integrate the different subsystems into a unified, dynamic layout. While this was corrected during the next phase, trucks maintained this structural transparency, as if to show that there was nothing in automotive technology to be ashamed of, and that the automobile was a highly technical commodity in the first place [6-4]. As we will see in Chapter 10, the passenger car followed another trajectory, where the desire to sell large quantities to a lay middle class public was seen as a reason to hide the technicality of the automobile and to emphasize its commodity character.

Two Schools of Early Car Design

Initially, car development got split along two schools, the carriage school and the bicycle school, the latter following the recent developments in bicycle technology. This was for instance apparent in efforts to adopt bicycle steering systems, such as in Benz's first three-wheeler car of 1886 (Figure 6.4B), complete with caster: the stability enhancing principle of tilting the upper end of the wheel fork to the rear. It also was visible in making use of tube technology and led to a separate, light-weight car species, the *voiturette*, which competed with the ever heavier carriagelike cars for five, six, or seven passengers of the large aristocratic families and their friends. Heavier cars opted for a frame consisting of I- or U-shaped steel beams on which the different subsystems were suspended, including the open gear-wheel shafts as we have seen in the previous chapter. Whereas these heavier cars saw their iron wheel tires replaced by solid rubber versions, builders of the lighter cars started to experiment with pneumatic tires developed from bicycle technology. Pneumatic tires mounted on clincher rims (Michelin) were such a *cross-over* innovation [6-5].

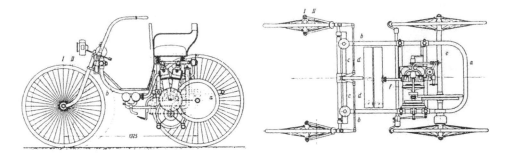

Figure 6.4A

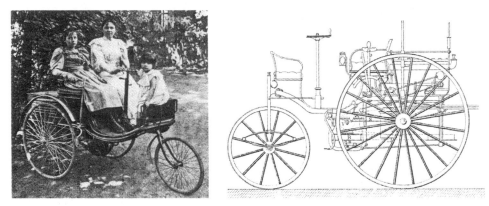

Figure 6.4B   **Figure 6.4C**

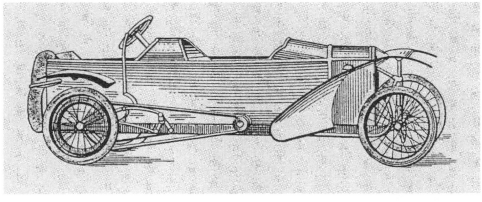

Figure 6.4D

Figure 6.4 Steering systems in early cars. A: Gottlieb Daimler's *Stahlradwagen* with tiller steering and bicycle spoke wheels; B: Karl Benz's three daughters on his *Motorwagen* with single bicycle wheel steering and caster; C: An earlier version of Benz's *Motorwagen* with rack-and-pinion steering; D: Bedelia cyclecar with central pivot steering ([6-2], p. 40–43).

By far the most important interface between the car and its user was the steering mechanism. Here, it can clearly be seen from how many different sources early designers got their inspiration. Karl Benz in one of the several versions of his first car avoided the steering problem altogether by using a single bicycle front wheel (see Figure 6.4B), whereas Gottlieb Daimler's *Motorkutsche* (Engine Carriage) even maintained central pivot steering. And although most pioneers started to apply the Ackermann principle, the British sporting car Bedelia (a cyclecar as a revival of the *voiturette*) had central pivot steering until after the First World War (see Figure 6.4D). The latter example shows that technology (*properties*) and *functions* have a very loose relationship, indeed: there is no one-to-one relation between changing car culture (and increasing speed desires) and something called technological "progress." To give another example: Daimler's *Stahlradwagen* (Steel Wheel Vehicle, see Figure 6.4A) from 1889 not only had bicycle-type spoke wheels, it also had so-called tiller steering (reminiscent of the steering of boats, but in fact also present in a doubled configuration in bicycle steering), which only around 1900 was succeeded by the steering wheel as standard.

Remarkably, an earlier version of Benz's *Motorwagen*, also with a single wheel in front (taken from a carriage and without a slanted position of the bicycle wheel applied later, see Figure 6.4C), had a kind of rack-and-pinion steering, which later would become the standard, but whether this design actually and directly influenced the later designs is not known. It is more likely that later reinventors took the idea from the state of the art as given in the handbooks and the overviews in the trade journals. The rack-and-pinion system allowed a torque multiplication, so the heavier wheels (than those of the bicycle and the *voiturettes*) could be turned with less effort, against the price of more turns of the steering wheel: early steering gears had a de-multiplication of about 4, meaning that four full turns of the steering wheel were necessary to turn the outgoing steering rod once. This was increased to about 10 during the inter-war years, and after the introduction of high-speed driving on freeways it again was increased to around 18. When the principle of steer gearing was adopted, tiller steering (with a de-multiplication of less than 1) was doomed, of course, because of the limited maximum allowable stroke of the tiller.

By then, around the start of the second phase at the beginning of the 1920s, a consensus among engineers and users alike had resulted in a victory of the heavier car school. The failure of the cycle cars (especially popular for a brief period in the UK) showed that motorists wanted a "real" car and not a four-wheeled cycle, so now a plethora of efforts were initiated to miniaturize the heavy car. This may have inspired the constant concerns about *simplification* (as we have seen in the case of the transmission, in the previous chapter) and the *reduction of unnecessary weight* (as we will see in the following section, on the steering box) that since then haunted the history of car technology.

6.3 Steering: Keep It Simple, Be Precise

The design challenge of the steering gear was to afford a sensitive transfer of manual effort at and from the steering wheel, and an insensitive transfer of road shocks in the reverse direction. The solution was the self-braking property (with high internal friction caused by gliding) of wormlike transmissions, such as the worm-and-nut configuration. Perhaps because of that, Henry Ford's planetary steering gear as applied on the Model T (1909) was not taken over by later car designers (Figure 6.5).

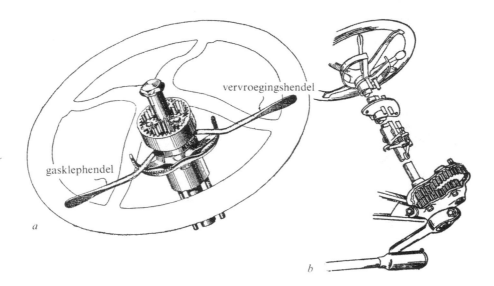

Figure 6.5 Steering gears with cylindrical wheels; a: Ford Model T; b: Chevrolet, 1927 ([6-2], p. 43).

Such sophisticated, "indirect" solutions came with a price, however (and again showing that technological development is not necessarily linear): they all necessitated extra rods between the steering box and the steered wheels, introducing extra spots of wear and tear and inaccuracy in a mechanism that increasingly was seen as pivotal to safety. The first adoption of the self-braking principle of the worm-and-nut transmission was the Marles steering box, designed by Henry Marles of the Marles Steering Company in 1913, who replaced the nut by a simple roll (Figure 6.6A). The same company introduced another variant, where the roll was replaced by a finger or lever; this system became known as the Ross steering box, after the Ross Gear & Tool Company (Figure 6.6B). Together with the Gemmer gear (called after the Gemmer Manufacturing Company of Detroit, with an adjusted Marles system) the Ross system dominated the market of sophisticated steering boxes, also in Europe as ZF took a license in 1932.

An evolutionary extension of the steering box family in another direction took place in the 1930s, when the Saginaw Steering Gear division of General Motors developed the original worm-and-nut principle into an even more precise reduction, the circulating ball steering gear. This Saginaw gear (also applied in precise guidance systems in tooling machines like lathes) was first applied to trucks, but went into passenger car production in 1940 (first on the exclusive Cadillac V12), and was then adopted by Daimler Benz, which equipped all its passenger cars with it (Figure 6.6C).

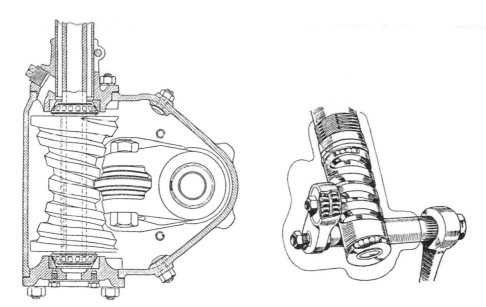

Figure 6.6A   **Figure 6.6B**

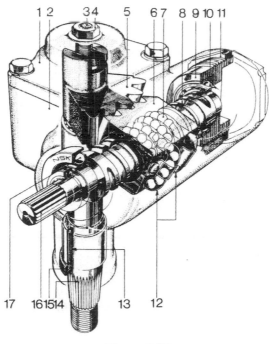

Figure 6.6C

Figure 6.6 Sophisticated steering boxes as adaptations from the original worm-and-nut principle of high-friction stearing gears. A: Marles steering gear; B: Ross steering gear from ZF; C: circulating ball steering gear from NSK. *Sources:* A,B: ([6-2], p. 45); C: ([6-6], p. 351).

Rack-and-Pinion Steering

A credulous evolutionist might now have thought that this development of seemingly ever increasing perfection would end in the victory of the most sophisticated principle of the rotating ball principle. Quite the opposite occurred, however: just like what happened in the ZF transmission case, car technology's evolution moved back to the simple rack-and-pinion system, of which the historical records do not even contain an inventor's name; like the simple spur gear wheel, it forms a part of the anonymous reservoir known as state of the art.

This example of the Pluto Effect was no doubt enabled because the simple system could benefit from the same improvements in materials and production methods as the more expensive versions, but perhaps also because this "direct steering gear" (as engineers know it nowadays) did not need vulnerable couplings between extra rods to the steered wheels: the rack propelled the steering rods at the wheels directly. In other words, the rack-and-pinion combined the functions of steering reduction and steering rod, a nice example of *convergence* of properties and the general trend of *compactification*.

Revived initially in light vehicles with independent wheel suspension (see Chapter 7), where it enabled a large wheel turning angle at relatively small angles of the steering wheel, the rack-and-pinion system has meanwhile reconquered much of lost terrain. As if no indirect gearing ever existed, modern handbooks now provide the usual shortlist of advantages: easier to produce, a lighter and stiffer construction, and turning with less friction ([6-6], p. 353, [6-3], Chapter 4, p. 45–64). In the process, the steering column was connected to the steering box through a Hardy disc (see Chapter 4), whereas the column itself was often (and still is) split in two or more parts, connected through universal joints ([6-7], p. 388).

It seems as if the steering gear problem, with the victory of the rack-and-pinion, was solved, if we look at the characteristic process of *black-boxing* taking shape in the handbooks of the last 50 years or so, where not many words are wasted to this component assembly ([6-8], Chapter 4, 239–268). It was as if engineers wanted to have their hands free to develop the entire steering system as a unit, introducing the safety steering column (which collapses under impact in the case of a frontal collision), power steering to increase the comfort of steering heavy passenger cars (all of them working as a hydraulic control system), and four-wheel steering as the zenith of sophistication. In the mean time, the rack-and-pinion type had taken over most of the advantages of the competition, especially their high precision (Figure 6.7), a mechanism we by now will immediately recognize as an example of the Pluto Effect.

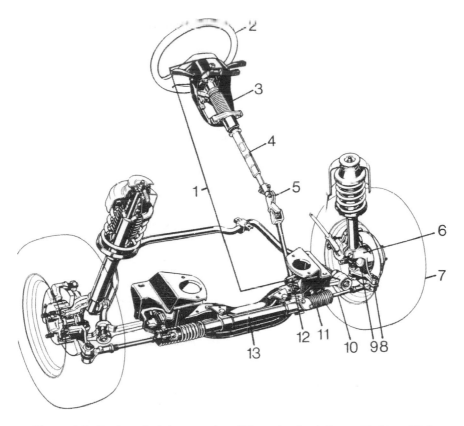

Figure 6.7 Rack-and-pinion steering (13) on the Ford Sierra ([6-6], p. 304).

But there is another reason why the rack and pinion may have won the battle of the steering systems. For this, we must delve a bit deeper into the intricacies of automotive steering theory.

6.4   Steering the Car, Theorizing Dynamic Behavior

What the rack-and-pinion type of steering gear also took over from its indirect competition was the smaller tooth distance at both ends of the rack, just like the worms in the more sophisticated systems had been given a different pitch at their extremes. This clever solution was introduced to satisfy the new, long-range and high-speed use on the highways and freeways built from the 1920s onwards on both sides of the Atlantic Ocean. Under such conditions, the reductions were much too small, resulting in erratic steering behavior at high speeds. On the other hand, when cornering in town traffic and especially when parking the car, a faster moving of the steered wheels is needed, so higher de-multiplications were introduced at the extremes (resulting in a larger angle of the steered wheels with a given angle of the steering wheel). Now that the car developed into a multifunctional vehicle driven in busy city traffic as much as at straight freeways, the steering gear had to be citylike at low speeds, whereas it had to be much more indirect at high speeds.

Such designs were the result of careful testing and theorizing about the car's dynamic behavior, and it is this accumulation of engineering knowledge that increasingly started to define what car technology was (and is) all about: cars are not so much new because of their components (although there are components that were radical innovations when they appeared in the car, such as the CV joint in the wheel drive shafts, or radial tires, as explained in Chapter 7) but because of their *ensemble* of systems and subsystems cooperating under dynamic conditions of driving, cornering, braking, accelerating, and, indeed, colliding with other objects.

Oversteer and Understeer

The change of tire knowledge toward a more scientific approach happened along two trajectories, in and outside the tire factories. In the mid-1920s, new scientifically educated engineers started to join the "old rubber dogs" in the factories—chemists in the first place, but also mathematicians, such as physics and mechanics specialist Albert Healey at the British Dunlop Rubber Company, who started systematic research in the plasticity, the resilience, and the abrasion resistance of rubber compounds. It was Healey who, as a commemorative volume of Dunlop claims, published "the world's first technical paper on the pneumatic tyre," titled "The Tyre as a part of the Suspension System," presented to the Institution of Automobile Engineers in 1924 and published in 1925 ([6-9], p. 55) Remarkably, scientification went hand in hand with approaching the car as a system.

The second trajectory took place outside the tire factories. It was the engineers in the test facilities and development departments of the car manufacturers, and the government laboratories who started to investigate car accidents resulting from tire skidding, and they made the decisive step to a truly systems approach by including the road surface into the picture. Such engineers found out that the Ackermann principle is a principle characteristic for a static vehicle and thus applies only to the car when it does not or hardly moves. As such, it was important to realize low-friction cornering in horse-drawn carriages, and it still is important if basic adjustments to the steering geometry have to be done, for instance in the garage.

But when the car is driven around a corner, the four small contact surfaces between the tires and the road, surfaces through which all the energy of driving and cornering has to be transmitted, deform, thus producing cornering forces that enable to steer the car through the curve. These engineers discovered the *slip angle* of the wheels, the phenomenon that the vehicle path does not coincide with the steered direction. They started to redraw the Ackermann diagram for dynamic conditions. In the UK, it was Maurice Olley of the local General Motors branch, who in 1937 introduced the concepts of oversteer and understeer after careful testing backed by sophisticated mechanical calculations. Olley found that a slip-free cornering was only theoretically important (a condition he called neutral steer), and that mostly one of the axles carries a heavier load than the other, resulting in a larger cornering force and a larger side slip angle. If the load is largest on the front wheels, these wheels take a wider sweep than the rear wheels, so the driver has to constantly correct this by introducing an extra angle at the steering wheel, a situation of *understeer*. If, however, the rear axle is heavier (because of luggage or extra passengers) the car tends to drive into a spiral, and it is this instable *oversteer* condition that car engineers learned to avoid by careful fine-tuning of the load distribution over the axles. They did so by varying the inflation pressure of the tire, and it took them until after World War II before the entire tire community understood the intricacies of this dynamic behavior. The Hillman Imp, for instance, with a large part of the engine weight resting on the rear axle, had a standard front tire pressure that was half of that of the rear tires ([6-9], p. 69–71). After the war, insights in the steering characteristics of the tire became standardized, routinized, and *black-boxed*, as can be seen in the introduction of the so-called Gough plot (proposed in the 1950s by Dunlop engineer Eric Gough ([6-9], p. 86).

Scientification: Steering Error

Research in Germany further revealed the importance of the geometry of the transmission system between the steering wheel and the steered wheels. The position of the latter is defined by the difference of the wheel angles, and this difference is called theoretical *toe-out on turns*. "Faultless steering" occurs if a steering system always obeys the primary steering law, according to which the steered wheels, in order to perform pure rolling, have to have a common turning center. It was Wunibald Kamm (whom we will meet extensively in Chapter 12), working in the German car industry before the Second World War, who showed that such faultless steering only occurs if the center lines through the steered wheels cross at a straight line, the so-called Kamm line (Figure 6.8).[2]

Kamm also showed that the Ackerman steering trapezium does not follow this Kamm line (except for one particular point), meaning that in nearly every position of the steered wheels the real toe-out on turns differs from the theoretical one. This difference is called angle of error, and the curve that can be drawn by connecting all line crossings is called the *error curve*; this curve indicates the amount of deviation from the ideal steering situation. In the course of developments of the best steering gear, the direct rack-and-pinion option appeared to make it easier for the manufacturer to realize a steering behavior close to the ideal situation, because of the triangular position of the steering rods. Indirect steering boxes necessitate a quadrangle, which makes realization of the close-to-ideal steering more difficult ([6-6], p. 312–315, 353–354).

---

2. In fact, the Kamm line is the result of the intersection of the center line of the outer wheel and the mirror image of the heart line of the inner wheel (mirrored in the line BD in Figure 6.8).

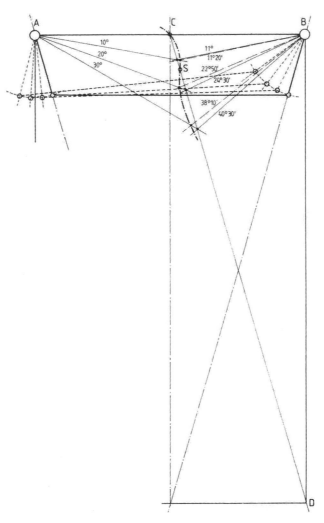

Figure 6.8 Steering theory to get a grip on the car's dynamic behavior. The Kamm line CD of theoretically ideal steering, the actual steering error curve (in bold, interrupted) and the only point of theoretically correct steering (in point S) ([6-6], p. 312).

6.5 Conclusions

Like the gear wheel, the rack-and-pinion steering box evolved into a typical automobile component from relative anonymity. During the inter-war years, it joined many other, less anonymous components (such as the carburetor, the battery ignition, and the CV joint) and formed, in close "cooperation" with them, an automotive structure in the full flow of getting adjusted to a dynamic context of use.

This context was ambiguous and a real challenge for manufacturers and engineers: it formed the reason why Henry Ford called his Model T universal and a family car. The new buyers during the 1920s and 1930s had developed a combined practice of city driving and high-speed coasting at the new highways and freeways. The properties to *afford* this dual use profile were not so easily adjusted. On the one hand engineers developed technologies that allowed motorists to drive fast. The overdrive was such a technology, a fixed reduction in the final drive (integrated later as the highest step in gear boxes) that kept the maximum engine speed (and through that its fuel consumption) relatively low. Motorists and manufacturers alike had to learn the hard way that freeways should be used with care: dozens of motorists stranded along the road with blown-up engines and boiling radiators when the Pennsylvania Turnpike between Harrisburg and Pittsburgh opened in 1940. Similar things happened during the opening of the first *autobahnen* in Germany, when Dutch motoring journals warned their readers not to step on the gas too enthusiastically as soon as they had reached the smooth concrete slabs that seemed to invite to speed up and race [6-10].

A close-reading study of the *Journal of the Society of Automotive Engineers* reveals that engineers in this phase did their utmost to enable this type of high-speed drive while at the same time not jeopardizing the car's ability to drive in busy city traffic, with its stop-and-start and very variable speed and steering angle characteristics. The struggle to either go to four steps in the manual transmission (in Europe) or to introduce a fully new automatic transmission (in the United States) is an expression of this dilemma, which ushered into the *divergence* of two separate car cultures. In some cases this dualism even was embedded in one and the same component, such as in the tooth distance gradient in the rack-and-pinion steering gear: the middle position was used at high speeds and was reduced in transmission ratio to prevent cars on the freeway from disappearing from the road at the slightest turn of the steering wheel, while the increased ratios at the rack's extreme ends were used in city traffic.

It was the gradual process of *scientification*, the intertwining of engineering skills with sophisticated knowledge of the car's dynamic behavior, that after half a century of development set the automobile clearly apart from its predecessors and companions, most notably the horse-drawn carriage and the bicycle. This process followed trends that are meanwhile familiar to those who have read the previous chapters (such as *variation* and *divergence*, and *compactification*). This chapter also showed how technological development is not linear, returns to earlier solutions, bifurcates, and follows trajectories that each need to be explained in their societal contexts. *Scientification* is such a context, which governed engineering in an increasing way.

References

6-1. Wilhelm Treue, *Achse, Rad und Wagen; Fünftausend Jahre Kultur- und Technikgeschichte* (Göttingen: Vandenhoeck & Ruprecht, 1986).

6-2. Gijs Mom, "Het rijdend gedeelte in historisch perspectief," in M. Arkenbosch, G. Mom, and J. Nieuwland, *Het rijdend gedeelte; Band A: historie, theorie, banden en wielen, besturing* (Deventer, 1989) (Part 4A of *De nieuwe Steinbuch; de automobiel;*

handboek voor autobezitters, monteurs en technici onder redactie en coördinatie van drs. ing. G.P.A. Mom), 25–71.

6-3. Jan Norbye, *The Car and Its Wheels—A Guide to Modern Suspension Systems* (Blue Ridge Summit: TAB Books, 1980).

6-4. David Gartman, *Auto Opium: A Social History of American Automobile Design* (New York/London: Routledge, 1994).

6-5. G. R. Shearer, "The Rolling Wheel—The Development of the Pneumatic Tyre," *Proceedings of the Institution of Mechanical Engineers* 191 (1977), 75–87.

6-6. M. Arkenbosch, G. Mom, and J. Nieuwland, *Het rijdend gedeelte; Band A: historie, theorie, banden en wielen, besturing* (Deventer, 1989), (Part 4A of *De nieuwe Steinbuch; de automobiel; handboek voor autobezitters, monteurs en technici onder redactie en coördinate van drs.ing. G.P.A. Mom*).

6-7. E. Blaich e.a., *Internationales Automobil-Handbuch; Umfassendes Lehr- und Nachschlagewerk für alle Gebiete der Kraftfahrt* (Lugano: J. Kramer, 1954).

6-8. Giancarlo Genta and Lorenzo Morello, *The Automotive Chassis; Volume 1: Components Design* (n.p.: Springer Science+Business Media, 2009).

6-9. Eric Tompkins, *The History of the Pneumatic Tyre* (n.p. [Birmingham], 1981).

6-10. Gijs Mom, *Atlantic Automobilism: The Emergence and Persistence of the Car, 1895–1940* (New York and Oxford: Berghahn Books, forthcoming).

Chapter 7
Wheel Suspension: Who Will Absorb the Shocks?

7.1　Introduction: The Suspension as System

Like the gear wheel, the prop shaft, and even the steering box, the wheel suspension's rods, springs, and ball joints are low tech compared to the complexity of a combustion engine or an automatic transmission. On the other hand, sophisticated production and calculation techniques are necessary to adapt a coil spring to its multiple, dynamic tasks within the wheel suspension, as we will see in this chapter. Such components benefit as much from state-of-the-art knowledge of materials as the more complex component groups do. Nevertheless, it is the arrangement of the components (and the theory behind it) that makes the suspension into a truly automotive technology and which, conversely, defines automotive technology in the first place.

In its *functional drift* from supporting the carriage passenger cabin upon its wooden framework to the separation of the unsprung weight (the part that moves with the road's unevenness, so to speak) from the sprung rest of the car, the suspension's evolution was governed by one big question and characterized by two trends. The question was which of the parts (including the passengers) had to absorb the shocks coming from the road when driving. The trends that documented the multiple answers to this question were fine-tuning through *functional split* and the insight that the sprung weight had to be as compact and light-weight as possible. Nearly all changes in the suspension can be understood from those two perspectives. They had to be incorporated in a constant effort to guide the change of angles and positions of the wheels under the influence of dynamic situations such as cornering, braking, accelerating, and the need to always keep contact with the road. This effort aimed at reconciling two seemingly irreconcilable characteristics of dynamic car behavior: comfort and safety.

Only during the second half of the 20th century did a third trend join the other two: the impact of the *scientification* of vehicle dynamics on the positions and angles of the wheels. As a follow-up on the beginning scientification of the tire from the mid-1920s as described in the previous chapter, we will dedicate some lines to this trend at the beginning and end of this chapter, but will spend more time on it in Chapter 12. Nowadays, engineers call these wheel positions and angles camber, toe-in and toe-out, kingpin inclination, kingpin offset radius or scrub radius, caster angle and caster (castor if you are British) and wheel offset. All these positions and angles have to be controlled by rods and other guiding elements during the multidirectional movements of the wheels and the body.

Figure 7.1 shows how a recent handbook of automotive technology explains the complex movements a car body and its wheels undergo, movements that have to be moderated, adjusted, and smoothed in order to compensate for the loss of comfort more than a century ago, when slow-riding, horse-drawn carriages were abandoned in favor of the automobile. Figure 7.2 shows the major wheel positions and angles, nowadays part of the routine vocabulary of the automotive engineer.

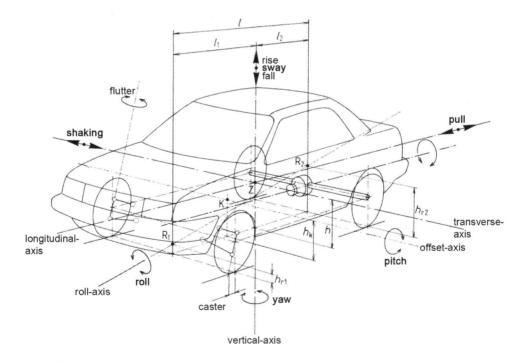

Figure 7.1 Movements of the car body and its wheels
(current terminology) ([7-1], p. 168).

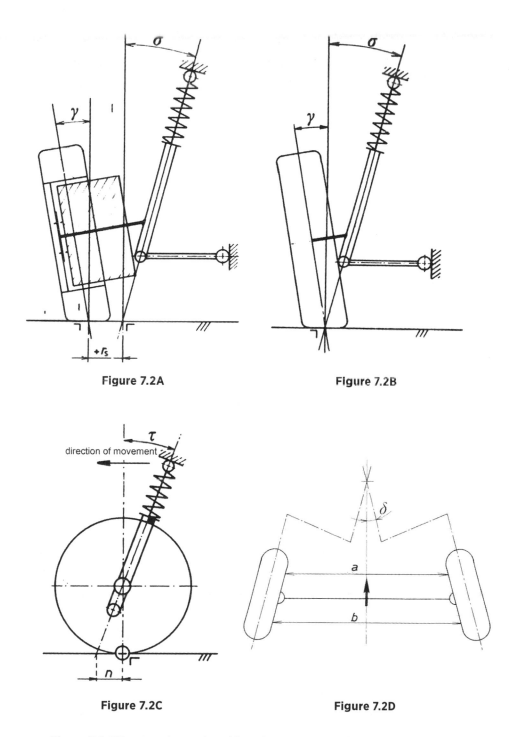

Figure 7.2A

Figure 7.2B

Figure 7.2C

Figure 7.2D

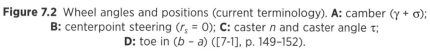

Figure 7.2 Wheel angles and positions (current terminology). **A:** camber ($\gamma + \sigma$); **B:** centerpoint steering ($r_s = 0$); **C:** caster n and caster angle τ; **D:** toe in ($b - a$) ([7-1], p. 149–152).

Co-development

Told like this (as so often happens in engineering courses), it is as if engineers and manufacturers knew what was at stake, and that they only had to find the technical means to fulfill these requirements. But this romantic idea (according to which some engineers are "ahead of their time" because their ideas were only realized or taken into account later, ideas that stem from hindsight) is rather misleading: while the properties of the suspension were developed, the knowledge about their functionality was co-developed as well. For instance, as we will see in the following sections, it was only after the introduction of the high-speed tires during the 1920s and 1930s, and especially after the radial tire was introduced during the post–WWII phase that the theory about tire and car dynamics developed into what we now define as true engineering science. This should not surprise us, because cars that drive 50 km/h need different suspension systems *and* different knowledge than cars that can drive four times as fast.

Projecting our insights into the past would strengthen the prejudice that previous generations of engineering were more primitive. They were not. Their engineering knowledge was as true as ours, even if we are able to show their errors and faults (as future generations will do with our knowledge). Also, most engineers' knowledge during, say, the first phase of suspension theory was fully *adequate* (a much more useful term than primitive or ahead of its time), although in this chapter and the next we will also encounter examples of a lack of knowledge that truly hampered some developments considered necessary by some contemporaries. We will see that these errors were the result of ignorance, which on its turn was related to the state of (largely empirically generated, and not very much scientifically reflected) knowledge of automotive engineering.

In this chapter we will first give an overview of the general development of wheel suspension, including its main component assemblies, such as the spring, the damper, and the wheel guidance subsystem (Section 7.2). We will then focus on one crucial component group, the tire, to show how a seemingly simple controversy about the importance of comfort can lead to a radical change that shook the supplier world of tire manufacturers to its very foundations and triggered a genuine turnaround in vehicle handling (Section 7.3). As usual, we will close this chapter with conclusions (Section 7.4).

7.2 How to Guide the Wheels? Big Problems, Many Solutions

Suspension is a feast of forces and the paradise for engineers who like to calculate dimensions of parts on the basis of these forces. When analyzing the confusing multitude of solutions to the wheel suspension problem (an indication of its shear complexity), some general trends and distinctions are helpful.

The first distinction is between the driven axle and the nondriven axle, where the former requires extra care because of the driving forces working on the wheel positions as well. When the driven wheels also become the steered wheels (in the case of front-wheel drive) the dance of the forces under dynamic conditions becomes even more complex. In order to master this complexity, engineers have the inclination to conceptually split these forces into functional groups and assign them to separate parts, such as guidance rods, springs, and dampers. This development we called *functional split* in the previous section, but parallel to this split, and enabling it, was a *split of properties*, as in this case new technologies are introduced to enable the same functions.

On the other hand, as soon as a new part is introduced, a tendency to efficiency thinking continuously incites engineers to have the single component play a combined role (for instance, by having the damper guide the wheel as well), and the dance can start all over again. The history of car suspension, in other words, is a history of mechanical force management that follows a cyclic trajectory of *simplification* and *complexification*, *split* (bifurcation) and *recombination*. The development is far from linear.

Take the standard Panhard configuration with its engine and transmission in the front and the car's rear wheels driven through a propeller shaft (see Chapter 4): this configuration was equipped with rigid axles suspended by semielliptic leaf springs (in the shape of the lower half of an ellipse) that had to take up all forces, causing frequent spring fractures ([7-1], p. 53–60). One solution to alleviate the forces upon the springs was to have the load partially taken up by a torque tube, which we encountered in Chapter 4 as the bulky alternative to the Hotchkiss drive. Despite the attractive lightness of the latter, however, the torque tube was applied in passenger cars until the 1970s. Another solution was the De Dion axle, where the forces were taken up by the chassis itself, at the same time reducing the unsprung weight considerably. This solution, too, was applied until well after WWII, when the cry for independent suspensions had already been heard for decades.

Indeed, the following shift from rigid to independent suspensions, which started at the same time as efforts to change to front-wheel drive (see Chapter 2), is a clear example of a *split of properties* as in the latter case the movements of one wheel do not interfere with the movements with the other wheel on the same axle, allowing car designers to fine-tune the wheel movements more precisely. This was done because the rigid front axle showed disadvantages that started to appear only when cars were able to perform high-speed behavior, and when brakes were added to the wheels, which on top of that were equipped with soft tires (see Section 7.3). Such wheels could slip into *shimmy*, a vibration around the vertical center line that is transferred through the steering rods to the other wheel and then leads to incontrollable vibrations in the steering wheel.

The reluctance to abandon the rigid axle sprung from the insight that the dynamic stability at the reigning car speeds was easily realized under all springing conditions: the wheel angles such as toe-in (the front of the driven wheels being closer together than the back, resulting in parallel wheels when driving) and camber (if positive: the upper end of the wheel pointing outwards, thus avoiding conditions leading to wheel shimmy) and the distance between the contact surfaces between the tires and the road were kept constant as a matter of course. The reluctance to change the rigid axle was also visible in the shaping into a curve of its middle part under the engine, such that the usually high center of gravity of rigid-axle cars could be lowered.

Proponents of the rigid axle competed with the higher comfort realized by the independent wheel suspensions by mobilizing the Pluto Effect, for instance by lowering the internal friction of the leaf spring through lubrication. The resulting loss of wheel guidance was then compensated by special rod systems such as the Panhard rod, the Watt linkage, and even triangle and four-point guidance systems ([7-2], p. 164). The split into guidance and springing functions even led to the introduction of rigid axles with coil springs. And although the attractiveness of the rigid axle for engineers is testified over and over again in automotive history (for instance by the reintroduction of semi-independent suspensions after WWII), this could not stop the slow but steady emergence of the independent solution, or better: solutions. Figure 7.3 indicates that even then the options engineers had in terms of guiding rod configurations seemed to be limitless, a nice example of evolutionary *variation* in a situation of large mechanical complexity combined with high uncertainty about exact mechanical behavior and its theory.

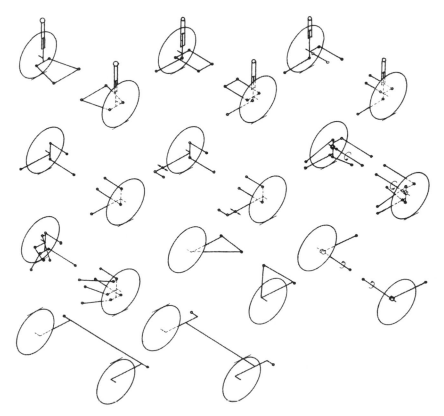

Figure 7.3 Ideal typical variation of independent wheel suspension systems ([7-1], p. 552–553).

Independent Wheel Suspension

As such, independent wheel suspension has pre-car roots: the Obéissante steam car of Bollée from 1873 was equipped with it. But the reintroduction in the car from the second half of the 1920s was inspired by the shimmy problem unknown to Bollée. The price to reintroduce this principle was high: many wheel guidance systems in the 1920s did not live up to the requirements of exact wheel guidance and reliability, while manufacturers hesitated, because some cars with rigid axles and very stiff springs (such as Bentley, Bugatti, Delahaye, and Maserati) showed a high grade of dynamic stability.

Over the course of about half a century, the (double) transverse wheel guidance configuration began to dominate over all other options, such as the longitudinal trailing systems, the swing axles (German *Pendelachse*), and all kinds of semi-independent systems. Such configurations *converged* into the double wishbone suspension (after the shape of the guiding arms reminiscent of the shape of the central bone in the chicken chest) or A arm, and it was the *lock-in* of this transverse guiding principle that then opened a new fine-tuning playing ground for suspension specialists ([7-2], p. 93).

It was here, within the car factory, that a new type of engineers started to join the classic chassis engineers who worked on a trial-and-error basis. The new, mathematically educated engineers emancipated axle kinematics into a mature body of engineering knowledge, laying the basis for a scientific adaptation of the wheel guidance to the increasing speeds and forces. As we will see in Chapter 12, it was American aeronautical engineers like William F. Milliken and Leonard Segel (with predecessors such as the British Frederick W. Lanchester and the German Wunibald Kamm) who for the first time in a systemic way applied sophisticated calculation and simulation techniques to the car's suspension system. These insights, Milliken and his co-researchers claimed, had to be brought into automotive engineering from outside, by aeronautical engineers who were much more used to a mathematically driven problem-solving attitude.

This happened quite late in the car's history, shortly after the Second World War. Although the American manufacturers considered themselves champions of the comfortable ride, the German car industry was known for its precision engineering in this respect, generating solutions such as Porsche's Weissach axle and Daimler-Benz's multilink suspension. Several versions of the latter, put one after the other since the introduction of Mercedes' transverse system on its front axle as far back as 1929, give a nice overview of the development of this type of suspension, which really could break through after the introduction of the coil spring and the telescopic damper.

But it was British Ford engineer Earl MacPherson (1891–1960) whose MacPherson strut (patent application in 1947, granted 1953) became nearly universal on front-wheel drive cars from the 1980s. That happened because it combined scientific insight with simplicity and a low-cost perspective, resulting in a very compact configuration, thus leaving more space for the passengers. As so often happens, the history of automotive technology hides a predecessor, in this case in the form of the Lancia Lambda from 1923, with its vertical wheel guidance system to which Lancia remained faithful until (and including) its model Aurelia (1950–1959) (Figure 7.4) ([7-2], p. 215–217), [7-3].

The introduction of the MacPherson strut took the form of a European invasion of the United States: first applied to the English Ford models Consul and Zephyr (1950), it was taken over by Peugeot (model 404, 1960), Honda (around the same time), while even Volkswagen gave up its famous longitudinal suspension with trailing arms for the new option in 1971, before it appeared on the Opel Rekord of 1977 and only then was adopted in the 1980 models of General Motors' X cars (Chevrolet Citation, Buick Skylark).

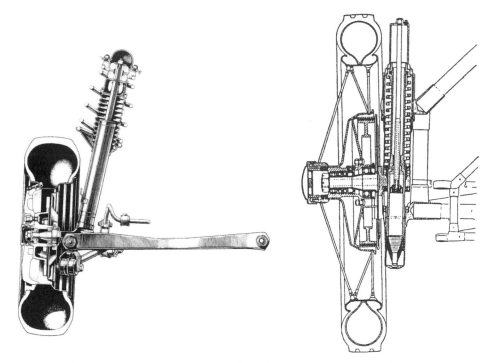

Figure 7.4 MacPherson strut as favorite suspension system since the 1980s (here applied by Honda) and its predecessor, the vertical guidance system on the Lancia Lambda of 1923 ([7-4], p. 504, 51).

For driven rear axles the diagonal or semitrailing wheel suspension was often preferred (pioneered by BMW on its 700 model of 1961), whereas semi-independent rear suspensions were revived by Volkswagen as late as the 1970s. This was not as strange as it sounds, because the post-war habit to add so-called stabilizers to the independent wheel suspension systems (which connected the suspension by a torsion bar alleviating body roll) had a similar effect ([7-2], p. 100). As so often in the history of automotive technology, developments took a (non-linear) backward turn, as it were, decreasing the "independence" of the independent suspension.

Springs and Dampers

In the course of these long-term (and certainly not linear!) developments another *split of properties* took place: while the leaf springs, as we saw, also took on a wheel guiding role, and the contact surfaces between the blades of multiple-leaf springs provided a form of damping through friction, engineers tried to get a grip on the very complex interplay of forces and speeds by pulling these functions apart. As a result, they then could be fine-tuned within their own specialism of springing, damping, and guidance. But once dampers were introduced, they were sometimes given extra wheel guidance functions, as we have argued previously. In other words, as so often happened in the last century of automotive technology, part of engineering effort went into paying the price of an earlier choice in favor of one special advantage, which generated a host of disadvantages that had to be remedied.

The compact coil spring, for instance, lacked any lateral stability, which then had to be taken over by guiding rods, complicating the suspension system as compensation for the initial simplification. Another example of this is the strut of the MacPherson system, which was damper and guiding rod at the same time. By the end of the century, leaf spring proponents struck back by introducing plastic (epoxy resin) blades reinforced with glass fibers, carbon fibers, or Kevlar (Du Pont) and Twaron (Akzo), mobilizing the Pluto Effect to defend (or revive) the leaf spring solution ([7-4], p. 451 453).

On top of that, the coil spring's lack of internal friction (an advantage, initially) necessitated a whole new subfield of *damping technology*, which started with friction-type damping mechanisms and went through options like inertia-type damping (such as could still be found as late as the 1940s and 1950s on the Citroën 2CV) as well as the hydraulic lever damper (Figure 7.5). But here, too, one option got *locked-in* and was further developed into the telescopic hydraulic damper: the friction, so to speak, was delegated to a separate element, a nice example of *functional drift*.

American Monroe was one of the first to propose the telescopic damper in 1932, soon becoming the largest damper supplier in the world, with a production of 50 million in the 1980s. It was introduced on the Hudson a year later. Since then, the evolution of the damper is one of fine-tuning of asymmetric characteristics: the in-bound behavior has to differ from the out-bound movement. *Properties* were adjusted to allow *functions* to be realized to the full: friction dampers could already be made asymmetrical (such as in the case of the Gabriel snubber), and this was later re-created in telescopic hydraulic dampers through the different dimension of valves inside the damper piston for the outward and inward strokes (an example of the Pluto Effect). The electronically controlled version (replacing the manual control for adjusting the damper characteristics) was developed in Japan and introduced on the Mitsubishi Galant of 1984.

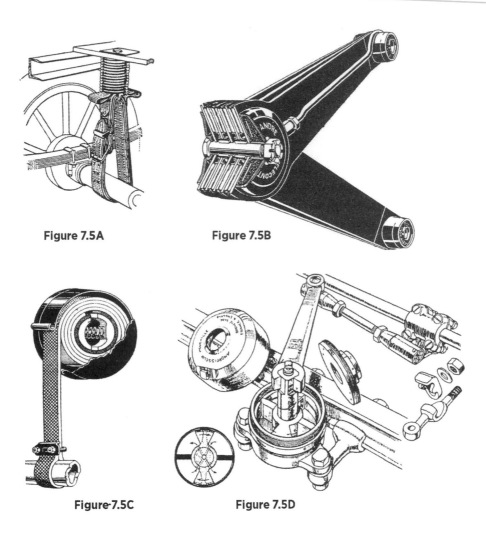

Figure 7.5A   Figure 7.5B

Figure 7.5C   Figure 7.5D

Figure 7.5 Damper options. **A:** leather strip shock absorber with spring;
B: multiple-plate friction damper; **C:** Gabriel snubber with brake pads;
D: hydraulic lever damper (Houdaille, 1921) ([7-1], p. 48–49).

In the area of *springing*, the Pluto Effect can be seen at work as well. While leaf springs
could be made progressive (requiring more force for the same amount of deformation)
through the addition of blades, coil springs realized the same effect through the gradual
change of the spring's body shape. The most sophisticated in the history of automotive
metallic coil springs was the mini bloc spring from Opel in 1978, where progessivity was
realized by all three possibilities: a variable wire diameter, a variable pitch, and a
variable coil diameter. It is a nice example of how production and design sophistication
can make a seemingly simple component into high-tech. The *split of properties* also led to
new spring types, such as the rubber spring (popular in the 1950s and 1960s) and the gas
spring (Figure 7.6).

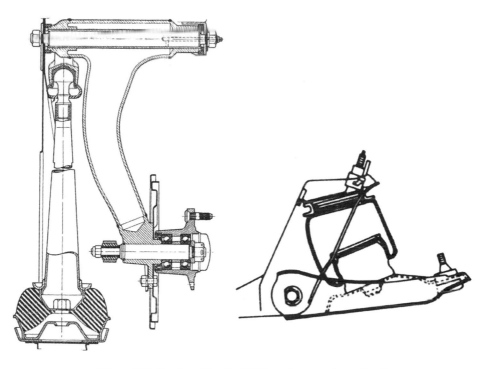

Figure 7.6 Rubber (Austin Mini) and gas springs (Ford
Lincoln Continental, 1984) ([7-4], p. 614, 477).

Citroën, which boasted to be especially innovation-prone, introduced in 1953 a hydro-pneumatic suspension system that *recombined* springing and damping functions, thus preluding on a tendency of the final years of the century, when electronically controlled suspension systems compensated for the separation of functions that had taken place half a century or so earlier. A decade after Citroën, Austin introduced its Hydrolastic system (on its 1100 model, in 1962) on the basis of gas springs, an effort to mimic the air springs that meanwhile had begun dominating the suspension development of trucks but could also be found on some passenger cars, such as the Ford Lincoln Continental from 1984. At the other end of the spectrum, in light cars where space around the wheels became even more restricted by higher comfort requirements for the passengers, manufacturers introduced simple metal replacements for coil springs, such as the torsion bars under several Renaults.

7.3 Who Should Absorb the Shocks? Defining the Role of Tires

All the changes described in the previous section literally and figuratively circled around the wheel, and especially the tire: the element that produces, while rolling, a small contact surface of perhaps 10 by 12 centimeters (three by four inches) between the car and the road through which all driving, braking, and cornering energy has to flow.

Students of the history of car technology have a case when they claim that the tire should be granted a major role similar to the internal-combustion engine.

When, about a decade after the first combustion engine cars equipped with massive rubber tires appeared on the streets, the pneumatic tire was taken over from bicycle practice (see Chapter 6), most of the basic options that would govern the following three decades were already known. The beaded tire (proposed by the British R.C. Wilson and American W. Bartlett, 1890) was used until the 1920s, when it was superseded by the wire-bead type, also proposed in 1890, by Charles Welch. This straight-side tire, as it was called, became standard in most American makes in 1916, with Ford's Model T as the exception. The tire with supple bead without wire was mounted on a clincher rim, allowing motorists to take it off with special spanners, until the sport of the car's unreliability started to become a nuisance for the less technically inclined motorists, and the Stepney rim was introduced, which could be more easily dismounted from the wheel as a unit of tire and rim. The next stage was the Rudge-Whitworth wheel, which could be dismounted in its entirety ([7-5], p. 439).

Indeed, the transfer to the much heavier vehicles (than the bicycle) with their much more complex wheel behavior asked such an amount of tolerance from the early sporting clientele that they turned the tire's short longevity into an asset. Repairing tires in the beginning was part of an extremely expensive sport and a token of pioneering machismo (if the task was not delegated to the chauffeur) (Figure 7.7). Electric taxicab fleets around the turn of the century were often failing because of the extremely high tire costs, despite the dozens of tire makes their managers tried out [7-6].

Figure 7.7 During the first phase, tire repair was seen as part of the sport of automobiling, although the actual work was often delegated to a chauffeur. *Source: Collection of the author.*

Elastic Wheels and Balloon Tires

Early tires had to be hard, containing air at high pressure, thus transferring the road shocks to the passengers without much springing action. The deflection (when passing the road surface) of the diagonally oriented textile fabric of the canvas within the tire caused heat-up, leading to frequent punctures as one of the quintessential aspects of early motoring. Before the ignition would take over as the *reverse salient* of automotive technology (the component causing the most frequent defects; see Chapter 1), the pneumatic tire played this role, until its unreliability became somewhat manageable, through a veritable crisis.

By the middle of the first decade of the new century the problems around the tire seemed insurmountable, and all kinds of *elastic wheels* were proposed (some handbooks counted 140 types) that instead took the deflection up in their spokes (Figure 7.8). That was true for the large deflections, however, as only the pneumatic tire seemed to be able to absorb the tiny bumps: "our tire drinks the obstacle," as Michelin advertised ([7-7], p. 219, 524–527).

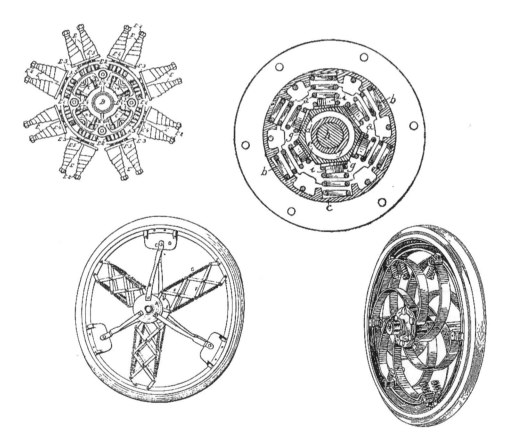

Figure 7.8 Elastic wheels proposed 1906–1909 as alternative to the lack of springing in the high-pressure tires ([7-1], p. 37).

It was electric vehicle technology that came to the rescue here. Because of their higher weight, special tires were developed for electric cars on the basis of a bicycle tire patent from 1892 by John Palmer, who proposed to avoid the crossing of the textile threads by making rubberized cords laid parallel in the canvas. Once mounted on electrics, such cord tires were soon (around 1914) taken over by combustion engine cars, and also appeared on trucks during the First World War, allowing a decrease of rolling resistance by one quarter.[1] If we combine this with the decision in 1910 by the American Goodyear company to add carbon black to the tire rubber (which decreased the wear and tear by a factor of three to five), the pneumatic seemed to be rescued for the car.

Not for long, so it seemed. Engineers at the tire suppliers observed during the early 1920s that users deliberately underinflated their tires in order to get a more comfortable ride when car speeds gradually increased, consciously accepting the higher costs of rising tire wear. These suppliers developed balloon tires with a stronger carcass, enabling a lower air pressure that led to the tire taking up more of the springing work of the total wheel suspension. Engineers from the car manufacturers fiercely opposed this innovation (they did not see this as an improvement), because they soon found out that such tires would increase the danger of shimmy, a "pernicious malady of mystery which is defying the cunning of near all car engineers," as one of them exclaimed [7-9], [7-10]. French race car designer Jean Albert Grégoire (see Chapter 4) also remembered how shimmy was a "catastroph," especially for designers of fast cars ([7-11], p. 41).

Within two years users had massively changed to the new tire type, forcing the car industry to start scientific research into the very elusive phenomenon of "ride comfort." As is often the case within the car as a constantly changing system of subsystems, the attack of one problem caused a chain effect, necessitating design changes in adjacent components up to the complete redesign of the suspension subsystem, a process that took nearly two decades in an atmosphere of utmost uncertainty [7-12].

In such an atmosphere, the number of alternatives seems to increase. Indeed, as we have seen, every manufacturer developed its own solution, mostly by making the geometry of the wheel guiding parts more exact and increasing the resilience of the springs. This gave American engineers the reputation of being the world's specialists in wheel guidance systems and laid the foundation for the "typical American" feel of very soft, "comfortable" springing [7-13].

The more exact configurations of the rods and springs enabled engineers to increase the tire pressure slightly, but a decade later the "terrors" of the user were repeated, when super balloon tires appeared on the market (also called doughnut or airwheel tires, in the latter case because their technology was derived from airplane tires), which again were immediately embraced by users [7-14], ([7-6], p. 440).

---

1. By far the best history of tire technology proper still is [7-8].

An American engineer in 1932 exclaimed that the new tires "absorb(ed) road shocks to an almost unbelievable extent" (which tells us something about the modest role played by the springs and the dampers in this period). At the same time, he had to acknowledge a "tendency to induce shimmy and tramp" (the latter being a term for lateral vibrations of the wheel springs) which he saw as "inherent but no doubt will be remedied," as he optimistically, and rightly, predicted. He also mentioned fantasies of his colleagues about "the possibility of eliminating springs if tires are made big and soft enough," testifying to the fact that as late as the early 1930s confusion about the exact role of every suspension component was still vivid among car specialists. But he was not convinced: "(I)t seems to me that a tire is fundamentally intended to envelop small road obstacles and absorb jars from irregularities, whereas the springs, aided by shock-absorbers, take care of the big upward thrusts of the tires, which have no snubbing [damping, GM] action." ([7-5], p. 441). He was right again, as we now know. Better formulated: we give him right, as current car technology is the result of a trajectory decided upon in the 1930s. In other words, it could have gone differently; there is no law of nature that forbids to use the springs, tires, and dampers for slightly different functions, and if there is, other properties could very probably have been developed.

From this perspective, the elastic wheel can be considered as a typical example of "a road not taken," as the academic discipline of the history of technology would call it, but it seems more adequate to grant this alternative a much more important role: through the Pluto Effect it threatened to take over the lead from rubber tire technology (through the "stealing," as it were, of the comfort function, realized in the terms of its own "vocabulary," through the deformation of metal), which on its turn triggered innovation within the mainstream technology, leading to the forced introduction of balloon tires.

Because of this development, tire pressure were reduced by two-thirds, and the walls of the tire were now taking up a part of the (high-frequency) springing function. This can be explained by comparing the tire on the road to a boat floating on water: "the tyre sinks onto the road surface, until the contact area is such that the reaction of the road, measured as contact area x inflation pressure, is equal to the load carried." Here, again, the nonlinearity of autotechnical evolution strikes us: by removing the internal friction from the tire, car comfort deteriorated, which had to be compensated within the subsystem, by introducing separate dampers ([7-8], p. 48, 50).

Tread Profiles and Materials

Meanwhile, tire technology entered a phase in which it became a part of the general aesthetic appearance of the car, leading to innovations, of which the technical function-ality was not always clear. This can be shown in the endless variation of tread profiles. At first they were placed *upon* the tire circumference (meant to grip behind cobble stones, like horse's shoes) and their pattern seemed to follow aesthetic rather than technical norms. But the lateral tread incisions (called *sommeren* in Germany after Robert

Sommer, who proposed it in 1932, or siping in the United States after J. F. Sipe, who already took a patent on it in 1922) were clearly technically inspired, although American manufacturers started to apply this procedure only as late as the 1950s (Figure 7.9).

Figure 7.9 Transverse tire tread incisions according to the Sommer patent ([7-1], p. 39).

Here, a more scientifically based insight in the function of tread patterns started to evolve during the first half of the 1920s, around the same time that scientification and mathematization in tire and suspensions technology emerged ([7-8], p. 56, 61). It is in this phase, too, that systematic research into the characteristics of high-speed tires was set up, while the growing resistance against the car in an ever more urbanized society also made tire and car engineers focus upon the noise produced by tires. By carefully (and asymmetrically!) redesigning tread patterns, the screaming *tyre sing* was changed into a "mutter" ([7-8], p. 63–64).

The 1930s and WWII are further known to have incited research into the material problem of the tire. With increasing tensions between Western motorizing countries with and without access to colonial rubber plantations, the need to find (partial) replacements became a matter of national security and led to the introduction of synthetic rubber types (such as German Buna, a mixture of butadiene and sodium [*natrium* in German], the latter used as a catalyst for the polymerization of the former) and later to the substitution of cotton by rayon and nylon. Higher-tensile-strength materials for the carcass (like rayon and nylon) enabled the tire walls to be made thinner, thus improving the heat release and through this the high-speed performance of the tire. Such improvements also benefited the constant reduction of the tire's rolling resistance (see Figure 9.1), which on its turn decreased the energy consumption of the car as a whole. By the car's centenary in 1986, a 7.5 kg tire contained about 3 kg of rubber (of which 1.2 kg synthetic), nearly 2 kg of soot and other fillers, more than 1 kg of chemical additives, and 1.25 kg of steel ([7-15], p. 450).

Indeed, while as late as 1929 the last passenger car on massive rubber tires appeared on the market (a British Trojan) and ten years later Mercedes introduced the last car (the Nürburg model) on wooden wheels, the future seemed to lay in the low-pressure bias-ply tire, in which the cotton cords were replaced by rayon and nylon shortly before and after WWII, respectively. Meanwhile, the tire profile (the relation between height and width of the tire section) changed considerably toward an ever broader footprint and a small (more speedy) height (Figure 7.10).

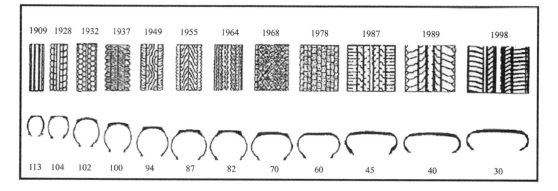

Figure 7.10 Development of the tire profile (height divided by width, in percent) over ninety years ([7-16], p. 175) *(courtesy of Springer)*.

Indeed, around this tire, which after the war, in the United States largely began to be *black-boxed* as a mature technology on which nobody seemed to pay much attention, a near-monopolistic market was built of suppliers (dominated by Goodyear, Firestone, Uniroyal, BFGoodrich, and General Tire in the United States), who annually made small improvements and increasingly were put under pressure by the car manufacturers to further decrease their prices. Firestone's thousand person R&D lab, for instance, introduced cosmetic changes every year (raised lettering, a new tread shape), so much so that it increased the number of tire types between 1968 and 1972 from 4,000 to 6,700 ([7-17], p. 435–436).

This went so far that by the early 1960s engineers called the lightweight, low-cost tires "four-dollar rags," as their sidewalls were meanwhile thinned by halving the number of plies (layers of bias fibers) to two. As a result, the tire suppliers earned their profit mostly in the vast replacement market. In this situation, in the midst of the period of exuberance after the Second World War, a second European invasion in the United States took place, just at the moment that the building of the American Interstate Highway network allowed motorists to drive much faster than before the war. The new car culture led to problems of heat-up and safety, which incited Wisconsin Democratic Senator Gaylord Nelson to put car safety (and the role the tire played in this) on the national agenda, as we will see in more detail in Chapter 10 [7-18], [7-19].

The Radial Revolution

What follows is a story of business espionage, cut-throat competition among tire suppliers, and the clash of two automotive cultures (a European and an American), leading to a redefinition of car comfort and handling. It was the French tire producer Michelin that started in 1946, in deep secret, research on a radically new type of pneumatic tire, based on a steel belt built into the rubber circumference of the tire. Michelin was especially prone to look for steel as a reinforcement material for passenger car tires, because it had already done some work on steel cord in truck tires and the incentive to transfer this knowledge to the passenger car became visible when Citroën's heavy front-wheel drive cars came on the market.

The choice for steel was scientifically inspired: tire wear is a function of misaligned running, so "if we make a tyre which produced its cornering power at a smaller misalignment, or slip angle, then it should wear lower" ([7-8], p. 86–87). The new steel-belt tire, called radial tire by Michelin's Italian competitor Pirelli (which later developed a tire with an alternative belt made from plastic, the Cinturato), did away with the bias plies and *split the functions* of carrying (in the belt) and comfort (in the sidewalls) (Figure 7.11.).

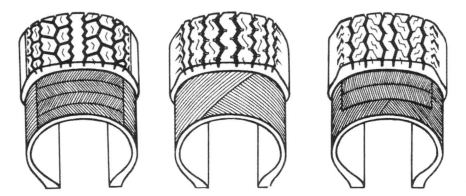

Figure 7.11. The radial cord tire, the diagonal (bias ply) tire, and the belted bias ply tire ([7-1], p. 232).

It became an immediate success, certainly after the introduction of the tubeless version in 1953 (the tubeless principle itself already introduced in bias-plied versions by Dunlop in 1935 and Goodrich in 1948). The Michelin X was presented at the Paris car show of 1949 and appeared on the Citroën (partially owned by Michelin) 2CV and 11 CV Traction Avant. By the 1980s, Michelin had obtained about one-quarter of the European market ([7-15], p. 443).

The tire looked "weird as hell," American engineers observed, but American customer organizations soon found out that their longevity was nearly double the life of the bias versions (which was, at best, 15,000 miles). It was mail order company Sears that jumped to the occasion and soon was selling 1 million imported Michelin radials a year, acquiring a market share of one-quarter, but car manufacturers remained skeptical. The reason for this was the different car culture in the United States, where the entire wheel suspension had been adapted (meaning: made very precise and direct) to the bias-ply tire to compensate for the tire's slow response to steering commands. As usual, they tried to mobilize Pluto, for instance by introducing polyester and fiberglass belting (in the Goodyear Polyglas of 1967), but the Firestone 500 Steel Belt radial led to the largest product recall in American automotive history so far, because the supplier had decided to manufacture it with adjusted bias tire production equipment. In 1978, consumer groups and the U.S. National Highway Safety Administration (one of the results of Senator Nelson's initiative) forced Firestone to a costly $150 million action of recalling 8.7 million tires. In 1988, Firestone was taken over by a Japanese manufacturer, Bridgestone, a move up to then unheard of in the history of American automotive manufacturing ([7-17], p. 432, 443, [7-20], p. 44).

Some manufacturers understood that the transfer of a foreign technology required *translation* work, and that the harsh handling and the increased sound production were not intrinsic to the technology (as many erroneously assumed). Instead, it could be (at least partly) eliminated by mounting special rubber bushes, driveshaft isolators, and other radial-tuned changes derived from "numerical analyses of skid and rolling resistance (and) wet traction" ([7-18], p. 36). Also as usual, new *properties* do not come without adjustment of the *functions*, so drivers (especially the well-trained among them) had to learn that the sensation of "running on rails" (meaning: very precise steering) came with the price of a sudden, unannounced and violent breakaway in cornering. Such characteristics were typically engineered out by adjusting the surrounding suspension system, to begin with in the new Citroën DS with its oleo-pneumatic system in 1955 ([7-8], p. 88).

In 1967, market leader Goodyear introduced a belted bias tire, accompanied by an advertising campaign that cast doubt on the radial's advantages. On the other hand, some American car manufacturers in that year made the radial tire optional, while dealers struggling with quality problems of the American-made radials ordered Michelins as replacements. From the moment that Ford made the new technology standard on its Lincoln Continental Mark III in 1970, market penetration went even faster than in Europe. Automakers massively switched to radials in fall of 1972: whereas 11% of GM cars and 26% of Fords had radials in 1973, two years later these percentages were 86 and 90, respectively (Figure 7.12) ([7-18], p. 37-38, [7-17], p. 442).

Figure 7.12B clearly shows how the Pluto Effect works: in a classic substitution game between bias and radial tire, a competing technology (bias belted) tries to capture the advantages of the newcomer, helps trigger interest for the innovation, but ultimately has to give in to the new technology. In this rare and exceptional case, the Pluto Effect turned into a Sailing Ship Effect, because the mainstream technology could not cope anymore with the new requirements and expectations.

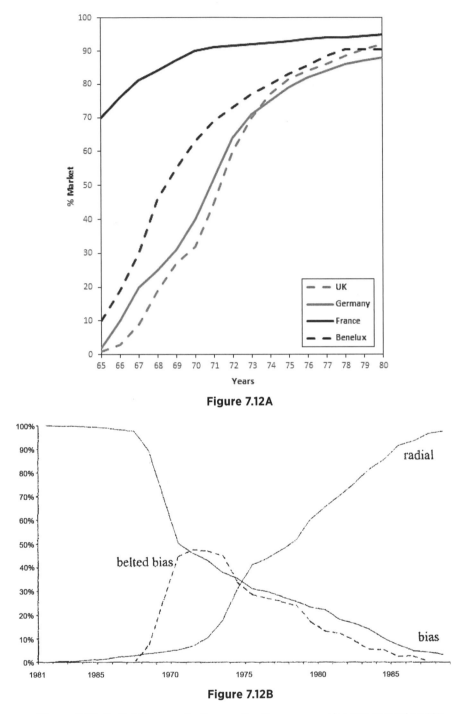

Figure 7.12A

Figure 7.12B

Figure 7.12 Radial tire penetration on the European replacement market (A) and on the American car market (B) (*A: sketch inspired by [7-19], p. 475, Figure 1, with permission from Oxford University Press; B: reprinted from [7-17], p. 441, with permission from Cambridge University Press*).

7.4   Conclusions

The radial's coup showed how we cannot approach car technology as a pile of things: it is a highly complex, ever changing system. It is a system bigger than the car as an artifact: because of the radial invasion, four of the five American tire manufacturers were taken over by foreign companies and a further oligopolistic phase took hold of this sector. In other words, changes in the car's (internal) structure have repercussion in the car's (external) system. In Europe, similar concentration tendencies were visible, of which Japanese tire manufacturers, after the Second World War, took advantage to acquire access to European and American markets.

The radial tire case meanwhile became a classic business school book example: whereas Firestone, thriving on routine and recruiting management on the basis of loyalty and a "uniform mindset," was taken over, Goodyear "emerged as one of the three global powers in the tire industry."[2]

Technically, the business concentration wave led to a *convergence* of automotive driving and car handling culture, with a more systemic dynamic steering approach based on high-grade steel-belt tire technology and the promise that at last the tire as a nuisance for lay motorists would be a thing of the past. For this, not only the suspension had to become truly automotive (emancipated from earlier carriage and bicycle technologies), but automotive engineering had to enter a phase of scientification. Meanwhile, dynamic tire design (in conjunction with the design of the entire suspension) belongs to the frontier of scientific car engineering.

Since then, car manufacturers have tried to get rid of the last "old-fashioned" character-istic of tire technology: the necessity to always carry a spare tire. They considered the introduction of a smaller, limp-home wheel as a step in that direction. Another step into the direction of the ideal of an absolutely safe tire is the electronic monitoring of the tire pressure, warning the motorists of a sudden pressure drop. "Green" tires with extra low rolling resistance are another contribution to further *black-boxing* of the tire in the overall automotive set-up: developed for electric vehicles, as usual in a world governed by the Pluto Effect, they were taken over quickly by mainstream car technology in an ongoing effort to keep abreast with the electric in the competitive run for the customer's preference.

The history of the car tire illustrates the width of the automotive system in another way. This case showed how the consumers triggered change. They were supported by suppliers: American suppliers in the balloon tire case, European suppliers in the radial case. While suppliers are often credited as being the real innovators in the history of car technology, the consumers, in both cases, voted with their feet (and their wallet) in favor of more comfort and lower costs, and the engineers had to follow, mostly by trying to

---

2. ([7-20], p. 50). Sull explains the failure of Firestone by referring to the strong "pull of the past" ([7-20], p. 52), which modern management should overcome. This book argues just the opposite: the more we are pulled to and embedded in the past, the better we can look into the future.

maintain and improve the road-holding characteristics. A clearer example of what we called *co-construction* in Chapter 1 can hardly be found.

In this chapter, we have seen how the constant efforts to get a grip on the handling of the car revealed a new trend of the *split of properties* and *functional split* and *functional drift* (such as the delegation of the guiding functions of the leaf spring to special rods), triggering opposing trends of *complexification, simplification,* and *compactification*. We have also witnessed the beginnings of a merger of aeronautical knowledge with its automotive counterparts, an example of mutual influence (a reciprocal crossover, as it were) between the two realms of engineering we will also encounter in the next chapter.

References

7-1.   Gijs Mom, "Het rijdend gedeelte in historisch perspectief," in M. Arkenbosch, G. Mom, and J. Nieuwland, *Het rijdend gedeelte; Band A: historie, theorie, banden en wielen, besturing* (Deventer, 1989) (part 4A of *De nieuwe Steinbuch; de automobiel; handboek voor autobezitters, monteurs en technici onder redactie en coördinatie van drs. ing. G.P.A. Mom*), 25–71.

7-2.   Jan Norbye, *The Car and Its Wheels—A Guide to Modern Suspension Systems* (Blue Ridge Summit: TAB Books, 1980).

7-3.   Jan Norbye, "Straps, Springs and Swivels—The Story of Suspension," *Automobile Quarterly* 3 No. 4 (Winter 1966), 412–419.

7-4.   M. Arkenbosch, G. Mom, and J. Nieuwland, *Het rijdend gedeelte; Band B: veersysteem, wielgeleiding, remsysteem, diagnose en uitlijnen* (Deventer, 1989), (Part 4B of *De nieuwe Steinbuch; de automobiel; handboek voor autobezitters, monteurs en technici onder redactie en coördinatie van drs.ing. G.P.A. Mom*).

7-5.   Burgess Darrow, "Pneumatic Tires—Old and New," *Journal of the Society of Automotive Engineers (Transactions)*, 30 No. 5 (November 1932), 438–444.

7-6.   Gijs Mom, *The Electric Vehicle: Technology and Expectations in the Automobile Age* (Baltimore: Johns Hopkins University Press, 2004).

7-7.   Gijs Mom, *Geschiedenis van de auto van morgen; Cultuur en techniek van de elektrische auto* (Deventer: Kluwer Bedrijfsinformatie BV, 1997).

7-8.   Eric Tompkins, *The History of the Pneumatic Tyre* (n.p. [Birmingham], 1981).

7-9.   O. M. Burkhardt, "Wheel Shimmying: Its Causes and Cure," *Journal of the Society of Automotive Engineers*, 16 No. 2 (February 1925), 189–191.

7-10. "Shimmy, Balloons and Air-Cleaners," *Journal of the Society of Automotive Engineers*, 15 No. 6 (December 1924), 482.

7-11. Jean-Albert Grégoire, *Toutes mes automobiles; Texte présenté et annoté par Daniel Tard et Marc-Antoine Colin* (Paris: Ch. Massin, 1993).

7-12. "More Flexible Springs Needed," *Journal of the Society of Automotive Engineers*, 16 No. 1 (January 1925), 10–11.

7-13. Tore Franzen, "Suspension Types Will Be Developed for Each Country," *Journal of the Society of Automotive Engineers (Transactions)* 33 No. 4 (October 1933), 347.

7-14. B. J. Lemon, "Judging Super-Balloon Tires," *Journal of the Society of Automotive Engineers (Transactions)* 31 No. 4 (October 1932), 403–411.

7-15. Olaf von Fersen (ed.), *Ein Jahrhundert Automobiltechnik; Personenwagen* (Düsseldorf: VDI Verlag, 1986).

7-16. Heinrich Huinink, "Die Entwicklung des Reifens; Der Reifen als Bindeglied zwischen Fahrbahn und Kraftfahrzeug," in Harry Niemann and Armin Hermann (eds.), *Geschichte der Strassenverkehrssicherheit im Wechselspiel zwischen Fahrzeug, Fahrbahn und Mensch* (Bielefeld: Delius & Klasing, 1999) (DaimlerChrysler Wissenschaftliche Schriftenreihe, Band 1), 169–183.

7-17. Donald N. Sull, "The Dynamics of Standing Still: Firestone Tire & Rubber and the Radial Revolution," *Business History Review* 73 (Autumn 1999), 430–464.

7-18. Tim Moran, "The Radial Revolution," *Invention & Technology* (Spring 2001), 28–39.

7-19. Donald N. Sull, Richard S. Tedlow, and Richard S. Rosenbloom, "Managerial Commitments and Technological Change in the US Tire Industry," *Industrial and Corporate Change* 6 (1997), 461–501.

7-20. Donald N. Sull, "Why Good Companies Go Bad," *Harvard Business Review* (July–August 1999), 42–52.

Chapter 8
Stopping the Car: How to Generate and Distribute Braking Energy

8.1 Introduction: Braking Levels Within the Car Structure

When the Dutch automobile club, on the occasion of parliamentary debates on motoring legislation, organized a demonstration in a seaside resort near The Hague in 1908, motorists and journalists alike were more astonished about the braking power than the propulsion power of the new contraptions. With horse culture in mind (its subject having a will of its own), they admired the quick braking times and short braking distances of the cars on display, an admiration that becomes all the more understandable when one realizes that the engine turned at a more or less constant speed, giving the feeling of a horse out of control ([8-1], p. 66, [8-2], p. 124).

Although braking, like steering and suspension, has its roots in pre-car technologies such as bicycles and horse-drawn carts and carriages (see Chapter 6), like these other *functions* of the chassis it acquired its automotive character through being a part of the car's structure tuned to its dynamic *properties*. The systemic aspects of braking are multiple: it influences the behavior of the suspension and of the drive train at one level, the behavior of the car as a whole at the next higher level, and even at the still higher level of the automobile system, where it forms a part of traffic safety (see Chapter 10) ([8-3], p. 60–69).

In the course of this process, the brake component assemblies themselves migrated from the transmission to the wheels (where the brake force had to be generated between the tires and the road), while their design shifted from external drums (such as cord and band brakes; Figure 8.1) to internal drums, no doubt because the brake shoes and their linings were better protected against road dirt and the weather.

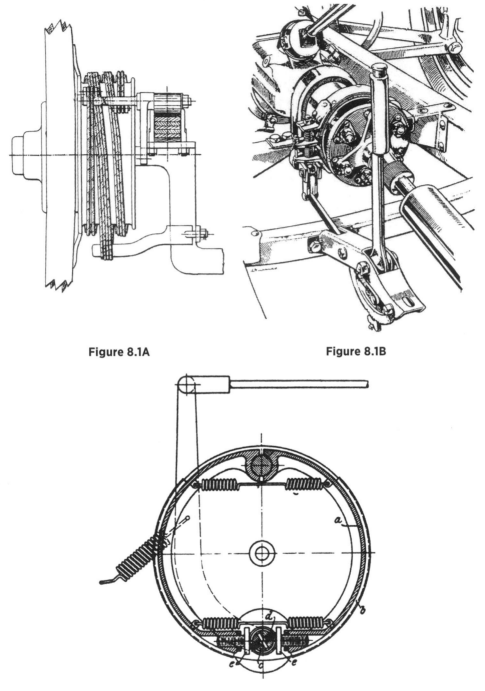

Figure 8.1A   **Figure 8.1B**

Figure 8.1C

Figure 8.1 Early external and internal drum brake types. A: cord brake as wheel brake; B: band brake as transmission brake; C: simplex internal drum brake. *Sources:* A, B: ([8-3], p. 61); C: ([8-4], p. 424).

Louis Renault, as early as 1902, patented the basic configuration of the internal drum brake which, at least in Europe, became universal before WWI on rear wheels. Renault's simplex type (as later engineers would call it, consisting of two semicircular braking shoes, its linings pressed against the inner wall of the turning drum by the motorist's brake pedal, thus generating braking energy) was also applied on Ford's Model N from 1906. Around the same time, the British Herbert Frood acquired a patent on heat-resistant asbestos as lining material, after he had conducted real waterwheel-driven test bench experiments since 1897. Whereas the basics of getting a wheel to stop were pretty much in place by the beginning of motorization, the development of the braking function followed a meandering route, but the systems approach was never far away.

In this chapter, then, we will first (in Section 8.2) investigate the adoption of the braking system or the way the motorist's foot pressure on the brake pedal was distributed and enhanced in order to exert braking energy under the wheels in such a way that the car would not skid. In the next section (Section 8.3) we will describe the emergence of the wheel brake proper, showing a nice example of ongoing variation on the simplex theme, followed by a radical innovation, the disc brake, after WWII. We will close this chapter with conclusions (Section 8.4).

8.2   The Braking System: How to Distribute Braking Energy over the Wheels

The story of the brake system is a story of diffusion of one man's idea, the Scottish mining engineer Malcolm Loughead, living in California, who invented a four-wheel hydraulic distribution system in connection with hydraulically operated internal drum brakes. It is also a story of many anonymous engineers who developed Loughead's idea further in a true evolutionary process of incremental steps, branching in different directions but never far away from the original idea, until a radical innovation shifted braking technology in a somewhat different direction, although the basic principle of hydraulic brake actuation was never abandoned.

Lockheed (as his anglicized name was spelled) was not the first: Fiat (1906), Itala and Maybach (1908), and Tatra (1909) proposed similar ideas. But Lockheed took a patent in 1917 on a wheel brake (slave) cylinder and in 1920 on a foot-pedal central (master) cylinder, which he licensed to many companies in motorizing countries. Among them were the Hydraulic Brake Company (HBC) in the United States (in 1930 acquired by Bendix Aviation), Automotive Products (AP) in the UK, Alfred Teves (ATE) in Germany, and the Société Française des Freins Hydrauliques Lockheed in France, in 1962 merged into Ducellier-Bendix-Lockheed-Air Equipment (DBA).([8-5], p. 133–147), ([8-6], p. 404–407), [8-7].

This invasion of an (Anglo-)American idea into the European continent (the reverse of the invasion of the radial tire idea from France described in the previous chapter) was a concerted effort to overcome the common practice of mechanical brake control, often on only two wheels. Popular history has it that the habit to brake only one axle was a heritage from bicycle technology, where the same hesitation to brake the front wheel existed out of fear of a turn-over (a header).

A study of the handbooks, however, reveals that it was not that simple. Arnold Heller, for instance, who published a German handbook for engineers as early as 1912, which went beyond the usual set up of a catalogue of technologies, expressed clearly the dilemma of early engineers. According to him, "normal" cars (as opposed to "traffic and sport cars," the former probably referring to taxicabs, buses, and the like) could well be braked on the rear axle alone. The alternative, brakes on the front wheels as well, would jeopardize the car's steering behavior too much. Heller (with a PhD in mechanical engineering) knew quite well that "on very steep streets the front wheels are subject to larger wheel loads and hence would be better suited to be the braked wheels than the rear wheels," but he did not mention the transfer of wheel loads from the rear to the front during the braking process ([8-4], p. 422).

The case of automotive braking is a nice example of a problem initially not well understood by the technical community.[1] Marcel Guillelmon, vice president of Renault's sales division in New York, had to show his audience at an SAE meeting in 1923 a toy car that, placed on an inclined plane with the front wheels locked, did not skid while it did turn nearly 180 degrees when the rear wheels were locked. Once accustomed to all-wheel braking, Guillelmon assured his audience, driving in a car equipped with only rear wheel brakes "gives one the same feeling of uncertainty and danger that would be felt if one were to ride on a railroad train at express speed and knew that the train was not equipped with airbrakes on all its wheels" ([8-9], p. 71).

"(M)ost accidents," one engineer opined without giving any evidence, "result from poor brakes" ([8-9], p. 72). That may be true, but Guillelmon pointed out the importance of culture (in this case the acquisition of skills and routine): rear wheel brakes feel dangerous only after one has had the experience of four-wheel brakes. People often seem to feel the limits of the existing technology only as soon as an alternative is at hand: disadvantages (and advantages as well), we conclude again, are social constructs. In France, brake specialist Henri Perrot told American engineers at another SAE meeting a year later, four-wheel braking was introduced from 1920 onwards, "first with the large cars, then with the medium and small cars and, in 1923, with the application of four-wheel brakes to taxicabs, jitneys and chars-a-banc."

---

1. The following paragraphs are based upon [8-8].

But Americans remained reluctant, also because they realized that changing the brake system would necessitate to "revise our front spring-suspension entirely" ([8-10], p. 102, 106). Again, we see here the slowly spreading insight under engineers that the car was a system, in which changes in one part had repercussions throughout the entire complex. By 1923, only 1.3% of American cars were equipped with a mechanically operated four-wheel brake system, but this percentage increased nearly tenfold during the following year and reached 44.3% in 1925 ([8-11], p. 90).

This is not to say that brake theory was well developed during these years: Rolls-Royce advised its customers in the 1920s that a car, when braked, needed twice the car's width on the road. Around the same time, the British top make introduced a model that distributed extra brake force during braking to the rear wheels ([8-3], p. 63).

Mechanical Brake Actuation

Mechanically operated systems came in two versions. The first version worked with rods, gearwheels, and even differentials to equalize the energy distribution over the wheels of the same axle (patented, for instance, by Henri Perrot in 1910 and expanded into a four-wheel system by Cadillac in 1922 after Perrot had emigrated to the United States). The second version was a cable system with less friction (introduced around 1930 after ideas from the British E. M. Bowden, who had devised it for barbers and dentists in 1896). When other makes within the General Motors family wanted to introduce the rod system, Bendix took a license and marketed, from 1924, its Bendix-Perrot system. While the rod system used differentials to equalize the braking forces between the wheels on one axle, the Bowden cable system worked with a balance lever to get the same effect.

Many brake actuating systems were combinations of rods and cables. Mechanically operated four-wheel systems, when applied in high-end makes such as Darracq, Hispano-Suiza, Isotta-Fraschini, and Metallurgique, had up to 50 joints, 20 bearings, and 200 parts, a challenge of maintenance and repair, and a source of inaccuracies that started to become a serious safety issue once car speeds increased during the 1920s (Figure 8.2).

The first European car with a hydraulic four-wheel system was the British Triumph 13/30 of 1925, followed a year later by the exclusive Adler Standard 6 in Germany. But Mercedes chose to stick to the mechanically operated system up to its model S, of which production ceased only in 1941. Volkswagen even used a cable brake as late as the 1950s ([8-5], p. 102), ([8-2], p. 63). The principle still survives in current car technology as a parking brake system, on rear drum brakes, where it at the same time functions as an emergency system if the hydraulic system would fail.

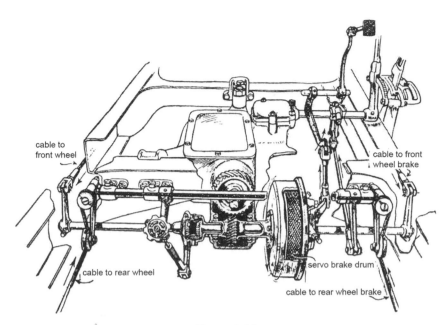

cable to front wheel

cable to front wheel brake

cable to rear wheel

servo brake drum

cable to rear wheel brake

Figure 8.2A

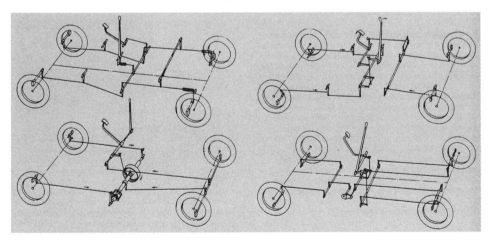

Figure 8.2B

Figure 8.2 Mechanically operated brake systems. A: with rods and gear wheels as two-wheel system; B: with rods and cables as four-wheel systems ([8-3], p. 62), ([8-6], p. 97).

Mechanical versus Hydraulic Brake Actuation

Hydraulics were certainly not seen as a clear "improvement." That is a hindsight construction. Users and engineers had to be convinced of the "superiority" of hydraulic braking. The German supplier ATE set up advertising campaigns, and Adler asked the famous Clärenore Stinnes, who circled the globe in the end of the 1920s in a model Standard of this

German make, to write a letter of recommendation. In 1921, an American Duesenberg with such a system won the Le Mans race, while hydraulic braking appeared in a "normal" Chrysler of 1924. According to the official company history of ATE, it was a survey of 5,000 brake inspections conducted by the German automobile club in 1937 which unmistakably proved the system's "far superior" characteristics "in every respect." Its efficiency (0.8 to 0.9), ATE claimed, was twice that of the mechanical system. But while by 1938 nearly all German makes had a car with such a system on offer, the American car industry stuck to the double option of a mechanically and a hydraulically operated parallel system until well in the 1940s. European Ford switched to hydraulics as late as 1950. Also late were Renault (until 1947, when the 4CV appeared) and Riley (until 1952) ([8-3], p. 64).

By then, the first efforts to separate the brake circuit in a front and rear subsystem had been undertaken, probably in racing car technology, and taken over by Jaguar in 1953 and Aston Martin in 1955. This was an expression of increasing public concern about safety, while it at the same time revealed the engineers' uncertainty about the reliability of nonmechanical systems: leakage seemed scarier than breaking (Figure 8.3).

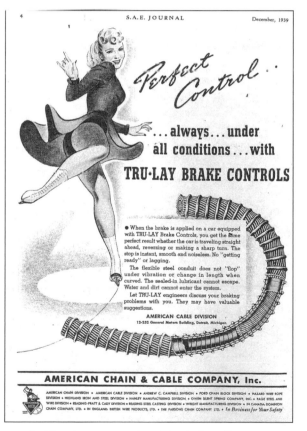

Figure 8.3 Brake control has been an underestimated continuous concern in the history of automotive technology, as can be observed in this advertisement that tries to rescue mechanical control against the threatening alternative of hydraulic control ([8-12], p. 4).

Scary, too, was the tendency of the brake fluid (initially a mixture of water and alcohol, then a mixture of alcohol and castor oil) to absorb air and release this in the form of bubbles at the moment of braking, causing a collapse of system pressure. Fluid formulas with a higher boiling point (such as diacetonal alcohol) were introduced, until in 1946, the Society of Automotive Engineers specified brake fluid on the basis of glycol ether with a minimum boiling point of 150° C (later increased to 180° C).

The diagonally separated system, meant to maintain a balanced braking effort without skidding, appeared on the Saab 95 and 96 of 1963. Equipped with tandem master cylinders (two cylinders, one after the other, each operating a separated partial system), the diagonally separated brake actuating system was only made possible through the 1960 patent of German physicist and car engineer Fritz Ostwald of the so-called negative (or inboard) scrub radius. The scrub radius is the distance between the point where the vertical steering axis touches the road and the center point of the contact surface between wheel and road, causing a torque around this center such that it counteracts the inclination of the car to break out. Although the negative scrub radius (negative offset or over-center point-steering) already was applied to the Oldsmobile Toronado of 1966, it was only used to counteract acceleration forces; the Audi 80 of 1972 was the first car in which Ostwald's insight was deliberately applied.

For a short transition period, there were two opposing schools, one (led by General Motors) emphasizing the advantages of a positive radius (on rear-wheel drive cars), the other (led by Mercedes) a proponent of the negative version. Since then, this stabilizing trick can be found in every car in the world, again an illustration of the systemic character of the car and the interdependence of brakes and tires under dynamic conditions, as well as of the cyclical sequence of technical *divergence* and *convergence* ([8-3], p. 64), ([8-5], p. 333, 337), [8-13], [8-14], ([8-15], p. 103).

8.3 Drums and Discs: Substitution and Coexistence

Lockheed's system was based upon the internal brake drum, of which the wedges to spread the brake shoes mechanically were replaced by hydraulic slave cylinders. His system was seen as beneficial not only because it allegedly produced less friction in the mechanical connections, but also because the size differences between the piston surfaces within the master and slave cylinders allowed for an easy multiplication of the forces exerted by the motorist's foot on the brake pedal.

Whether mechanically or hydraulically operated, however, with increasing car speeds the motorist's braking energy had to increase as well. Some of the variations proposed to solve this involved increasing the effective brake lining and drum friction surfaces (by mounting two parallel shoes, or by adding an extra drum per wheel), but this was obviously limited by the space within the wheel hubs and by the insight that the unsprung weight should be kept low. Engineers soon introduced so-called servo systems by positioning and connecting the brake shoes such that they reinforced each other's pressure against the drum surface. Bendix proposed a duo-servo brake in 1927 (with two self-reinforcing brake shoes, effective in both directions of the drum), which soon was applied on the rear axles of several American cars. Lockheed developed a duplex system for the front wheels (with two slave cylinders per drum), and even drum brakes with three self-reinforcing shoes were applied. Because wheel brakes with such systems were difficult to synchronize (causing skidding during braking), engineers later preferred to shift the servo effect outside of the drum by introducing so-called power brake systems, with trailing instead of self-reinforcing shoes in the drum brakes, a nice example of *functional drift*, leading to *functional split* (the servo function transferred from within the drums to an outside location, resulting in the separation of the braking and the reinforcing function).

Power Braking

Just like most other aspects of hydraulic braking, power braking was prefigured in a fully mechanical form, for instance in the Rolls Royce Phantom (1933) equipped with a friction clutch that connected the transmission's exit shaft with the brake master cylinder when the brake pedal was operated. Such friction servos were used by Bentley until 1965. Nonmechanical energy for such systems was often tapped from the engine's intake system, such as the most well known of all, the Dewandre vacuum brake servo (proposed by the Belgian Albert Dewandre in 1924) and its American competitor, the Mastervac servo by Bendix (Figure 8.4). Together with the Hydrovac (a constructive combination of vacuum servo and master cylinder), the Bendix systems were built into more than 1 million military vehicles during WWII.

Hydraulic brake assist systems were introduced as early as 1927 (Fiat, Delage, equipped with four-wheel systems), but the real breakthrough of this alternative appeared with the Hydro-Booster from Bendix (Lincoln Continental, 1977, the hydraulic energy tapped from the power steering pump), although the Citroën DS-19 from 1955 included it into its overall hydraulic system, which also controlled the spring-and-damper systems. In Europe, the hydraulic brake boost system appeared on the BMW 7 series, from 1977. Even electric systems appeared on the do-it-yourself market in the United States in 1934 (8-6], p. 408–410), ([8-3], p. 65–67), [8-16]. We have seen earlier (see Chapter 6) that heavy trucks followed a different path: that of the more bulky but cheaper air pressure braking system.

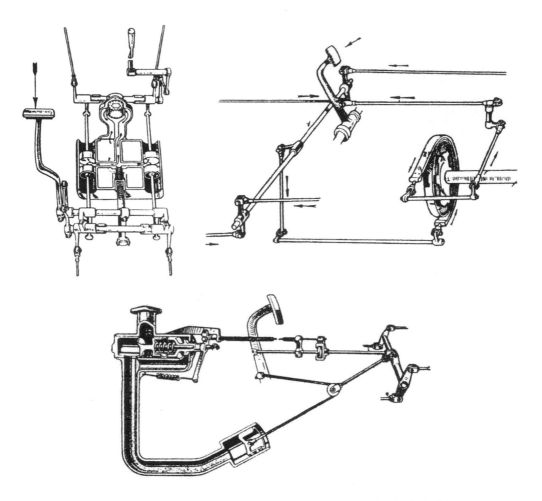

Figure 8.4 Power brake systems: hydraulic (Fiat), mechanical (Ballot) and on engine vacuum (Dewandre, Ballot) ([8-3], p. 66).

Increasing braking forces also caused changes in brake lining material. Metal wires embedded in the asbestos to enable heat dissipation were introduced as early as 1909, whereas Ford used cotton textile soaked in asphalt until the 1920s. Asbestos-free linings became available from 1938 in Germany (in the context of efforts to develop a self-sufficient economy) and around the same time in the United States (because of lower costs). Only after WWII, with increasing environmental and health concerns, asbestos was replaced by plastic fibers. By then, in order to avoid the need to drill the linings and rivet them to the shoes, brake linings started to get glued to the pads, a practice initiated by Chrysler, Chevrolet, and Crosley in 1949 ([8-5], p. 136, 138–139), ([8-6], p. 400), ([8-3], p. 62). Next to the brake fluids with higher boiling point, brake drums were mounted with cooling fins or made from aluminum (around a cast-iron friction ring), but the fear for fading, the sudden drop of friction characteristics through a chemical recomposition by the temperature rise in the linings, remained very real. This fear increased when cars could be run on freeway speeds by the 1950s, and automatic transmissions diminished the braking effect of the engine when the accelerator pedal was released.

The Disc Brake Revolution, Interrupted

It was the disc brake that promised to solve this problem. As usual, the disc brake had its predecessors, most notably car pioneer F. W. Lanchester's patent of 1902 (Figure 8.5). Alternative options were proposed, such as brakes in which the entire disc surface acted as friction surface (applied in the multiplate brake of 1911 in a Metz car, or even in a Chrysler of 1949).[2]

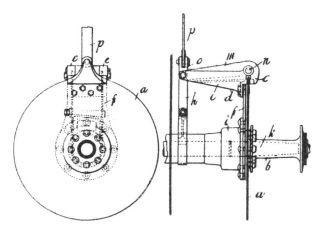

Figure 8.5A

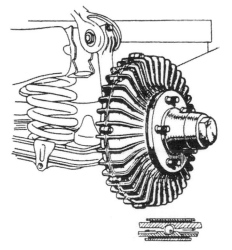

Figure 8.5B

Figure 8.5 Early disc brakes; A: Lanchester, 1902; B: Chrysler, 1949 ([8-3], p. 67–68).

---

2. For an extensive overview of disc brake history see ([8-6], p. 410–419),([8-5], p. 289–307).

Before the disc brake's breakthrough on front wheels after WWII, they appeared on military vehicles and aircraft during the war. Because of its higher costs, this was a typical innovation starting at the top of the car pyramid. The D-type Jaguar that won the 24-hour race at Le Mans in 1953 was equipped with the new brake type, because it could approach corners much faster than the competition. BMW had disc brakes on its V8 model (502, 3.2 L Super) in 1959.

Partial disc brakes (such as the one in Figure 8.5A) allowed better cooling: they were preferred because disc surfaces within the wheel hubs were not large enough to produce the same braking energy at the same hydraulic pressure as a comparable drum brake, so brake pressures had to be increased, causing larger heat build-up ([8-5], p. 262). This is a nice example of a solution causing new problems.

In Europe, the transition to disc brakes was rather quick, with the Renault R4 as one of the last in 1986. Of the 24 makes having front-wheel disc brakes in 1963, only one was American. The heydays of the disc brakes in the United States started only in the second half of the 1960s, when the disc with fixed caliper (having two cylinders on both sides of the disc, stemming from aircraft technology) was superseded by a simpler and much less expensive system with a floating caliper, with only one slave cylinder (Figure 8.6). Chevrolet applied the new system in 1969. The incentive to shift to a simpler and more compact system increased when center point steering, with the vertical steering axis meeting the center of the contact surface below the tire, enhanced steering comfort as this made the turning of the wheel easier.

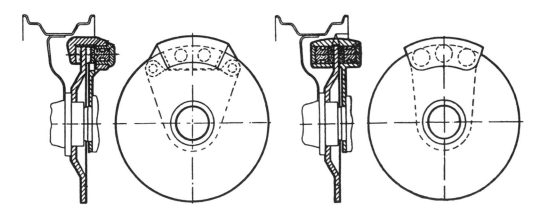

Figure 8.6 Disc brakes with floating (left) and fixed caliper ([8-5], p. 288) (*courtesy of Erik Eckermann*).

The disc brake solved the fading problem, but efforts to control its temperature behavior continued, such as adding internal ventilation or replacing the cast iron of the disc by carbon fibers. The latter solution was applied in Formula 1 racing cars in 1985, which cut the disc's unsprung weight in half ([8-6], p. 419). Contrary to the general expectation of a quick universalization of the disc option, however, this seemingly linear development trajectory was interrupted as the industry settled on a hybrid system consisting of discs on the front wheels and drums (in which a parking brake could easily be installed) in the rear wheels, connected with the master cylinder through a diagonally separated system.

Mechanical ABS

It was this configuration that formed the basis of the worldwide attempts to automate the braking process in the case of certain dangerous situations. The idea to prevent the braked wheels from being locked (jumping from rolling to sliding friction, unable to counteract the inertia forces of the vehicle anymore) had been formulated nearly 150 years before by a train braking specialist from Westinghouse: "In order to reach the maximum deceleration power of a train the wheels may never be allowed to slip. The pressure of the brake blocks on the wheels should in fact be diminished just before slipping starts. The best solution would be to start with a very high pressure between block and wheel as soon as the brakes are deployed, and then lower it gradually until the train has come to a full stop."[3]

The prolific transitory phase of the inter-war years saw several inventors (such as Werner Mom in the United States, in 1932) ([8-6], p. 422) trying to develop a viable anti-lock braking system. As early as 1920, French top-car producer Gabriel Voisin experimented with a system in which every wheel propelled a small pump; the brake pressure collapsed as soon as the wheel stopped turning. In 1938 a system for trains was developed.

It was Dunlop's fully mechanically controlled Maxaret system introduced 1952 for aircraft, which was also tested on a Ferguson racing car in 1961. Later, fully mechanically operated systems were proposed and developed (such as the SCS Lucas-Girling system), but antilock brake automation only succeeded when car electronics emerged by the end of the 1960s, as we will see in the next two chapters.

8.4   Conclusions

The braking of a car is a nearly ideal example of technological evolution: incremental changes on several levels of the car's structural hierarchy along a variety of pathways, interspersed with punctuated [8-17], [8-18] occurrences of *radical change*.

At the basic level, the evolution of the wheel brake proper shows an early *lock-in* of the internal drum brake. For a while two different cultures coexisted: in Europe this brake type was near universal on the rear wheels before the First World War, whereas in the United States the external drum brakes close to the transmission enjoyed some popularity.

---

3. Engineer Douglas Galton, quoted in ([8-3], p. 69).

Once locked in as a concept, a *variation* of solutions emerged during the 1920s, circling around the problem of brake force multiplication within the brake, leading to duplex and servo brakes, until, during the 1930s, the problem was delegated toward the next-higher system level, where power brakes were developed. It freed the drum from self-reinforcing brake shoes, which were very difficult to synchronize between wheels.

Power brakes witnessed their heydays after the Second World War, when they were combined with disc brakes, a radical innovation stemming from railway and aviation technology, which by the end of the 1960s started to become universal on front-wheel brakes, while the rear wheels often kept using the drums, because they allowed an easy combination with the parking brake system and because they were cheaper.

At the system level, meanwhile, the shifts went from two-wheel to four-wheel configurations, and from a mechanical (rods, cables) to a hydraulic set-up, according to a pattern proposed by Malcolm Loughead in the 1910s and early 1920s. Heavier vehicles opted for air braking, cheaper but bulkier. When power steering emancipated from its mechanical set-up, braking sometimes borrowed from the overall hydraulic system, but the most popular system was pneumatic, borrowing its energy from the engine. Electric systems were also proposed.

The preference, especially in the United States, to combine hydraulic and mechanical systems reveals a remarkable mechanical bias among automotive engineers, a fear for leakage rather than for break. This conservatism of mechanical engineers of the chassis engineering type haunted the industry for at least two decades and had to be overcome by external force: by the cross-fertilization of automotive and aeronautical engineering, as we saw in the previous chapter, a process we will study in detail in the next chapters.

The fear for leakage also points at a discourse around brakes that became increasingly important: the safety discourse. It led to efforts to decrease the risk of skidding further by developing automated antilock braking systems. This trend of *automation* could thrive on a parallel trend of *homogenization* of the braking system, where the disc brake and the hydraulic actuator became mainstream, and another trend of *convergence* of American and European culture to one universal, homogenized system. Other trends supported this as well, such as the trend to *compactness* as well as a *trend toward low costs*.

However, this case also shows how technological evolution is not so much about substitution of old by new, but often about coexistence of old *and* new: drums and discs, mechanical (parking) and hydraulic brake actuation, vacuum and electric power braking each play their roles in modern braking systems. Homogenization does allow for *hybridization*, allowing remnants of an earlier trend or variation to be revived and coexist with what most engineers consider the real innovations. That the "old" technologies show at least as much innovative creativity in adjusting to the new, threatening technology, is often forgotten.

This case also showed the importance of the *Americanization* debate, flaring up time and again among historians of technology. In this debate, the consensus has meanwhile emerged that this process has been much more complex than a simple one-way scheme of influences from America (as the term implies) would suggest. It was a difficult interplay in two directions, full of *translations* (adaptations of foreign technologies to the local circumstances) and crossovers, testified by the invasion of "American" hydraulic brake technology (à la Lockheed) as well as by the opposite phenomenon of the invasion of the radial tire from Europe.

Now that we have given an overview of the most basic technologies that emerged within the car during the 20th century (increasingly introducing the systemic and scientific aspects as we went through the chapters) , we will be able to zoom out a bit more and take a wider context into consideration. We will do so in the following eight chapters. In preparation to these chapters, the following table gives a wrap-up of the trends we have identified so far in the concluding sections of this and the previous chapters (Table 8.1).

Table 8.1 Overview of basic trends in the history of automotive technology	
Trend	**Definition**
variation (of properties)	the increase of different (sets of) properties enabling the (more or less) same function(s)
speciation	(or bifurcation) the emergence of new species as a result of the process of variation
homogenization	the decrease of variation of properties enabling (more or less) the same function
divergence (and convergence)	a special form of variation at the population level: structured speciation (bifurcation) on another continent, in another country, within another vehicle type (f.i. trucks, racing cars) (respectively its reverse)
complexification (& simplification)	the tendency of artifacts to combine more (& less) components (and split [& combine] the properties) to enable the same general function in a more sophisticated way
compactification	decrease in size of an artifact with a constant set of properties
automation	the tendency of self-control of machines, resulting in a shift from (human) skills to (technical) properties
black-boxing	the tendency to require less technical knowledge to understand the properties of an artifact (example: the piston, nowadays bought from the shelf)
cheapening	purchase price decrease though simplification, or through deliberately not choosing a more sophisticated solution (examples: rack-and-pinion steering; drum brakes on rear axles). Cheapening can be part of the manufacturer's strategy of price differentiation between market segments.

References

8-1.　Gijs Mom and Ruud Filarski, *Van transport naar mobiliteit; De mobiliteitsexplosie (1895–2005)* (Zutphen: Walburg Pers, 2008).

8-2.　Gijs Mom (met medewerking van Charley Werff en Ariejan Bos), *De auto; Van avonturenmachine naar gebruiksvoorwerp* (Deventer: Kluwer Bedrijfsinformatie, 1997) (Conam-reeks nr. 3).

8-3.　Gijs Mom, "Het rijdend gedeelte in historisch perspectief," in M. Arkenbosch, G. Mom, and J. Nieuwland, *Het rijdend gedeelte; Band A: historie, theorie, banden en wielen, besturing* (Deventer, 1989) (Part 4A of *De nieuwe Steinbuch; de automobiel; handboek voor autobezitters, monteurs en technici onder redactie en coördinatie van drs. ing. G.P.A. Mom*), 25–71.

8-4.　Arnold Heller, *Motorwagen und Fahrzeugmaschinen für flüssigen Brennstoff; Ein Lehrbuch für den Selbstunterricht und für den Unterricht an technischen Lehranstalten aus dem Jahre 1912* (Moers: Steiger Verlag, 1985) (Reprint of 1922 edition, 1st ed.: 1912).

8-5.　Erik Eckermann, *Dynamik beherrschen; Alfred Teves GmbH; Eine Chronik im Zeichen des technischen Fortschritts* (Frankfurt am Main/Stuttgart: Alfred Teves GmbH/ Motorbuch Verlag, n.y. [1986]).

8-6.　Olaf von Fersen (ed.), *Ein Jahrhundert Automobiltechnik; Personenwagen* (Düsseldorf: VDI Verlag, 1986).

8-7.　M.R. Clements, *King of Stop and Go—the story of Bendix: A history, 1919–1963 in South Bend, Indiana* (n.p., n.y. [South Bend: Bendix Aviation Corporation, 1963]).

8-8.　Gijs Mom, "Constructing the State of the Art: Innovation and the Evolution of Automotive Technology (1898–1940)," in Rolf-Jürgen Gleitsmann and Jürgen E. Wittmann (eds.), *Innovationskulturen um das Automobil; Von gestern bis morgen; Stuttgarter Tage zur Automobil- und Unternehmensgeschichte 2011* (Stuttgart: Mercedes-Benz Classic Archive, 2012) (Wissenschaftliche Schriftenreihe der Mercedes-Benz Classic Archive, Band 16), 51–75.

8-9.　"Four-Wheel Brakes," *Journal of the Society of Automotive Engineers*, 13 No. 1 (July 1923), 70–72.

8-10.　Henri Perrot, "Four-Wheel Brakes," *Journal of the Society of Automotive Engineers*, 14 No. 2 (February 1924), 101–106.

8-11.　J. R. Cautley and A. Y. Dodge, "Development of a Modern Four-Wheel Mechanical Braking-System," *Journal of the Society of Automotive Engineers*, 17 No. 1 (July 1925), 87–90.

8-12.　*Journal of the Society of Automotive Engineers* (December 1939).

8-13.　Erick Eckermann, "Fritz Ostwald—Protagonist des negativen Lenkrollradius; Sein Demonstrationsmodell: ein Technik-Füllhorn," *Automobil Revue* (1 August 1996), 19, 21.

8-14. Colin Campbell, *Automobile Suspensions* (London: Chapman & Hall, 1981)

8-15. Jan Norbye, *The Car and Its Wheels—A Guide to Modern Suspension Systems* (Blue Ridge Summit: TAB Books, 1980).

8-16. Olaf von Fersen, "'Negativer Lenkrollradius'; Ein geometrischer Trick entpuppt sich als wichtiger Sicherheitsfaktor," *Automobil Revue* (23 March 1973), 25, 27.

8-17. James M. Utterback and Fernando F. Suárez, "Innovation, Competition, and Industry Structure," *Research Policy* 22 (1993), 1–21.

8-18. Connie J.G. Gersick, "Revolutionary Change Theories: A Multilevel Exploration of the Punctuated Equilibrium Paradigm," *Academy of Management Review* 16 (1991), Nr. 1, 10–36.

PART II
SYSTEM

Chapter 9
Automation: Driver Deskilling and the "Electronic Revolution"

9.1   Introduction: Trends in Automotive Technology

In the previous eight chapters, we have focused on the evolution of subsystems and their component assemblies. In the coming three chapters (and in a certain way also in the remaining five of Part II) we will zoom out and look at developments that cross the boundaries of the subsystems. This enables us to identify trends of a higher order (touched upon very briefly in the previous chapters), such as automation, safety, the emergence of an environmental consciousness around the car (dealt with in Chapters 9, 10, and 11, respectively) and scientification (Chapter 12). Such more general trends build upon the *basic trends* we have mentioned on several occasions but in a nonsystematic way during the previous chapters. Table 9.1 gave an overview of these trends. Other, less frequently mentioned trends in the previous chapters such as Americanization will play a role at the aggregate level and will be wrapped up in the final Chapter 16.

Trends are not linear: they come and go, emerge and fade away, bifurcate and counteract each other, but we cannot deny that overall, seen from hindsight, there exists some *path dependency* (and perhaps even some *teleology*) in their unfolding as long as one concentrates upon the typical artifact. The car indeed became more safe, got automated, and diversified while its components became more compact and more complex, and its design and production became more scientific, as we will see in Chapter 12. Indeed, from the perspective of the typical artifact, technological development differs from random biological evolution. It is at this level that debates among historians of technology rage over the question of the purposefulness of the technological path.

On a population basis, however, developments are less clear-cut: it is doubtful (or simply not true) whether global automobilism became safer or whether its environmental burden diminished. Even on the most basic level, the level of technical trends, the population of cars has evolved, on average, toward higher compression ratios in the combustion engines and lower rolling resistance values of tires, but maximum engine power and vehicle weight have first diverged (affording different car cultures in the United States and Europe) and then converged. They thus reflect a trend of a higher order: the homogenization of global car technology during the last 50 years (Figure 9.1; also see Figure 1.3 and Figure 2.6).

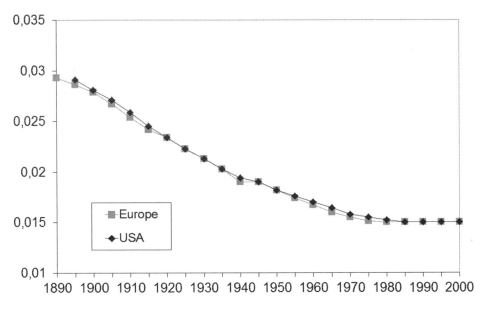

Figure 9.1A

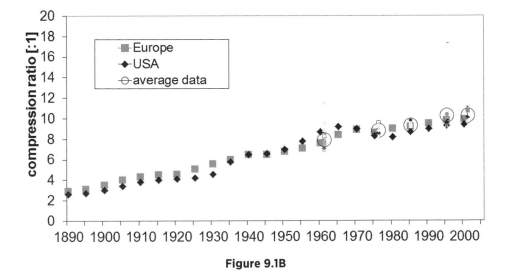

Figure 9.1B

Figure 9.1 Basic technical trends: tire rolling resistance (A) and compression ratio (B), both an expression of the engineering state of the art and, thus, not very different for Europe and the United States. *Source:* See Chapter 1, Figure 1.3 and n. 4.

Automation and Electronization

In the final, concluding chapter, we will come back to the issue of trends, distinguishing between basic evolutionary, automotive, and more general societal trends. Focusing, in this chapter, on the automotive trend of automation, three remarks are important before we enter in its development over the last century. First, just as most other trends, automation is not a linear process, but it develops in leaps and bounds and can even be redone. For instance, the governor that regulated engine speed disappeared (as we saw in Chapter 3) and full control of engine speed was given back to the driver, even during the process of his and her deskilling.

Second, automation, as we have seen on multiple occasions in the previous chapters, is not the same as electronization. Far from it: the tendency to deskill the driver has been as old as the car itself. Remember, for instance, the introduction of centrifugal and vacuum advance in battery ignition systems during the inter-war years, as shown in Chapter 2, or the hydraulic regulation of gear shifting in the automatic transmission, in Chapter 5. This trend struggled, so to speak, with that other tendency (weaker, it turned out) that knowledgeable users wished to keep control over their car. Such users deplored the manufacturers' urge to simplify the car's handling, making it (literally) foolproof, usable for the lay man and woman who wanted to use the car as any other commodity, such as a refrigerator or a washing machine. And yet, there still seems to be a bottom line that even the most deskilled lay driver does not yet wish to cross: the decision to seal the cyborg-like unity of car and driver in the form of the smart car and leave the major decisions in terms of safety and driving fun to the machine is still a bridge too far, as we will see in Chapter 16.

And third, not every automatic sequence in the car is the result of automation. The distribution of electric energy over the four spark plugs in order to create a spark in accordance with the working cycles in every cylinder is an automatic process, but it has never been done by hand, for the simple reason that the process is (and was) too fast in the first place. Likewise, the lambda sensor measuring the oxygen in the exhaust gases to control mixture formation electronically (see Chapter 11) is no automation either: it is an added, automatic function that would not have been possible without the introduction of that sensor. Nor can the force multiplication taking place in the hydraulic brake circuit, generated because of the difference in pressure surfaces of the master and slave cylinders (as we have seen in Chapter 8), be called automation: not only was the same multiplying effect accomplished before, when the braking system functioned purely mechanical (with rods and cables), but even then the effort to apply braking force (by hand or by foot) was not taken from the driver.

The borderline between increasing technical sophistication and automation is fuzzy indeed: how to judge, for instance, the shift from the thermometer measuring the slow temperature rise of the engine cooling fluid by a simple emergency light that urges the driver to stop? Although the stopping itself still has to be done by the driver, the skill of judging a sudden temperature rise is clearly simplified by the invitation to stop. Knowledge of engine thermal behavior is replaced by knowledge to obey a red blinking warning light. The case is comparable to the change from cranking the engine to deploying the starter motor, which in most cars still has to be actuated by turning the ignition key but does not necessitate us anymore to go out of the car and apply the hand crank. However this may be, automation takes place at the *interfaces* between driver and car. From the perspective of the dual nature of technology (see Chapter 1), automation describes the *shift from functions to properties*.

Driving Skills

This shift can probably best be understood through the work of American psychologist James Gibson, whose concept of affordance we introduced in Chapter 1 as the opportunity to act offered by the artifact and observed by the user. Gibson's experiments showed that the perception of the environment is "accompanied by proprioception of the active self" ([9-1], p. 195). The term proprioception, the sense of one's own body parts, already indicates that skills and affordances are closely related to our senses.

Automation, then, can be described as the process of unburdening some of the driver's senses to enable him or her to concentrate on those senses that really matter in the daily experience of car driving. Such senses are eye sight for observing the surroundings (traffic, landscapes) and the body's haptic capacities to feel the movement of the car (as feedback for safe steering, as the thrill of accelerating). Table 9.2 gives a taxonomy of functions and skills and their related properties, senses, and the automation that was deemed necessary in order to enable the driver to concentrate on his or her main practices. The table certainly does not cover the entire spectrum of skills, let alone that it gives an overview of their historical change: it is, considering the state of the art, a first tentative sketch of what should be studied in much more detail by historians of technology and of psychology.

The table also shows that several skills and interfaces we have been dealing with in the previous chapters have meanwhile disappeared: ignition advance, mixture strength formation, lubrication of the chassis, and the enriching of the mixture when starting the cold engine have all been taken from the users' hands, literally. In a broader perspective, the diagnostic listening to the sounds of the engine and the car while driving has been delegated to the mechanic in the garage, while the washing and polishing of the car body has been delegated to the car wash [9-2].

In some cases, driver skills nearly vanished in one subculture, but stayed on in another. The automatic transmission which broke through into a universal form of deskilling of gear shifting in the United States, followed by Japan but rejected by the majority of car drivers in Europe up to this very moment, is an example. Automation, in this case, was accomplished by entirely hydro-mechanical means, and electronization only fine-tuned this process, but could not drastically alter it anymore.

Table 9.1 Taxonomy of skills and related properties and senses

	Function/skill	Property/interface	Automation	Sense/cyborg organs
Basic driving skills	Steering	Steering wheel	Power steering	Hand
	Propelling	Accelerator pedal	Cruise control	Foot
	Gear shifting	Gear lever and clutch pedal	Automatic transmission	Hand/foot
	Braking	Braking pedal	Power braking/ABS	Foot
	Interior climate regulation	Levers and switches at dashboard	Climate control	Body skin
	Overtaking	Turning lights	Lane change assistance	
	Emergency braking		ABS	
Traffic skills	Navigating		Navigation system	Eyes and ears
	Cornering	Tire		
	Anticipating	Lidar	Adaptive cruise control	
	Parking		Automated parking	
Special driving skills	Mountain driving	Retarder		
	Night driving		Night vision assistance	
Social skills	Participating in city traffic	Claxon	Blind spot detection	
	Socializing in the car		In-car entertainment systems	

Nowadays, skills can also be bought, even if the driver would never have been able to develop them herself. This is for instance the case when a modern, fully electronically equipped all-terrain car offers the possibility to select General Driving, Grass/Gravel/Snow, Mud/Ruts, Sand and Rock Crawl, as is the case with the Range Rover of the fifth generation. According to some national legislators, some skills are better not combined, such as using the mobile phone and steering a car, because both make use of the brain's "working memory" [9-3], [9-4]. It would require another book to describe the development of these skills in conjunction of the changes in the technology. Because of a lack of studies in this respect, we will focus, in the remainder of this chapter, on the electronic "revolution" in general (Section 9.2), while we deal with car navigation as a case study in Section 9.3. We will close this chapter with conclusions (Section 9.4).

9.2   Postwar Automation, an Electronic Revolution?

Although, as we saw in the previous section, automation has a long pre-electronic tradition, it cannot be denied that the emergence of car electronics opened a vista upon an acceleration of the car's automation process [9-5]. But before the hopes of car manufacturers started to run high, so much so that they—around the centenary of the car, in 1986—began to use terms like *autronics* (coined by Alfa Romeo in 1984 ([9-6], p. 1)) to characterize the new trend, there was a long and especially hidden prehistory that goes back to the days before the Second World War. Remarkably, it was the electric vehicle that benefited first from the new technology, in the form of one of its two versions, *power electronics*. Like its second version, *signal electronics*, early power electronics was based on the electronic tube or valve, made of glass, and developed by American Lee De Forest for application in radio as a vacuum tube (see Section 9.3). His compatriot Peter Cooper Hewitt developed a gas-filled tube that could function as a one-way valve, applied in charging systems for electric vehicles.

It was the solid-state semiconductor technology, however, before the war in the form of metallic selenium rectifiers, after the war in the form of germanium transistors that enabled a miniaturization that triggered the breakthrough of electronics. This breakthrough, which can best be structured through the construction of different *generations*, which stand for different technologies dominating in different phases (Table 9.2), took place on the basis of silicon rather than germanium, because the military preferred the former as they could better operate at higher temperatures ([9-7], p. 130).

Three Generations of Electronization

According to this table, the *first generation* consisted of electronization of existing components. This was attractive for manufacturers because it got rid of moving mechanical parts, thus promising an increase of reliability and longevity, much desired in a phase of mass motorization. This generation emerged largely out of sight of the user, especially during its first phase, which started in 1958 when German supplier Robert Bosch GmbH introduced a so-called variode-regulator of the battery charging system and continued when Bosch introduced transistors in direction indicators in 1962.

The most covert conversion to the new technology happened when American suppliers in 1960 brought a three-phase alternative-current generator with electronic rectifier on the market as a replacement of the long-established direct-current type. It was quickly followed by similar systems from Japanese supplier Nippondenso (1961) and Bosch (1962).

Less hidden to the public were developments set in motion by Japanese car manufacturers who—as newcomers to global car production—used electronics as a token of their preparedness to innovate: in 1962 the Toyota Crown was equipped with an electronically controlled overdrive.

Table 9.2 Generations of car electronics

Generation: main characteristics	Phases and applications
First generation (1958–1975): replacing mechanical properties	**first phase (until end of 1960s):** from DC to AC three-phase generator with rectifiers on the basis of silicon diodes; Toyota Crown 1962 with electronically controlled overdrive **second phase (first half of 1970s):** Integrated Circuit (IC) enabled new functions: gasoline injection Bosch; Japan: electronic ignition with IC control (1973)
Second generation (1975–1985): integration, limited to engine management; emergence of microprocessor, on one chip; from analog to digital	**first phase (1975–1980):** microcomputer (microprocessor in combination with a fixed program memory) controlled ignition systems; electronically controlled carburetor in 1978; integrated engine control (gasoline injection and ignition system) 1979; trip computer 1979; self-diagnose system 1980; automatic transmission 1981; Bosch digital ABS (1978) and Motronic motor management (ML 1) 1979; LH1-Jetronic with heat wire (1981) succeeding analog L- and K-Jetronic **second phase (1980–1985):** **breakthrough; start of "electronic revolution"** 1980, 100% of sold cars in Europe equipped with electronic rectification and voltage regulation 1980: Siemens attacks Bosch monopoly
Third generation (1985–2010): development of electronics as a car-specified system	active systems; communication with outside world: navigation
Fourth generation (2010–now): integrating the interior into the car management system: comfort, warning systems	entertainment systems; seat adjustment system; lane change warning and other persuasive systems described in Chapter 16; smart car

The best example from this phase, however, is the electronization of the ignition system, up to then defect cause number one with its contact points and its delicate centrifugal and vacuum advance mechanisms. This process started, as in most other electronic systems, by the supplier industry in the beginning of the 1960s. It was to be found in nearly every American car in 1975, whereas European cars at that moment only were equipped with the new technology for 5%, a share which increased to one-third by the beginning of the 1980s. The reason for this was the independence of the European supplier industry (who turned out to be unable to deliver in time), whereas most

American suppliers (AC and Delco at General Motors, Motorcraft at Ford) were integrated into the car manufacturer.

The *second phase of the first generation* brought an expansion of the replacement possibilities through the introduction of the integrated circuit (IC), which at the same time enabled the creation of extra functions, hardly conceivable in mechanical or hydraulic form. Japanese firm histories have this phase start around 1965, while the European newcomer to the automotive field, Siemens, placed it in the beginning of the 1970s. The IC not only made the DC generator and (as we saw previously) the mechanically/pneumatically controlled ignition disappear within the next decade, it also enabled the electronization of mixture formation. Most well known became Bosch's Jetronic systems, such as the D-Jetronic (in the Volkswagen 1600E of 1967, its Japanese version marketed by Nippondenso as D-EFI in 1971) and the electronic, analog versions of the L- and K-Jetronic (1973), on the market in 1981 and 1982, respectively.

The *second generation* is about integration of digital functions within a microprocessor on one chip and covers about a decade (1975–1985). Like the first, it can be divided in two phases, the first (until 1980) characterized by LSI (large-scale integration) technology, which for the first time enabled a new type of car technology, instead of an electronized, improved existing technology.

American Bendix, for instance, announced in 1973 a fuel injection system based on microprocessor control, applied in the Chevrolet Cosworth Vega of the 1974 model year. Nippondenso proposed in 1976 a micro computer (a processor with a fixed program memory) controlled ignition system (called MISAR), followed by an electronically controlled carburetor in 1978 and a fully integrated engine control system in 1979. That same year Nippondenso marketed a trip computer with a self-diagnostic function, followed in 1981 by a micro computer controlled transmission. It was in this phase that Bosch appeared on the market with its ABS (1978, see Chapter 10). One year later, its motor management system Motronic marked the breakthrough of digital electronics at this company.

The most enthusiastic application of the new technology, however, took place in Japan, where both top models (Royal and Turbo) of the Mitsubishi Galant III (1983) appeared with a hitherto unsurpassed amount of electronically controlled systems: speed-controlled power steering, adjustable spring and damper characteristics, electronically controlled four-step automatic transmission, a variable spring characteristic of the engine mountings, an automated heating and ventilation system, variable windscreen wipers, automatic switch-off of rear window heating after 20 minutes, interior lighting extinguished five seconds after the doors are closed, centrally controlled door locks, an electronically controlled windscreen washer, and more. The proliferation of systems and gadgets in this phase was so chaotic that leading officers in the car industry started to worry about the reliability of car electronics (Figure 9.2).

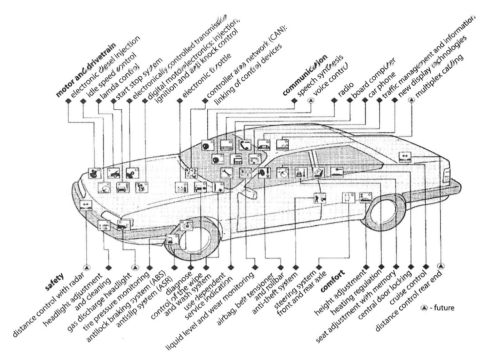

Figure 9.2 Car electronics of the second generation, in a chaotic representation by Bosch (*courtesy of Robert Bosch GmbH*).

The Failure of the Electronic Carburetor

How much the new technology influenced the market conditions can be shown by analyzing the efforts by German carburetor giant Pierburg, which tried to develop an injection system but failed. This failure, Pierburg's own jubilee history confessed, was due to a reluctance among leading officers (including Alfred Pierburg himself) to embrace electronics, part of a general adherence to mechanics among the German car manufacturers. This reluctance led to "one of the largest catastrophes the company has ever gone through" ([9-8], p. 117). Pierburg then joined forces with Bosch to found a common company to develop an electronically controlled carburetor. Introduced in 1983, the Ecotronic was applied on the BMW 316 and 518, based on Pierburg's 2BE carburetor. The most remarkable element was a lambda sensor in the exhaust system, measuring the amount of oxygen and controlling an automated choke.

But at least as important to explain Pierburg's reluctance to change to fuel injection, the jubilee publication argued, was what we have called the Pluto Effect: the electronically controlled carburetor could not be matched by competing injection systems in simplicity, they claimed (Figure 9.3). Coupled with the fact that Bosch did not allow competitors of its own injection systems to blossom, Pierburg vanished from the market as a major player in the mixture formation field, one of the most spectacular victims of the electronic revolution ([9-8], p. 140, 157, 166), ([9-9], p. 763). This chain of events deserves

179

a better study, as it provides one of the most fateful examples of a company leadership blinded by the Pluto Effect: nowadays, every student of innovation and competition between alternative solutions knows that while a new, threatening technology is in the lift, the "old" technology is also influenced by that effect. The effect is often even more misleading, because the improved "old" technology starts to grow first, followed by the new technology, and only in hindsight one can observe that the former was nothing more than a transition to the latter. If this happens, we call this phenomenon the Sailing Ship Effect (after the sailing ship that in the end had to give in to the motor ship), and it is understandable that the Pierburg staff thought that the carburetor would not behave like the sailing ship: it is an exception in an evolution governed by the Pluto Effect, which privileges the "old," mainstream technology.

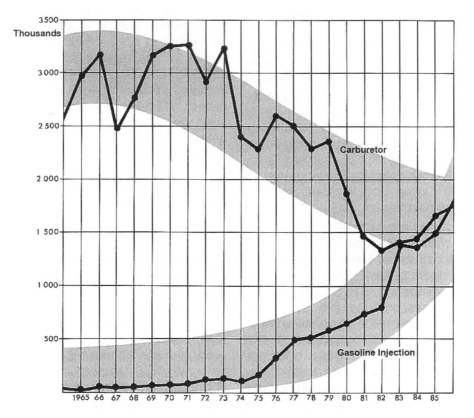

Figure 9.3 Competition between carburetors (*Vergaser*) and injection systems (*Benzineinspritzung*) in Germany between 1964 and 1986 (verticle axis, in thousands). The upswing of both competitors from 1982 must be due to the Pluto Effect: carburetors became electronically controlled as well, until injection appeared to be victorious (thus turning the Pluto Effect into a Sailing Ship Effect, where the mainstream technology succumbs to the new technology), probably due to environmentally inspired regulation ([9-8], p. 162) (*courtesy of Springer*).

Siemens Automotive as Newcomer

While Pierburg suffered, an even greater giant, Siemens, started to be interested in the new market. The largest electronics producer in Europe and the second largest company in Germany (behind Mercedes), Siemens was never much interested in the internal-combustionengine car (it had produced electric cars in the 1910s and 1920s ([9-10], p. 127–128)), probably the result of a market deal with Bosch. Siemens developed a prototype digital ignition and injection control system in 1973, but stopped this development a year later because of a negative internal market assessment. Only in 1980 did its interest revive. It tried to take over Pierburg, but this was blocked by Bosch. Then, from 1987, it started to invest heavily in car electronics, a commitment that was scaled by the takeover of the European company Bendix Electronics. The latter then evolved into a major competitor of Bosch under the name of Siemens Automotive.

Such skirmishes were the result of a sudden growth of the automotive electronics market. The turning point was 1980, in the middle of the second generation, when annual growth rates of 25% could be absorbed, four times faster than the general world electronics market. The growth was spectacular, but after a decade the car represented only 4% of this market, and more than half of that share was spent on engine electronics. By then, electronics represented 5% of the car's production costs, but optimistic prognoses predicted that by 2010 one-quarter of these costs were spent on electronics, a prediction which was not far off the mark. In 1991, General Motors boasted that the computation capacity of its cars had meanwhile become higher than that of the NASA Apollo space ship of 1969, and that the car's memory capacity had surpassed 40 kB. The car's electric power to enable this had meanwhile doubled to 2.5 kW, and in some cases even 4.5 kW.

A Crisis of Electronization

It was on the brink of the *third generation*, in the middle of this proliferation of functions and gadgets, that concerned leading officers of the German car industry, on the occasion of the centenary of the automobile in 1986, openly confessed that they faced a multiple electronics crisis [9-11]. First of all, suppliers started to have bigger economies of scale than car manufacturers, because they supplied more than one manufacturer. This led to a tendency of *homogenization* of technology, and the fear, among the car industry, that the character of each and every car make would be lost. The *democratizing* effect of car electronics, such was the fear, would make the car into just another example of consumer electronics, which was characterized by a low degree of brand loyalty and an emphasis on the utilitarian aspects of the commodity, like a refrigerator or a washing machine. Volkswagen, for instance, decided to keep the design of the main traits of software development in its own hands, a decision that later was followed by the suppliers toward their subsuppliers. Nowadays, for instance, Bosch has 6,000 software engineers working in India, but it keeps the steering of the design process in-house [9-12]. Software is the tool to integrate hardware ([9-13], p. 178).

A second element of this crisis was the explosion of defects, caused mainly by a comparable explosion of the number of connections. Apart from a technical fix in the form of solid state (which helped to overcome the connection problem to a large extent), the answer of the industry was a revival of the systems approach. This approach initially took the shape of a centrally controlled architecture, but soon it developed into a hierarchy of decentralized and mutually communicating subsystems. Thus, the automotive counterpart emerged in what in general technological development was called mechatronics. The new generation was "intelligent," eqipped with smart sensors, EEPROMS (electrically erasable programmable read-only memories) and clever testers, whereas "multiplexing" was proposed to solve the "chaos" of wiring (Figure 9.4 and Figure 9.5).

Clever testers became a necessity because electronic systems behave differently from mechanical systems: they tend to fail suddenly, without prior warnings to the skilled ear of a knowledgable motorist, failure that cannot be prevented by increased main-tenance. On top of that, electronic failures follow statistical patterns: combining many components with a small failure rate can result in a high overall failure rate that may directly affect car safety. Hence the interest among car electronics developers in military and aeronautical experience of self-test, feedback, and fail-safe systems ([9-6], p. 4–5).

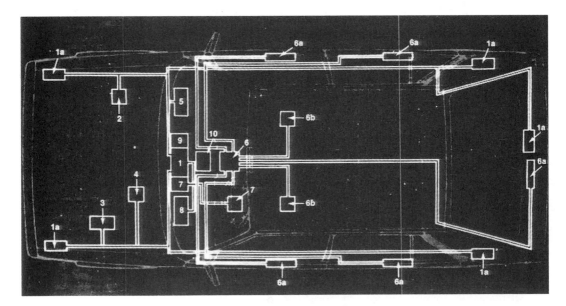

Figure 9.4 Multiplexing proposed by Motorola in 1983 ([9-5], p. 43).

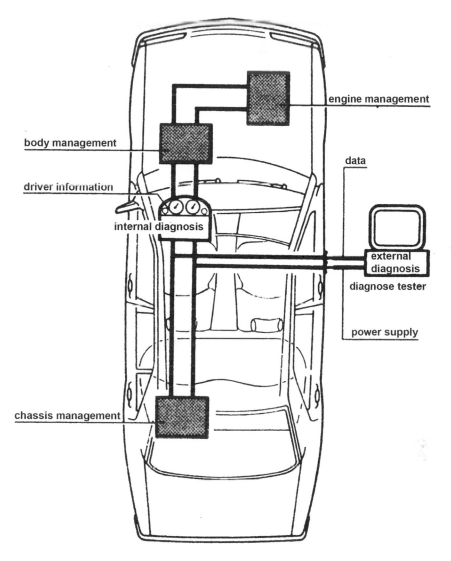

Figure 9.5 System electronics as proposed in the 1980s ([9-6], p. 6).

With the third generation, "active" technology became possible, such as Toyota's ECT-I ("electronic [sic] controlled transmission") introduced in the Lexus of 1990, or the motor management system of the Cadillac Seville of 1992, which regulated ignition timing and reduced gasoline delivery at the moment of gear shifting, or the extension of ABS with antislip control during acceleration (ASR). In general, in this phase the three major management systems of engine (mixture formation, ignition, exhaust gas control, anti-knock control, intake valve control), drivetrain (transmission control, clutch slip control, differential control in four-wheel drive cars), and chassis (ABS, ASR, adaptive springs) got connected into a vehicle management system (Figure 9.6).

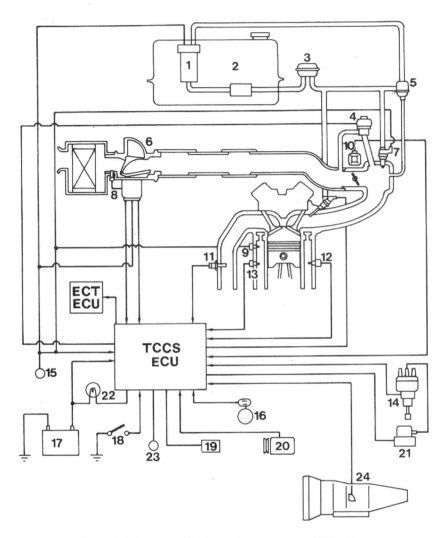

Figure 9.6 Motor and drivetrain management TCCS from Toyota, model year 1985 ([9-14], p. 196).

It was, however, the *fourth generation* of today, which allowed to integrate the car's interior into the system, optimizing comfort and entertainment systems, from the audio installation through parking support and seat and mirror adjustment systems, to a full thermal management system.

9.3   The Lay Motorist and the Navigation Revolution

When manufacturers, around the last turn of the century, started to equip their products with entertainment and comfort electronics, the feeling of a revolution spread among the users. They were truly lay users, as tinkerers were the first victims of the electronic revolution.

Entertainment electronics have a long tradition, although nobody in the early car radio days during the inter-war years would have called it such. In fact, the history of mobile communication goes even further back, as the experiments of radio pioneer Guglielmo Marconi (1874–1939) with a steam bus around the turn of the 19th century, and an American patent from 1908 on a wireless telephone in a horse-drawn carriage testify (Figure 9.7).

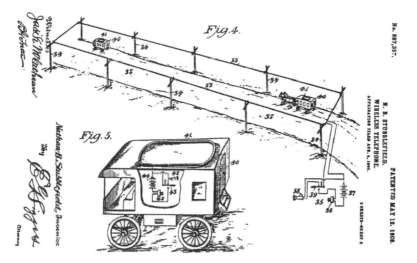

Figure 9.7 Patent on a wireless telephone in a horse-drawn carriage from 1908 ([9-5], p. 20).

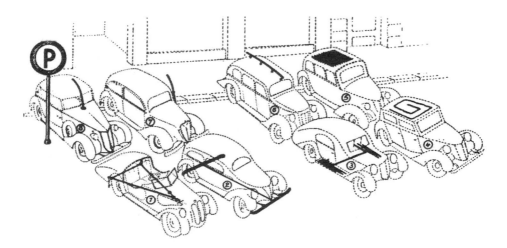

Figure 9.8 Variation of radio antennae ([9-15], p. 621).

Car radio became popular in the United States during the 1930s, and it seems that the antenna problem was one of the *reverse salients* (see Chapter 4) in this development (Figure 9.8). Another problem, also related to the connection of the car's internal structure to the car system, was the problem of interference caused by the spark ignition of the gasoline engine. Placing heavy resistances (up to 50 kilo-ohm) in series with the spark plug alleviated the problem considerably, but also increased fuel consumption by one-third, a problem that started to play a role when the radio migrated to Europe with its much less powerful engines: in Europe car radio caused a retarded spark, which led to burned exhaust valves and reduced engine power, reminding engineers once again that the car had a systemlike structure.

And while these and many other problems gradually became solved during the next three decades, the solutions resulted in separate cultures between the United States and Europe. There was a trend of custom design of radio sets; here was an emphasis on standardization. Conversely, there was a preference for medium wave lengths and VHF; here was a plethora of standards, from medium wave in the UK to long wave in France, to VHF in Germany, while the Treaty of Geneva in 1975 led to an overpopulation of radio stations per bandwidth. The radio evolved into much more than a simple device to get the outside world into the cocoon of the closed car body: when BMW in 1986 equipped its product with up to ten loudspeakers, it was quite clear that the car had become a "home on wheels."

The Fourth Generation of Electronization

In this wheeled home, the fourth generation brought new functions in the car that would not have been possible without electronics, because they did not mimic mechanical, hydraulic, or pneumatic systems as was the case during the previous generations. The best example of this is car navigation, the effort to deskill the driver in his or her ability to find their way in ever increasing traffic. Early one-way systems such as Blaupunkt's ARI (*Autofahrer-Rundfunk-Informationssystem*, car driver radio information system) were proposed as early as 1974, but the development of automotive communication systems received a decisive boost when French President François Mitterand in 1986 initiated a Europe-wide research program called Eureka meant to solve what was considered to be Europe's backwardness in electronic innovation. One of the Eureka projects was Prometheus (Project for a European traffic with highest efficiency and unprecedented safety), led by the German car industry, which developed prototypes of so-called convoy systems, where cars became an electronically controlled part of a "vehicle train," and the "intelligent car" had to "cooperate" with an "intelligent road."

In the United States comparable projects were initiated. They were part of an older dream of car use in which the human driver is regulated away (Figure 9.9). This, of course, is seen from the perspective of the knowledgeable user-tinkerer, and the industry claims that new generations, educated with social media as a matter of course, will look at this form of automation with quite a different attitude: they may compare it positively to their personal communication media and would see a smart car as just an extension of their daily habits and culture of staying wired. No wonder, then, that current cooperative research projects again are seriously considering the implementation of such systems, for instance the EU-sponsored SARTRE program (Safe Road Trains for the Environment).

Figure 9.9 Early examples of the driverless dream ([9-5], p. 46).

Another Eureka project was CARMINAT, a cooperation between Dutch electronics giant Philips, French car manufacturer Renault, and French telecom companies TDF and Sagem. Philips brought its CARIN (car information and navigation) system in, while Renault shared its ATLAS car driver information system and Sagem its navigation system MINERVE.[1] In Japan, Honda proposed the Electro Gyro-Cator in 1981 working on the basis of a gyro compass from aeronautics, a system that was pulled back from the market after 250 sold copies, because nobody wanted a system that needed a constant changing of a transparent map to be placed upon the screen.

Three Generations of Navigation Systems

During this *first generation* of traffic information and car navigation a shift toward so-called autarcic systems took place, when systems relying on a special infrastructure (for instance infrared beacons on every street corner in Bosch's EVA system [*Elektronische Verkehrslotse für Autofahrer*, Electric traffic pilot for motorists, 1982]) were superseded by systems working with electronic compasses, such as sensors of the magnetic field of the Earth in Blaupunkt's Travel-Pilot (1987).

In 1990, Honda brought a *second generation* on the market, working on the basis of a CD-ROM. In the United States, Chrysler proposed its CLASS system (Chrysler laser atlas and satellite system), but General Motors' ETAK system became the most successful, with its licenses to Japan (Clarion) and Germany (Blaupunkt's Travel-Pilot). Such *third-generation* systems based on satellite navigation (GPS, global positioning system) led to their breakthrough by the mid-1990s, at a moment that consumer electronics giants such as Sony and Pioneer started to get interested in the technology. One of the well-known examples is TomTom, the result of a complicated history of tiny start-ups, takeovers by large companies, mergers, and deals with the car industry.

---

1. MINERVE = *Média intelligent pour l'environnement routier du véhicule Européen*, Intelligent media for the road environment of the European vehicle; ATLAS = *Acquisition par télédiffusion du logiciels automobiles pour des services*, Long-range acquisition of automotive software for maintenance.

9.4 Conclusions

By now, software is the main source of automotive innovation. With a failure rate of one per million per year ("rebooting isn't an option for most automotive ECUs"), connecting up to 80 ECUs (electronic control units) in luxury vehicles, the architecture of car electronics is still partitioned along the main subsystems distinguished from the start of automotive technology: power train, engine, chassis, and body. But the system has meanwhile been expanded with a new subsystem: infotainment, emphasizing the fun character (as opposed to the transport character) of the car. While coding in the 1970s took place in Assembler, C has become the main programming language in the 1990s of an amount of code lines that meanwhile has surpassed one hundred million. Such technology is capable of performances that are hardly conceivable in hydro- or pneumo-mechanical terms: a current diesel engine injects "fuel quantities smaller than a pinhead up to seven times per stroke of the piston," which is 420 times per second at an engine speed of 1800 per minute, controlled electronically ([9-12], p. 93).

Although the electronic "revolution" started slowly, it gave rise to a redefinition of car technology on a scale and with a structural penetration that would not have been possible through those other two technologies (hydraulics and pneumatics) used to regulate and control the mechanical behavior of the car in previous decades. It was the hybrid combinations of these servo-systems that marked the real breakthrough of sophisticated control engineering.This does not mean, however, that electronics were ready to be harvested by the car industry; on the contrary: although the European car industry was faster than the American in applying the new technology, as late as the second half of the 1970s the production of Mercedes' ABS systems was delayed because of "the unreliability of production electronics" ([9-17], p. 130, 133). Some analysts even go so far as to assume that "the user's tendency to respond to an automated system with behavioral changes (…) may actually be counterproductive to what automation should achieve," mainly because of compensating behavior such as a somewhat riskier driving style ([9-18], p. 237, 238). Nonetheless, in the next chapter we will see how combinations of hydraulic and electronic regulation led to a newly conceived relationship between motorist and car and even passenger and car, thus laying the basis for a second "reinvention" of the car, after the "invention" of the speedy family car during the inter-war years.

References

9-1.   Edward S. Reed, *James J. Gibson and the Psychology of Perception* (New Haven/London: Yale University Press, 1988).

9-2.   Karin Bijsterveld and Stefan Krebs, "Listening to the Sounding Objects of the Past: The Case of the Car," in Karmen Franinovic and Stefania Serafin (eds.), *Sonic Interaction Design: Fresh Perspectives on Interactive Sound* (Cambridge, MA/London: MIT Press, 2013), 3–38.

9-3.   Bas van Putten, "Het beest doet alles zelf," *NRC Handelsblad* (December 8-9, 2012), 22.

9-4.   "Hoofdzaken," *De Kampioen* (December 2012), 52–54.

9-5.   Gijs Mom, "De auto-elektronica in historisch perspectief," in G.P.A. Mom and A.G. Visser, *Elektronica in de auto* (Deventer, 1986) (Part 9 of *De nieuwe Steinbuch; de automobiel; handboek voor autobezitters, monteurs en technici onder redactie en coördinatie van drs.ing. G.P.A. Mom*), 17–46.

9-6.   Gijs Mom, "Inleiding," in J. Kasedorf, *Automobielelektronica; principes en toepassingen* (Deventer: Kluwer Technische Boeken, 1987), 1–7.

9-7.   John Kenly Smith, Jr., "The Scientific Tradition in American Industrial Research," *Technology and Culture* 31 No. 1 (January 1990), 121–131.

9-8.   Günter Böcker, *Auf die Mischung kommt es an; Technik für die Mobilität: Erfinden - Entwickeln - Verwirklichen* (Meerbusch/Neuss: Lippert-Druck & Verlag/Pierburg, 1990).

9-9.   Holger Bingmann, "Chapter 14: Competence; Case C: Antiblockiersystem und Benzineinspritzung (Anti-Blocking System and Fuel Injection)," in Horst Albach, *Culture and Technical Innovation; A Cross-Cultural Analysis and Policy Recommendations* (Berlin/New York: Walter de Gruyter, 1994) (Akademie der Wissenschaften zu Berlin, Research Report 9) 736–821.

9-10. Gijs Mom, *The Electric Vehicle; Technology and Expectations in the Automobile Age* (Baltimore: Johns Hopkins University Press, 2004).

9-11. Hermann Scholl, "Elektronik im Kraftfahrzeug," in [VDI-Gesellschaft Fahrzeugtechnik], *100 Jahre Automobil; Tagung Fellbach, 17. und 18. April 1986* (Düsseldorf: VDI Verlag, 1986) (VDI Berichte 595), 307–312.

9-12. Jürgen Mössinger, "Software in Automotive Systems," *IEEE Software* (March/April 2010), 92–94.

9-13. Jonathan Coopersmith, "Old Technologies Never Die, They Just Don't Get Updated," *International Journal for the History of Engineering and Technology* 80 No. 2 (July 2010), 166–182.

9-14. J. Kasedorf, *Automobielelektronica; principes en toepassingen* (Deventer: Kluwer Technische Boeken, 1987).

9-15. *Automobiltechnische Zeitung* (1938) No. 23.

9-16. K. Tagami et al., "'Electro Gyro-Cator' New Inertial Navigation System for Use in Automobiles" (SAE Technical Paper 830659, 1983).

9-17. Ann Johnson, *Hitting the Brakes: Engineering Design and the Production of Knowledge* (Durham/London: Duke University Press, 2009).

9-18. Wiel Janssen, Marcel Wierda, and Richard van der Horst, "Automation and the Future of Driver Behavior," *Safety Science* 19 (1995), 237–244.

Chapter 10
Safety: From Shell to Capsule to Cocoon, from Danger to Risk

10.1 Introduction: Closing the Automotive Body

The definition of automotive safety from a technological point of view has been formulated during the 1960s, but as a concept, safety and security are much older, and should be placed in the context of the so-called risk society. The idea to change danger into risk came up in the 18th century when investors found that commercial shipping was a safe investment as soon and as long as they could insure the ships, thus making the chance of failure calculable.[1]

Applied to the automobile this insight has two consequences. First of all, safety is a characteristic of the automobile system rather than the automobile: without its risky context automotive safety cannot be understood. Second, the meaning of the concepts of safety and risk are deeply historical, which means that in every phase there was a struggle between different stakeholders to redefine them.

For most of its history, the car's main opposing stakeholders were those within and those outside: motorists and nonmotorists (pedestrians, cyclists) developed different perspectives on the car's benefits and drawbacks in every phase, and this led to different ideas how the safety issue had to be approached. Sociologists call such perspectives situational, as they seem to emerge in certain situations, irrespective of class background. The remarkable phenomenon occurs that one develops opposing attitudes toward the car depending on whether one is driving or crossing a street in front of a car.

In the earliest phase, the open-body automobiles were extremely dangerous for pedestrians and cyclists, but the motorists did not seem to bother as much. In an arrogant and

---

1. [10-1] It was the German sociologist Ulrich Beck who coined the phrase *risk society* (*Risikogesellschaft*): [10-2]. For an application of Beck's theory to the automobile system see: [10-3].

violent atmosphere of elite automotive "adventure," they used their vehicles as a type of "shell" (a more or less protective space, but open, like one half of a clam shell) to distance themselves from others. These others were sometimes angry peasants in the countryside, who tried to defend their traditional right to use the street for their social intercourse, and inhabitants of the southern peripheries of Europe and the United States and of the colonies. These groups saw the rich motorists in their extremely heavy automobiles as an intrusion of their space [10-4].

From Open to Closed Bodies

The introduction of the closed body had an enormous impact on automotive technology as well as its production and user culture, so much so that a British journal spoke of "almost a new motoring" ([10-5], p. 735). At first, the immediate effect on the drivers was a decrease in visibility, because the metal roof and small windows (in cold weather condensed by bad ventilation) were blocking the occupants' view. Although engineers successfully struggled against their sales departments to gradually increase the glass surface (as we know from the United States, where this has been investigated in detail), while ventilation and heating systems were added as well, soon a convertible version was introduced. Remarkably, these were more expensive, and, by the beginning of the 1930s they were especially in demand in the more variable climates of New England and the Midwest rather than in the sunny South or the Pacific Coast.[2] Around the same time, engineers were still complaining that the new mass-produced seats were often less comfortable than the old open-body ones. ([10-11], p. 474). The nearly 40% weight increase in smaller cars also necessitated further growth of engine power to propel the heavier load. Some students of the car's history even suggest that the shift to the closed body may have precipitated the demise of Ford's Model T, as this model's chassis was designed to carry an open body [10-12], ([10-13], p. 116). However this may be, the change to an all-steel closed body took decades to be completed. In Germany at the beginning of the 1930s, for instance, steel represented only 40% of the average car weight, and wood was still the dominant material for body construction ([10-14], p. 297).

It was the closing of the body during the 1920s that triggered a new understanding of automotive safety (Figure 10.1). This remarkably rapid process of "encapsulation" led to a new way of communicating with the outside world, through claxons, headlights, and brake and turning signals, and seemed to privilege the motorists above the nonmotorists when it came to surviving daily traffic. After WWII, this process went through a new phase of further cocooning, which made the former capsule turn into a cocoon. Now, the small, nuclear family was enclosed in a safety cage with deformable, protective front and rear parts that had the same effect as the silk cocoon has on the fragile butterfly. This cocoon now became part of a flow of other vehicles on the freeways. Adorned with sound barriers and often encased in concrete walls, the flow, rather than

---

2. Visibility: ([10-6], p. 237); Bad heating: ([10-7], p. 277); Revival of convertible: ([10-8], p. 680); New England: ([10-9], p. 299); Struggle: ([10-10], p. 31).

the car, became a second protective barrier.[3] At the same time, the number of casualties started to decrease. Why did this happen so late, and how did this happen? That is the governing question of this chapter.

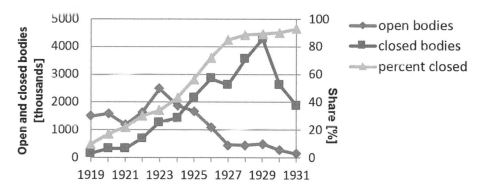

Figure 10.1 The closing of the car body in North America [10-16].

To answer these questions, we will follow the cocooning process along three consecutive cases. In Section 10.2 we will focus on the onslaught the encapsulation of the car provoked upon the nonusers, especially pedestrians in the city, and the cry among motorists to get rid of the dangerous "lunatics" in their midst. Then in Section 10.3 we give our attention to the post–WWII consumer movement against unsafe automobiles, and the response from the industry through the building of special safety cars. Section 10.4 is dedicated to the safety subsystem *par excellence*, ABS, and we will ask whether it really helped in the struggle to lower the amount of road casualties. We will close this chapter with conclusions (Section 10.5).

10.2 "Control the Lunatic!" The Hunt for the "Accident-Prone" Motorist

The emergence of the automobile has until recently been told as an uninterrupted success story. The initial resistance, especially among farmers, against the car has been described as temporary, which vanished as soon as these farmers started to discover the utility of the car for their own purposes. This misleading picture is the result of a questionable generalization of the American case. Recently, however, historical research revealed that even in the United States, as the country where car enthusiasm was the highest, the start of the second phase directly after WWI was characterized by a true struggle between a fledgling car lobby and a massive movement of concerned citizens, especially mothers, who protested against the amount of accidents and fatalities in the cities and the dangers their children were exposed to [10-17]. Persistence (the name we

---

3. My insight into the cocooning process of the car was considerably enhanced through participating in the Selling Sound project, together with Karin Bijsterveld, Stefan Krebs, and Eefje Cleophas, University of Maastricht. See [10-15].

gave to the second phase; see Chapter 1) was not a given for the automobile movement; it had to be conquered by the motoring community.

This motoring community responded threefold. In the first place the motorists themselves started to blame the adventurers in their own midst who gave the entire movement a bad name. Psychologists came to their rescue, defining accident-proneness as a special disposition among certain members of the population. They tried to develop lists of character traits that would predispose certain people to act dangerously. "Control the lunatic!" a poster of the Dutch touring club read (Figure 10.2). In some cases this hunt for the rotten apples among motorists, who in the majority did not consider themselves to blame at all, took on a racist flavor, for instance when American trial judges started sending many more black motorists than white motorists who were involved in accidents to re-education facilities. [10-18]. But the lunatics could not be found, and the hunt for them lost momentum once it became clear that most of the accidents occurred in everyday situations caused by average people.

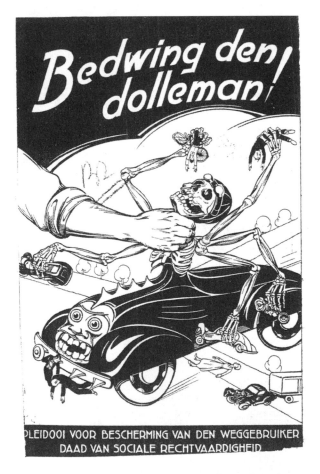

Figure 10.2 "Bedwing den dolleman!" (Control the lunatic!)
Source: Collection of author.

The second response came from the entire automotive community, including the car manufacturers, together with local police officers and politicians: massive campaigns were organized throughout American cities in the early 1920s to convince concerned citizens that the automobile was a good thing and that it was nonmotorists who were to blame for undisciplined behavior on the streets. When this did not help enough, even the American president warned against the onslaught on the roads. Such warnings, also made by many others, were often accompanied by a comparison to the atrocities of the recent war.

School children especially became the new target for re-education, a practice that quickly spread over the entire motorizing world. In the United States, the "three E's" became famous, coined by safety pioneer William Eno, standing for education, engineering, and enforcement [10-19].

On the technical side, there was a third engineering response necessary to really redefine the road safety problem.[4] This response came from a new stakeholder: the road engineering community, representing national governments who started to worry about the negative reactions from the public to the motoring movement. This community gathered in an international organization called PIARC (Permanent International Association of Road Congresses). On the initiative of the car and touring clubs as representatives of motorists and other road users, the issue of road signs was discussed during the first PIARC conference in Paris in 1908, but this was soon delegated to a special diplomatic conference held in October of the following year in Paris, where the signs were included in a convention on traffic (Figure 10.3).

Figure 10.3 The first road signs in the struggle for road safety, issued by the League of Nations, predecessor of the United Nations ([10-20], p. 1).

---

4. See for the following: [10-4].

The Choice in Favor of "Order"

By the time of the second PIARC conference in Brussels in 1910, it was clear that the "road problem" was redefined by the road builders as a "car problem." First the horse owners (in Paris 1908) and then the bicyclists (in Brussels 1910) were relegated to their own paths. Once established as primarily a matter of concern for motorists, the paved road became, during subsequent PIARC conferences, firmly embedded in a rudimentary automobile system, of which the highway formed the material spine. Soon, the United States became the example. From the painting of white stripes on the road to the struggle against automobile parking in cities by simply putting written notes on the windscreen (with the request to pay the fine at the police station), the European debate became increasingly colored by American examples. During this phase, just as in the more narrow world of automotive technology, an increasing *scientification* (and quantification) of the road problem became apparent, together with an increasing centralization of planning and financing of the ever-growing national road improvement projects [10-21]. But the technical problems seemed peanuts compared to the complex conditions that determined road safety. In this messy situation, the PIARC engineers chose order.

What did order look like in a road engineering perspective, and how was it realized? The participants of the PIARC conference held in Washington in 1930 witnessed how the management of traffic (defined as a flow of automobiles) emerged. With this, the definitive shift in the unit of analysis took place from the single vehicle to a systems approach. This shift was reinforced by the founding of road research laboratories in most motorizing countries (and most American states) during the second half of the 1920s and the early 1930s. These agencies then took up the issue of road construction technology and materials, as well as the testing of alternative solutions. Many countries completed the development of these networks well before the Second World War, a fact often neglected by historians of road construction because of their fascination for the more spectacular freeway projects. In many countries these huge improvement projects of paving, straightening, and widening existing roads were enabled by new taxes on fuel. This was a very efficient method because of the spiraling effect of mutually supportive increasing car registrations and increasing road building. Once established as a growth mechanism, a surprisingly rapid process (also for the contemporaries) of national and secondary road improvement all over Western Europe and North America was set in motion.

The last PIARC conference before WWII was held in The Hague in 1938. This meeting can be seen as the occasion where, under certain circumstances, special automobile roads came to be considered a safe solution to the road problem, mostly because of their radical separation of flows (between slow and fast traffic, and between traffic in opposite directions). This happened at a moment that the borders of the automobile system were not yet clearly demarcated. Inclined to find "technical fixes" to the road problem, PIARC engineers and related technocrats close to the national governments were not sure whether these fixes had to be applied at the level of the single vehicle or of the infrastructure. To the extent that vehicle speed was acknowledged as a constituent factor in road safety, the choice was clear: the speed of automobiles must be curtailed. On the issue of lighting, however, the opinions were still mixed. It was an open question whether automobiles should be equipped with lighting systems or if the entire road system should be illuminated. Also, during the first post-war conference in Sevilla (1923), a discussion took place about the question of whether mirrors for better and safer vision should be constructed inside automobiles only or also along the roads, and especially at crossings.

Road Accident Statistics

It was during the PIARC conferences in Munich (1934) and in The Hague (1938) that special sets of questions and conclusions were dedicated to the safety issue in an atmosphere of increasing anxiety about the "slaughter" on the roads on the basis of the first sets of statistics that became available. These statistics showed a general trend of steep increase for the 1920s. As we now know, the unsafe levels were extremely high compared to post–WWII values. The excellent French long-term statistics on traffic accidents reveal how the car fundamentally changed the societal safety experiences (Figure 10.4). So much had become clear by then: the risk society was a car society.

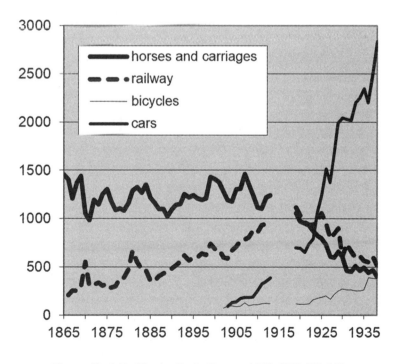

Figure 10.4 Traffic deaths in France, 1865–1939 [10-22].

The very rapid increase in traffic casualties in most motorizing countries may have been caused by the fact that the secondary road network was often ready earlier than the primary road network: because of their car-friendly surfaces and alignment taken over from the standardized characteristics developed for the primary network, such roads invited motorists to speed up, increasing the chance of violent encounters with fellow motorists (on crossings, from the opposite direction) and slower traffic.

What most worried many PIARC delegates was the alarmingly large share of pedestrians and cyclists among the deaths on the roads. In the UK for instance, cyclists were involved in one-third of all accidents, and they formed one-quarter of all road deaths. British pedestrians fared even worse: they formed 41 to 54% of all road deaths in 1933–1936. In 1936, about 60% of the French road casualties were nonmotorists. In cities, these figures were even more alarming. In Paris in 1932, cyclists and pedestrians accounted for three-quarters of all road deaths. For the UK the death toll of children since 1927 had reached 14,000 by 1937. In that year still half of the victims killed were pedestrians (and 40% of the wounded), while the remainder of the victims were motorcyclists (18%), bicyclists (18%), and nondriving car occupants (10%).

Inventing the "Death Rate"

From hindsight, it may be quite surprising to observe the optimism among PIARC engineers amidst these findings. American initiatives to organize special safety campaigns in selected cities suggested that something could be done against unsafe automobiles. Most American cities that started safety campaigns reported drastically diminishing fatality figures. Road safety, at last, seemed achievable, in a double sense: first by adjusting the infrastructure such that flows caused fewer injuries and deaths, and second (because the figures remained high, despite these measures) by adjusting the way people talked about the onslaught on the road. The observer in the *Atlantic Monthly* was certainly not the only one spicing his comments with a dose of sarcasm: "Eventually we may arrive at the comfortable stage in which the normal rise in casualties will not particularly annoy us, while a temporary drop would be hailed with genuine satisfaction" ([10-23], p. 730).

But this drop did not come, and so it was "created": in every country a careful statistical myth began to be constructed of a constantly declining "death rate" (Figure 10.5). This is not to say, of course, that road building engineers conspired to play down the lethal aspects of their profession, although one study on the British inter-war years claims that statistics may have been deliberately faked by the police to acquire more personnel ([10-25], p. 128). But it is remarkable how any signs (whether real or imagined) of a decrease in traffic risks were eagerly received by engineers, both inside and outside PIARC. It is no coincidence, either, that this desire for an optimistic reading of traffic fatality data occurred within the context of the same process of *scientification*, which had driven general road-building culture. Such an optimistic reading was achieved by normalizing the resulting graphs, dividing absolute accident, injury, and death numbers by some constantly increasing factor, mostly the number of cars or the number of vehicle-kilometers.

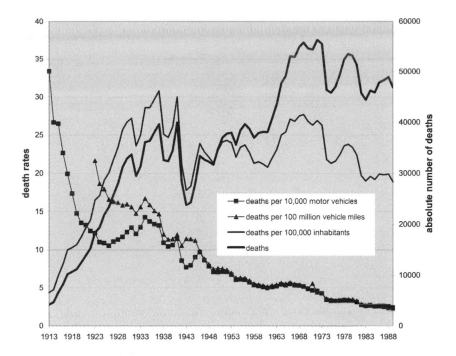

Figure 10.5 Mystification of road death statistics by constructing "death rates"; example: United States, 1913–1989 [10-24].

In the midst of this turmoil, the freeway appeared as a *deus ex machina*: this road type, based upon a radical application of the separation of flows principle, seemed to promise a solution to the safety problem. At the PIARC conference in The Hague (1938), there was quite some opposition against this idea. They saw it as an extremely expensive and undemocratic solution (undemocratic because it was authoritarian regimes like the German and the Italian who introduced the concept on a grand scale before WWII).

In other words, the openly and increasingly violent character of road traffic led to a redefinition of the road safety problem, from one that questioned the very basis of a future car society (at the beginning of the inter-war period) into a management problem of separated traffic flows at different speeds, drastically lowering the statistical chance of encountering a collision partner. Redefining the safety problem meant shifting (and at the same time reducing) the confrontation space between opposing interests from the entire society to the infrastructural part of the car system, reducing it from a social and cultural problem to a technical problem. By making the accident statistically invisible (and redefining it as a risk), the thrill of risk taking, which still formed the very core of the automotive adventure, could be rescued. Safe driving remained speedy driving.

Another factor that may explain the easy acceptance of the annual death toll is the statistical fact that during the 1930s, the share of nonusers among the accident victims diminished: motorists now started to kill each other instead of only women and children along the road. It is not without irony that the car industry only started to worry about safety when drivers and passengers threatened to dominate fatality statistics, proving that the industry as late as the immediate pre–WWII years still had some trouble to see its product as part of a system.

10.3 Protecting the Motorists: Active and Passive Safety

The motives to enclose the body in the 1920s were comfort and security rather than safety: protection against a hostile outside world in a capsule that insulated motorists acoustically and visually from their surroundings. The visibility problem emerged because defective ventilation caused misty windows. Suddenly, interface devices such as a claxon, lights, and mirrors became important and motorists had to learn new skills of anticipation and collision avoidance in busy urban street traffic.

Initially, it was not clear who was responsible in this domain. The car industry saw the road builders as the true safety guards, busy as they were in straightening blind bends, broadening narrow passages, and removing black spots where for unexplained reasons many accidents happened, and paving the roads with asphalt and concrete. The safety of the car itself was no item, at least not for many manufacturers. In 1930, General Motors president Alfred Sloan, who deliberately delayed the use of safety glass to keep costs down, declared: "it is not my responsibility to sell safety glass."[5]

Apparently, the motorists themselves, at least as represented by the car and touring clubs, felt a clear responsibility, shown, for instance, by their testing during the 1920s of devices to dip the blinding headlights (at that moment often fed by acetylene gas, burning very brightly) in case of an encounter with another car. Here, new skills had to be acquired as well, such as stopping when another car was approaching and switching off the lights. Like many other innovations, dipping systems emerged as mechanically operated devices (Figure 10.6) but soon were superseded by so-called double-filament light bulbs (1924), of which one filament was hidden behind a small metal panel enabling a dipped beam by using another part of the parabolic mirror of the headlight.

Here, an interesting *divergence* took place between a European and an American lighting culture, the latter characterized by an encased lighting unit (a sealed beam unit, 1940), which in Europe was considered to be too bright and blinding. Within Europe a divergence between cultures took place, too, as French lighting experts successfully proposed to use a yellow lamp as less blinding in their national car park ([10-29], p. 40-44). This makes clear that scientifically inspired engineering (as we will see in Chapter 12) not always leads to one single, best solution, but is subject to political and cultural power play. The history of vehicle lighting is an interesting but under-researched field, especially the post–WWII competition between several types of lamps (halogen, gas discharge, LED) and the resulting struggle between safety and aesthetics.

---

5. ([10-13], p. 150); costs down: ([10-26], p. 350). Safety glass consisted of a transparent foil in-between two glass plates (double glass), later followed by one layer of tempered glass, producing blunt splinters in a collision ([10-27], p. 53).

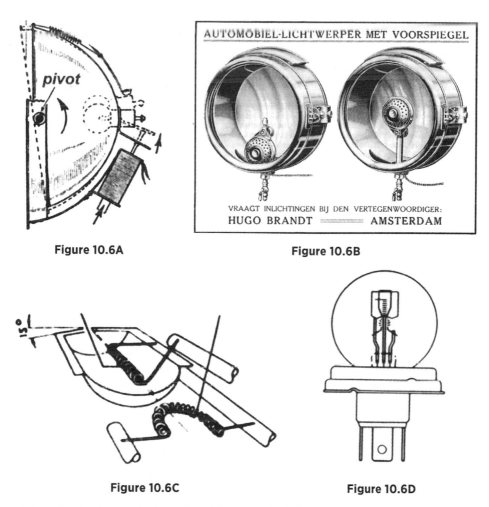

Figure 10.6A

Figure 10.6B

Figure 10.6C

Figure 10.6D

Figure 10.6 Dipping devices. A and B: Mechanical dipping devices to avoid blinding of the opposing motorist; C and D: Duplo light bulb with double filament, the one for the dipping beam behind a small, asymmetrical metal panel ([10-28], p. 41, 42, 371).

Safety Research

It was American lawyer Ralph Nader, who, in an atmosphere during the 1960s of increasing road safety anxiety, accused the American car industry of neglecting its customers' safety, using the Chevrolet Corvair as an especially dangerous example [10-27]. His accusations became a national issue (and his book, *Unsafe at Any Speed*, a national bestseller) when it appeared that the industry had spied in his private sphere. The Nader case is the culmination of a long gestation of road safety consciousness, on both sides of the Atlantic. It was accompanied, just as during the inter-war years, by invocations of war. In 1971, for instance, American Congress members observed "that we kill more people on the highways in a single year than we have lost in combat in Vietnam over the past 10 years" ([10-28], p. 7).

Well before Nader, medical research emerging from the aeronautical realm started to spill over to the automotive world, most spectacularly shown by biomechanic research on corpses and on live animals (for instance, at Wayne State University, Figure 10.7) [10-31]. In that second phase of automotive evolution (the inter-war phase of persistence), medical researchers started to investigate the bodies of (pedestrian) victims of car accidents, also in Europe. Others focused on the cars' occupants. One of the most famous was Hugh DeHaven, a freelance inventor who started his career by throwing eggs on a slab of half-inch foam, proving that it was the second collision (between the body of the occupants and the sharp edges of the car's interior) that caused the most harm, and something could be done about it. It was the beginning of a new engineering discipline (crashworthiness), first in aviation (accelerated during the war), after WWII also in automobiles, for instance undertaken in New York City's Cornell University Medical College ([10-32], p. 42, 47–48).

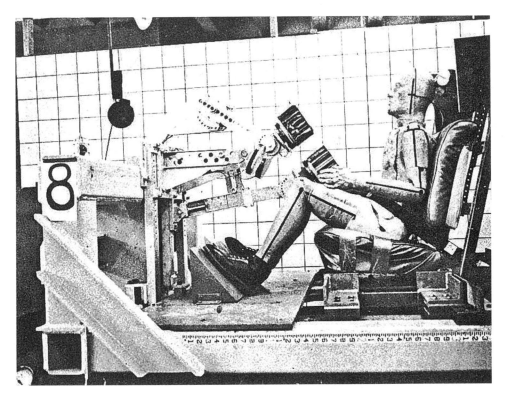

Figure 10.7 Sled test at Wayne State University ([10-30], p. 175) (*courtesy of SAE International*).

After the war, the German transport ministry became worried about the safety statistics when it realized that the relative number of injured per capita was higher in the United States, but their ratio between injured and deaths was lower. Mercedes, with a substantial stake on the American car market, joined the ministry's concern. This company had a reputation of safety research in the person of Béla Barényi, who as early as 1925 took a patent on a less dangerous steering column. That was about the same time that in the United States the Safety Stutz was proposed, a car with a very low center of gravity and equipped with safety glass. Mercedes equipped its products with safety door locks in 1958, offered safety belts as an option from 1957, and realized in 1959 its first car with Barényi's "safety cage with energy absorbing front and rear," laying the constructive basis for our so-called automotive cocoon. That year Mercedes did its first crash test ([10-27], p. 104–105).

Also, the research was accompanied by a redefinition of the safety concept, initiated by Italian Fiat engineer Luigi Locati, who in 1964 distinguished between active and passive safety, referring to the prevention of an accident and the alleviating of the accident's harmful consequences, respectively. When Mercedes took over this distinction, it added another distinction to the passive side, between interior and exterior safety, acknowledging at least the existence of an outside world (Figure 10.8).

Figure 10.8 Redefinition of automotive safety by Mercedes in 1966 ([10-27], p. 203) (*courtesy of Daimler AG*).

Safety Legislation and Regulation

The Stapp Car Crash conferences (called after the American safety pioneer Colonel John Paul Stapp) became the international intermediary of knowledge exchange, all the more so when American President Richard Nixon in 1969, in need of an electoral success and as a response to hearings held by the U.S. Congress on road safety, proposed to develop a special, very safe automobile. In a way reminiscent of the Man on the Moon program, and as such to be understood in the context of the Cold War, the Experimental Safety Vehicle (ESV) project between 1971 and 1974 was an international success of automobile propaganda, despite the failure of the project as such. It was accompanied by safety legislation, in the United States started in 1965 and prepared under the auspices of a new government agency: the NHTSA (National Highway Traffic Safety Administration), in Europe five years later.

Although specific safety cars were proposed before (for instance, the Rover 2000 in 1963), all manufacturers (also outside the United States) who wanted to be taken seriously as responsible actors contributed their safe prototypes to the ESV program, the best of which were at least 20–25% heavier than conventional cars (Figure 10.9). This led to a general skepticism about such aggressive cars "beyond any practical utility," although there were also "lightweight" ESV-prototypes proposed.[6] And although the project was abandoned because of the outbreak of the first energy crisis, the Pluto Effect reigned supreme: when Mercedes built the first airbag in its S class model at the end of 1980, the company considered this to be a belated result of the ESV project. In 1992 all Mercedes models were equipped with airbags ([10-27], p. 384). That was eight years after Lee Iacocca, then president of Chrysler, had rejected the airbag by referring to proposals to use its explosive power as a modern device to execute the death penalty, by slamming the head of the convict rearward and thus breaking the convict's neck [10-34].

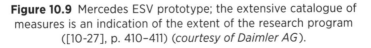

Figure 10.9 Mercedes ESV prototype; the extensive catalogue of measures is an indication of the extent of the research program ([10-27], p. 410–411) (*courtesy of Daimler AG*).

---

6. For instance, Fiat in 1974 proposed a small ESV that was 150 kg heavier than its 600 kg standard model 500 [10-33]. Utility: ([10-27], p. 351).

The airbag itself was also the product of a *divergence* of safety cultures on both sides of the ocean, despite the increased exchange of knowledge. Patented as far back as 1903 by Frenchman Desiré Liebau, the safety belt had to be defended by a nervous car industry against motorists who saw it as a device to attack their personal (bodily) freedom. That is why initially the airbag in the U.S. was presented as an alternative to the safety belt. In the end both devices had to be mobilized to keep motorists from being injured and killed by the notorious second collision of the motorists' bodies against the inside of the car [10-35].

10.4 Antilock Braking ABS

Part of the active safety reservoir of technologies, the history of the antilock braking system starts shortly before the Second World War, in the midst of a shift among leading research engineers toward the problem of *vehicle dynamics* (see Chapter 12). By then, research had shown that the friction coefficient of a sliding tire is lower than that of a rolling tire, and that loss of friction between tire and road would also prevent the tire from resisting lateral forces and thus would jeopardize steering stability. Although, as so often in technological evolution, ideas to prevent the wheels from locking when braked were proposed earlier (see Chapter 8), it was the engineer from the German brake manufacturer ATE Fritz Ostwald who in a master's thesis at the Technical University of Munich designed an electro-mechanical antilock brake for the first time. His patent from 1942 became part of the war booty of the U.S. Army, which, according to one student of ABS's history, explains the ten-year lead in this domain of the United States and the UK immediately after the war ([10-36], p. 771–772).

It was, indeed, a military source from which the first antilock system sprang. During the war, especially in the Pacific, acute tire and rubber shortages occurred because of aircraft skidding on runways, causing blowouts. The Aviation Division of the British Dunlop Rubber Company was one of the stakeholders that developed an antilock system. Called Maxaret, the device was retrofitted on a 1950 Morris 6 (and later on the Rolls-Royce Silver Shadow of 1966 and the racing car Jensen FF). Comparable developments took place in the United States, where aviation suppliers Bendix and Kelsey-Hayes applied antiskid devices to Chrysler and Ford cars, respectively. Called Sure-Brake (on the Chrysler Imperial of 1971) and Sure-Track, the systems did not enjoy a large demand. From an engineering point of view, their basic technical property, a dependence on the engine's vacuum, appeared a *cul-de-sac* when emission control systems appeared to need the same energy source, thus lowering the available negative pressure potential ([10-37], p. 27-30, 117).

The American historian and philosopher Ann Johnson, who performed the most thorough study of the development of these systems, convincingly showed that what we now call ABS was not derived from aeronautics, for two reasons. First of all, cars are not planes or trains (on which antiskid systems were also proposed before the war):

apart from a shortening of the braking distance, car drivers also needed to maintain directional control while braking. Second, when Robert Bosch in Germany around 1975 took over Bendix's affiliated company Teldix (probably to prevent AEG or Telefunken from doing so), which had developed an electronically controlled antilock braking system, it decided to redevelop the system from the ground up, as a fully digital system. This happened during 1972–1974, exactly the phase in which the experimental safety vehicles were being built (see Section 10.3). Nonetheless, aeronautics always shimmered in the background, if only because of the high-speed hydraulic valve developed by Teldix and the much less vehement skepticism toward electronic control in aeronautical engineering circles. Whether tributary to aeronautics or not, when Bosch and Teldix during the 1960s teamed up with Daimler-Benz, a powerful actor emerged that made the device into a typical engineer's concern, just like four-wheel drive: no marketing studies had been done showing that motorists were waiting for these systems. ([10-36], p. 774).

After nearly 20 years of development (including some years waiting for the energy crisis to subside, and some problems with the reliability of electronics, as we saw in the previous chapter), the second generation ABS was marketed at a political price. It was deliberately priced three to four times cheaper than it should have been, considering the development costs at Bosch alone of 50–80 million DMark ([10-36], p. 788, 791). Offered as an option on the Mercedes S class of 1978, ABS was subsequently applied, according to mutual agreements, by BMW and Audi.

In 1986 1 million devices had been sold. A year later the Bosch device had conquered three-quarters of the European market against a dozen or so competitors. The most important one was ATE, which had started its development directly with a third generation system (integrated in the braking system) and used this advantage to occupy the majority of the American market for a while (it appeared on the Lincoln Continental Mk VII of 1984 for the first time).

In 1986, about 10% of the European market had ABS (in Germany, 20%), whereas in the United States and Japan at that moment market penetration amounted only to around 2%. ([10-38], p. 773–774), ([10-36], p. 797). But then it went fast: in less than a decade the market became saturated (Figure 10.10) and by the end of the 1980s the Bosch technology (including the Japanese licensees) governed three-quarters of the world market. In 2004, ABS became standard on all cars built in the EU ([10-37], p. 157). Apart from the fully hydromechanically designed Stop Control System (SCS) from Lucas Girling, all systems were electronically controlled, which was of strategic importance for supplier companies, as these systems really helped trigger the "electronic revolution" of the 1980s, as we have seen in the previous chapter. Soon after the market success of ABS, the Bosch concern could show a comparable diffusion graph for its successor, ESP (Electronic Stability Program) (Figure 10.10B) [10-39]. Car handling had at last become a commercial challenge.

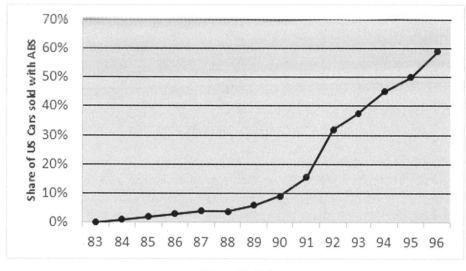

Figure 10.10A

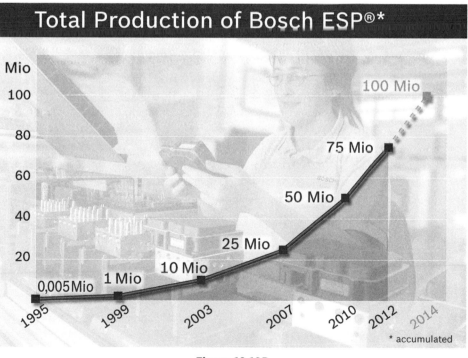

Figure 10.10B

Figure 10.10 Diffusion of ABS on the American market and total production of Bosch's ESP. *Sources:* A: ([10-40], p. 249, Fig. 3); B: [10-41] (B: *courtesy of Robert Bosch GmbH*).

10.5 Conclusions

The history of safety thinking and safety devices seems to support Alfred Sloan's cynical dictum that safety doesn't sell. Whether he was right in claiming that he is not responsible for selling safety devices remains to be seen, however. Ralph Nader would not have agreed, and neither do representatives of the consumer movement in his wake. Even the industry itself has meanwhile gained a different insight, considering the fact that ABS was introduced on the market without much concern about the users' wishes. What exactly was the reason for motorists' hesitancy to buy a "safe car" has not been researched very well, but the remainder of the ABS story suggests an answer.

Although ABS was developed into a standard system in Germany with the blessing of the German TüV (*Technische Überwachungsverein*; Technical Control Society), which had testified that it would prevent 7% of all accidents and could alleviate the consequences of another 10 to 15%, later research into the effects of the system were less convincing. Ann Johnson shows how psychological research during the 1980s in Germany and Canada coined the phrase *risk compensation* for referring to the phenomenon that motorists who knew that they had ABS braked later and drove faster (keeping the stopping distance the same as without ABS) ([10-37], p. 158–159), ([10-38], p. 773). Research by an Australian car club as late as 2004 confirmed most of these findings [10-42].

On top of that, such research could not find convincing evidence that ABS caused fewer accidents to happen and alleviated accidents' consequences. Instead, the number of fatal crashes of single cars (without a collision partner) were higher, a riddle still not completely solved. It seems that ABS was a solution to a problem that did not exist, at least not in the form as envisaged by suppliers, car manufacturers, and their engineers. The development aim to make sure that nothing would change for the driver did not come through: the ABS Education Alliance, set up by the main stakeholders (Bosch, Teves, Kelsey-Hayes, and Bendix), warned that in order to be effective the system had to be actuated by "keep(ing) hard continuous brake pressure." Motorists (at least those knowledgeable motorists who applied this procedure) had to "unlearn" to brake on skiddy pavement in a pumping manner and had to get used to the pulsating feel in the braking pedal when ABS was on (instead of suddenly lessening the braking force, which might have caused the single cars to veer off the road). ([10-37], p. 162–163).

ABS confirms what we have been saying throughout this book, with the systems character of the car's structure in mind: even the slightest change in the *properties* implies that in the long run the *function* has to be adjusted as well: "smuggling" ABS in without asking the drivers to adjust their braking habits is not a good strategy. Just like for every other innovation, and especially for those at the interface between driver and car, new skills have to be acquired whenever technical properties are changed. A better example of the cyborg-character of the modern car plus driver can hardly be given. Would the new, highly sophisticated dynamic stability control systems like ESC (in which ABS is integrated) perform the trick where ABS failed? The industry claims it does [10-43]. Indeed, expectations run high when it comes to the promises of automotive automation and the power of driver assistance (Figure 10.11). Are we witnessing a collective fantasy here? Only the future can tell.

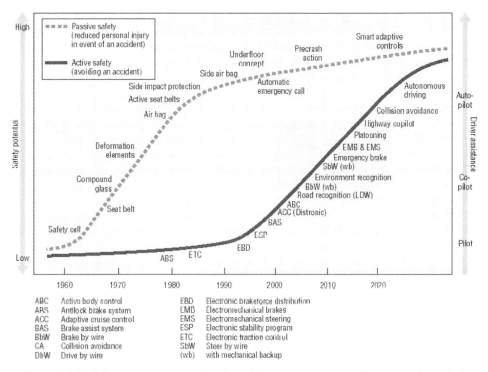

ABC   Active body control
ARS   Antilock brake system
ACC   Adaptive cruise control
BAS   Brake assist system
BbW   Brake by wire
CA   Collision avoidance
DbW   Drive by wire

EBD   Electronic brakeforce distribution
EMB   Electromechanical brakes
EMS   Electromechanical steering
ESP   Electronic stability program
ETC   Electronic traction control
SbW   Steer by wire
(wb)   with mechanical backup

Figure 10.11 Safety systems conceived as driver assistance, leading to a reformulation of driver skills (reprinted from [10-44], p. 92, with permission from IEEE).

Such innovations, together with measures by road safety authorities, further incite the competition going on nowadays between countries for being the safest in terms of road accidents.

References

10-1.   Peter L. Bernstein, *Against the Gods; The remarkable Story of Risk* (New York/ Chichester/Brisbane/Toronto/Singapore: John Wiley & Sons, Inc., 1996).

10-2.   Ulrich Beck, "Risk Society and the Provident State," in Scott Lash, Bronislaw Szerszynski, and Brian Wynne (eds.), *Risk, Environment and Modernity: Towards a New Ecology* (London/Thousand Oaks/New Delhi: Sage, 1996), 27–43.

10-3.   Jörg Beckmann, *Risky Mobility; The Filtering of Automobility's Unintended Consequences* (Copenhagen: Copenhagen University, Sociological Institute: 2001).

10-4.   Gijs Mom, *Atlantic Automobilism: The Emergence and Persistence of the Car, 1895–1940* (New York and Oxford: Berghahn Books, forthcoming).

10-5. "Where cars are better," *Autocar* 59 No. 1667 (14 October 1927), 735–736.

10-6. "The Passenger Car of the Future," *Journal of the Society of Automotive Engineers*, 5 No. 3 (September 1919), 236–241.

10-7. E. W. Goodwin, "Automobile Body Design and Construction," *Journal of the Society of Automotive Engineers*, 2 No. 4 (April 1918), 271–278.

10-8. Hermann A. Brunn, "Trends in Body Design," *Journal of the Society of Automotive Engineers*, 22 No. 6 (June 1928), 679–683.

10-9. J. W. Frazer, "Bodies Considered from the Car Buyer's Viewpoint," *Journal of the Society of Automotive Engineers*, 31 No. 1 (July 1932), 294, 299.

10-10. H.M. Crane, "How Versatile Engineering Meets Public Demand," *Journal of the Society of Automotive Engineers* 41 No. 2 (August 1937), 358–392.

10-11. Richard M. Bach, "Design and Style as Selling Factors," *Journal of the Society of Automotive Engineers* 22 No. 4 (April 1919), 468–474.

10-12. "Weights of 1921 Cars on Which Kansas Bases License Fee," *Automotive Industries* 45 No. 7 (18 August 1921), 330–331.

10-13. Sally H. Clarke, *Trust and Power: Consumers, the Modern Corporation, and the Making of the United States Automobile Market* (Cambridge: Cambridge University Press, 2007).

10-14. Christian Kehrt, *Zwischen Evolution und Revolution; Der Werkstoffwandel im Flugzeugbau* (Karlsruhe: KIT Scientific Publishing, 2013).

10-15. Karin Bijsterveld, Eefje Cleophas, Stefan Krebs, and Gijs Mom, *Sound and Safe: A History of Listening Behind the Wheel* (Oxford/New York, etc.: Oxford University Press, 2014).

10-16. Malcolm M. Willey and Stuart A. Rice, "The Agencies of Communication," Chapter IV in *Recent Social Trends; Report of the President's Research Committee on Social Trends, Volume I* (New York/London: McGraw-Hill Book Company, Inc., 1933), 167–217.

10-17. Peter David Norton, *Fighting Traffic; The Dawn of the Motor Age in the American City* (Cambridge, Massachusetts/London: The MIT Press, 2008).

10-18. Daniel M. Albert, "Primitive Drivers: Racial Science and Citizenship in the Motor Age," *Science as Culture* 10 No. 3 (September 2001), 327–351.

10-19. William Phelps Eno, *The Story of Highway Traffic Control 1899–1939* (n.p.: The Eno Foundation for Highway Traffic Control, Inc., 1939).

10-20. Juan Agustin Valle, "Police de la circulation" (report no. 77 of PIARC conference in Washington 1930).

10-21. Bruce Seely, *Building the American Highway System; Engineers as Policy Makers* (Philadelphia: Temple University Press, 1987) (Technology and Urban Growth series, eds.: Blaine A. Brownell, Mark Rose, e.a.).

10-22. Jean-Claude Chesnais, "La mortalité par accidents en France depuis 1826," *Population* (French edition) 29 No. 6 (November – December 1974), 1097–1136.

10-23. Seth K. Humphrey, "Our Delightful Man-Killer," *Atlantic Monthly* 148 (1931), 724–730.

10-24. Clay McShane and Gijs Mom, "Death and the Automobile: A Comparison of Automobile Ownership and Fatal Accidents in the United States and the Netherlands, 1910–1980," (unpubl. paper presented at the ICOHTEC conference, Prague, August 22–26, 2000).

10-25. Howard Taylor, "Forging the Job; A Crisis of 'Modernization' or Redundancy for the Police in England and Wales, 1900–39," *British Journal of Criminology* 39 No. 1 (Special Issue 1999), 113–135.

10-26. Jeffrey Robert Yost, "Components of the Past and Vehicles of Change: Parts Manufacturers and Supplier Relations in the U.S. Automobile Industry," (unpubl. diss. Case Western Reserve University, May, 1998).

10-27. Heike Weishaupt, *Die Entwicklung der passiven Sicherheit im Automobilbau von den Anfängen bis 1980 unter besonderer Berücksichtigung der Daimler-Benz AG* (Bielefeld: Delius & Klasing, 1999) (Wissenschaftliche Schriftenreihe des DaimlerChrysler Konzernarchivs, Band 2, eds.: Harry Niemann and Armin Hermann).

10-28. Gijs Mom, "De elektrische installatie in historisch perspectief, " in H. de Boer, Th. Dobbelaar and G. Mom, *De elektrische installatie* (Deventer, 1986) (Part 8 of *De nieuwe Steinbuch; de automobiel; handboek voor autobezitters, monteurs en technici onder redactie en coördinate van drs.ing. G.P.A. Mom*). 17–51.

10-29. Laurel L. Cornell, "The Problem of the Yellow Light and How Mobility Kills Older Drivers," (paper presented at the 10th anniversary conference of T²M, Madrid, 15–18 November 2012).

10-30. Theodore P. Wright, "Automotive Safety Research," in Theodore P. Wright, "Articles and Addresses of Theodore P. Wright; Volume IV: A Post-retirement Miscellany," (Buffalo, NY: Cornell Aeronautical Laboratory, Inc., Technical Services Department, of Cornell University, 1961) (typescript) 159–179.

10-31. Lee Vinsel, "Bodies at Unrest: Impact Biomechanics as a Regulatory Science" (paper presented at the Annual conference of the Society for the History of Technology SHOT, Copenhagen, 29 September 2012).

10-32. Amy Gangloff, "Safety in Accidents: Hugh DeHaven and the Development of Crash Injury Studies," *Technology and Culture* 54 No. 1 (January 2013), 40–61.

10-33. "How Fiat Designed Three Lightweight ESV's," *Automotive Engineering* 82 No. 2 (February 1974), 46–47.

10-34. Lee Iacocca, *Iacocca: An Autobiography* (New York: Bantam, 1984).

10-35. Jameson M. Wetmore, "Engineering with Uncertainty: Monitoring Air Bag Performance," *Science and Engineering Ethics* 14 (2008), 201–218.

10-36. Holger Bingmann, "Chapter 14: Competence; Case C: Antiblockiersystem und Benzineinspritzung (Anti-Blocking System and Fuel Injection)," in Horst Albach, *Culture and Technical Innovation; A Cross-Cultural Analysis and Policy Recommendations* (Berlin/New York: Walter de Gruyter, 1994) (Akademie der Wissenschaften zu Berlin, Research Report 9), 736–821.

10-37. Ann Johnson, *Hitting the Brakes; Engineering Design and the Production of Knowledge* (Durham/London: Duke University Press, 2009).

10-38. M. Arkenbosch, G. Mom, and J. Nieuwland, *Het rijdend gedeelte; Band B: veersysteem, wielgeleiding, remsysteem, diagnose en uitlijnen* (Deventer, 1989) (Part 4B of *De nieuwe Steinbuch; de automobiel; handboek voor autobezitters, monteurs en technici onder redactie en coördinatie van drs.ing. G.P.A. Mom*).

10-39. Daan Gerrits, Harm Gijselhart, Tsvetan Balyovski, Johan van Uden, and Jan van der Vleuten, "Anti-Blocking System" (student report for the course Cars in Context, Eindhoven University of Technology, 2012).

10-40. Francisco Veloso and Sebastian Fixson, "Make-Buy Decisions in the Auto Industry: New Perspectives on the Role of the Supplier as an Innovator," *Technological Forecasting and Social Change* 67 Nr. 2&3 (June/July 2001), 239–257.

10-41. "EU regulation enters into force in November 2011: ESP compulsory in all new car models" (Bosch press release, Stuttgart, 2011).

10-42. David Burton, Amanda Delaney, Stuart Newstead, David Logan, and Brian Fildes, *Effectiveness of ABS and Vehicle Stability Control Systems* (Noble Park Noth, VI: Royal Automobile Club of Victoria (RACV), April 2004).

10-43. Isabel Wilmink e.a., *Socio-economic Impact Assessment of Stand-alone and Co-operative Intelligent Vehicle Safety Systems (IVSS) in Europe* (Delft: TNO, 2008).

10-44. G. Leen and D. Hefferman, "Expanding Automotive Electronic Systems," *IEEE Computer 2002*, 92.

Chapter 11
Environment: Discovering the Other

11.1   Introduction: Inventing the Environment

If we look at the history of automotive technology from an environmental point of view, we may observe some reluctance among motorists and manufacturers alike to approach the world beyond the car with some form of empathy. At the turn of the 20th century, the environment was the world outside to be conquered. The aggressive connotation of this term was important, as the world consisted of "nature" to be opened for the motorists touring through the countryside. Early grand touring took place by invading the rather closed cultures of peoples and their habits in the south, whether in Europe or the United States. Other people to be "conquered" were opponents who had to be convinced that car driving may be an elite pastime but it had a right to exist. An in-depth analysis of the motorists' motives on the basis of novels and poetry (a surprising share of early car pioneers also belonged to the literary avant-garde, especially in Europe) reveals that the automotive adventure had narcissistic, even aggressive traits [11-1]. It was mainly a pastime of white, upper-middle class, and aristocratic men.

A more optimistic read sees the car emerge out of a horse-dominated urban quagmire of environmental degradation. However, this seems to be a projection of our current environmental consciousness upon a time, when car manufacturers used this argument as propaganda against the horse. Indeed, cities smelled badly and horse dung had become a problem of local hygiene, but only after a well-organized horse economy had started to collapse: well before the car appeared in the city, the horse and carriage was already retreating, mostly because of spatial problems in the crowded inner cities and because of increasing horse and carriage (maintenance) prices. When, on top of this, artificial fertilizer was developed by a burgeoning chemical industry, the logistics of dung, hay, and fodder flow to and from the neighbouring agricultural countryside began to crumble [11-2].

It was the environment, also consisting of nonusers backed by local governments, that before the First World War alerted motorists of the nuisance they were causing. In

Germany, for instance, urbanites complained about smelling and smoking exhausts, but car proponents (not a few of them engineers) responded that this was caused by old-fashioned, badly adjusted engines that would soon disappear from the market, thus automatically solving the problem. They were right: the black smoke (from too rich fuel-air mixtures) and blue smoke (from lubricants leaking past the piston into the combustion chambers of the cylinders) vanished from automotive culture as soon as more precise production techniques were adopted and ignition and carburation systems were more carefully adjusted ([11-3], p. 150).

Health Aspects of Early Automobilism

This is not to say that physicians weren't aware of the invisible negative effects of the automobile on human health, but they were a minority compared to the car proponents who saw the car as a harbinger of public health because it reduced dust, flies (from horses), as well as "direct infections of stablemen and veterinarians." ([11-4], p. 284). Medical doctors and hygienists knew about the poisonous properties of carbon monoxide and started to warn users and mechanics to be careful with running car engines in their garages. They also experimented with exhaust gas treatment in the laboratory during the 1930s ([11-5], p. 15, 19, 106).

Manufacturers had to start thinking about CO poisoning once they decided to seal off the passenger cabin; American insurance companies warned them in 1933 that leaks in the cabin made more than half of all cars contain CO and "7 per cent. [sic] of them contain quantities that may cause collapse." ([11-6], p. 150).

Medical doctors also were the pioneers of protest against the dust plague. In European Monaco, a League against Road Dust (*Ligue contre la poussière sur les routes*) was set up, which appeared to be instrumental in raising local governments' consciousness about the potential harmful effect of road dust. This was all the more important because already during the last decades of the 19th century dust had been connected to the notorious miasma, bad, unhealthy air containing dangerous germs.

Although some touring and automobile clubs initiated driving tests of specially equipped cars (a famous Dutch make for instance marketed a "dustless Spyker" with a specially made "dust-free" plating under the chassis), the usual reaction of the motoring community was to point at the unpaved roads. They again got their way: because cyclists and pedestrians were the primary victims of this nuisance, national governments in all motorizing countries set up extensive road-paving programs to satisfy a growing part of their walking, cycling, and motoring middle-class constituency. As long as road paving was not completed, sprinkling (horse-drawn) carts and (motorized) cars made the road surface wet in the warm and dry seasons, especially in the cities. New pavement materials (such as asphalt and concrete) had to be applied, as the vacuum created directly behind the car's rolling rubber tires sucked the dust from between stones or clinkers of the more traditionally paved roads ([11-7], p. 9).

For many observers, the car seemed to become environmentally benign once it became embedded in a system of roads and regulations, and the remaining protesters against congestion and other nuisances were easily drowned out by the urban indignation about road safety (see the previous chapter). So much was clear: cars, it seemed, were not bad individually (so the individual owner was not to blame, really); they only caused a problem if they were plentiful. In this sense, automobilism from its very beginnings had a clear collective flavor, despite its attractiveness as a vehicle to celebrate the new individualism of modern society.

What was not so easily solved away, however, was the acoustic nuisance the car was responsible for. No wonder, then, that we will dedicate the next section (Section 11.2) to this nuisance, before we focus on the re-emergence of environmental consciousness around the car's exhaust in the post–WWII era through two case studies: the controversial claim of the diesel engine as a clean engine (Section 11.3) and the emergence of environmental regulation, including the struggle between two competing principles to abate tailpipe emissions, lean burn versus the catalytic converter (Section 11.4). As usual, conclusions close the chapter (Section 11.5).

11.2 Engineering Car Noise While Closing the Body: Liberating Vision from Sensual Interference

The noise problem should be seen in its context of increasing sensitivity among the middle classes for the consequences of urban growth: congestion, smell, and, especially sound [11-8], [11-9]. It took the founding of many local committees against street noise to convince manufacturers and motorists alike that they somehow had to try to decouple car driving and noise production.

In one respect noise abatement was accomplished through a change of attitude: pre–WWI pioneers often opened a special valve in the exhaust to decrease the counter-pressure in order to create better scavenging within the cylinders and thus generate more power. Car drivers and especially motorcyclists then developed the skill to drive with an open exhaust to show off their "powerful" driving style, although nobody up to then had given any evidence that it would really help, in a technical sense. In most countries the intervention of the national government was needed to amend road legislation and make this form of sporting driving illegal. At the same time, making unnecessary noise was increasingly seen as uncivilized and, in European class societies, as something characterizing the lower classes.

But even with their exhaust valves closed, cars during the 1910s and 1920s, when their numbers started to count, had a really negative impact on their (mostly urban) surroundings, as we can observe when going through the first decade of *The Journal of the Society of Automotive Engineers*, which started to appear in 1917 [11-10]. One engineer estimated that mimicking the quiet drive of the electric vehicle, left behind a decade ago, cost half of engineers' time, noting: "Our goal, as engineers designing an automobile engine, is to create powerplants that shall not be heard during operation and are free from vibrations." Slowly, engineers started to see noise (and vibrations in general) as waste, a sign of loss of power [11-8], [11-11]. In other words, "noise" (or sound as nuisance) is a social construction. Soon, the struggle against noise expanded to the entire vehicle, most particularly the howling of gearwheels (11-12], p. 388).

Sound and Comfort Engineering

Although advertisements in European trade journals revealed a similar discourse on that side of the ocean (Figure 11.1) [11-14], U.S. specialists from Bell Laboratories as early as 1925 came to SAE International meetings to explain the intricacies of noise production and measurement. Around the same time engineers and scientists started basic applied research laying the groundwork for a new engineering attitude toward vibrations and resonance. The "snapping sounds," the "slapping impacts," and the "oil swishes" in the transmission and the squeaking brakes were attacked along three routes: by more precise engineering avoiding production variation, by sound absorption, if the first solution was not effective enough, and, much more preferred, by changing the sound [11-15], ([11-16], p. 461), [11-17], ([11-18], p. 105).

Figure 11.1 Opel advertisement from 1930s Germany emphasizing the importance of silence: "the miraculous quiet running of the small Opel" ([11-13], p. 20).

Absorption materials filtered high-pitched noises out, giving light cars a "heavier feel." By then, "noise studies" had effectively abated disagreeable sounds to such a degree that some engineers feared it might jeopardize safety as motorists "unconsciously run faster until the increased speed of the car reaches the noise level to which [they] are accustomed. (..) Noise reduction," they concluded at a Noise Symposium, "tends to reduce the feeling of vibration, even though the vibration itself has not been reduced" ([11-19], p. 267), [11-20].

Through sound engineering, car engineers started to understand that their work was as much about the soft, psychological sides of their hard technology. But the closing of the car's body caused a new noise problem: as "sounding boards," body panels amplified vibrations from the propulsion system and other chassis subsystems, as well as from the road, while windows rattled in doors, and panels squeaked when moving relative to each other ([11-21], p. 630).

Such problems made acoustic engineering into one of the essential elements of riding-comfort engineering, a decade-long effort to translate and convert the driver's experiences and those of the passengers into a comfortable "capsule." Crucial in this development was the co-evolution of measuring devices, a difficult endeavor because sound did not behave in a linear way, but changed logarithmically, as was rediscovered in 1923 in relation to noise interferences in telephone lines. The (deci)bel, called after acoustic and telephone pioneer Alexander Graham Bell, was introduced in the course of the 1920s and became "pretty much universal" by 1930 ([11-9], p. 159–161).

But sound engineering was much more than juggling with logarithms: even the sound of shutting the door was an object of study, as potential buyers preferred a "soft" sound rather than a "tinny" one ([11-22], p. 487). After a decade of research and practical experience, body panels became coated with "noise deadener" (often asphalt) as a matter of routine, while components were increasingly rubber-mounted, notably the entire engine and transmission, and the body on the chassis (Figure 11.2) ([11-23], p. 249), [11-24].

Figure 11.2 Advertisement in a French journal for Silentbloc, a famous rubber mounting to combat noise transfer to the interior of the body. I thank Dr. Stefan Krebs, Maastricht University, The Netherlands, for providing me with this illustration [11-25].

Eliminate Sound to Enhance Vision

A gradual shift from technology to appearance and esthetics in both the industry's marketing and buyers' preferences accompanied the closing of the body. By the end of the decade, when the body had slowly been made "rattle-proof and leak-proof," appearance had advanced before comfort in perceived consumer desires (only then followed by performance, reliability, economy, and durability) ([11-26], p. 682). While the classic chassis engineers (especially the engineers who designed the engine and transmission) remained skeptical, new body engineers tried to reconcile the wish for comfort with the increasing complexity of the car. One such a new engineer opined: "(M)any engineers haughtily turned a cold shoulder to the idea [of extending engineering into the body, GM]. Only upon the insistence of management did they give it any consideration" ([11-27], p. 437), ([11-28], p. 107). Car engineering, up to then mostly an affair of mechanical engineers specialized either in engines or transmissions, began to be split into specialisms, a process that accelerated through the parallel *scientification* of the profession, as we will see in Chapter 12.

The "emancipation" of the body engineer can nicely be followed through the annual SAE International meetings: while it organized its first session on the body in 1914, a separate division as part of the Detroit Section was founded in 1928 ([11-29], p. 25–26). By the beginning of the 1930s body engineers had started to speculate about the need to develop unit constructions allowing the body to be further lowered and to be given a streamlined shape (rounded corners, slanted windscreen) for a new popular type of use: long-range touring. Specially designed coloring of the body enhanced the "speedy" look ([11-30], p. 38).

However, wind resistance resulting from the higher speeds introduced a new source of noise, while at these speeds the sound of the tires also started to become noticeable in concert with the increased engine sounds. Each technical refinement created a new noise issue; engineers seemed to be heading toward an insurmountable noise barrier. This became all the more acute when, at the end of the 1920s, car radio made a rapid entrance. But despite all these measures, Americans were equipping themselves to go on long tours with the family in a car interior that in 1937 still showed a sound and noise level of 82 decibel at 60 mph (2 dB more than a large orchestra) ([11-21], p. 629), ([11-31], p. 369).

By the end of the 1930s leading officers of SAE noticed that "appearance" was such a hot item that car owners "will accept a certain degree of discomfort" ([11-32], p. 141). Noise abatement and sound engineering meanwhile had developed so far that engineers openly started to fantasize about an "ideal car [that] should rapidly traverse an ordinary road in such a manner that the only indication of motion would be the sight of the passing landscape." This quote says it all: thinking from the point of view of the driver, engineers had been trying to eliminate sound to enhance vision. And just like in the case of safety (as discussed in the previous chapter), they had focused on the driver and passenger rather than on outward noise. The cocooning effect was completed when engineers appeared able to seal off the capsule against odors and noxious gases from outside, including the car's own exhaust ([11-32], p. 778).

Car manufacturers after WWII would pay dearly for this narcissistic approach of the noise issue. Governments would start imposing noise limits for cars in an effort to clean up the environment of all forms of harmful waste (Figure 11.3). Did it help? Not in terms of complaints, as nowadays cars are still mentioned as one of the major sources of complaints about noise, next to children ([11-9], p. 244–245). But who wants to get rid of children?

Unlike sound, however, many of the other waste products were visible, as we shall see in the next section.

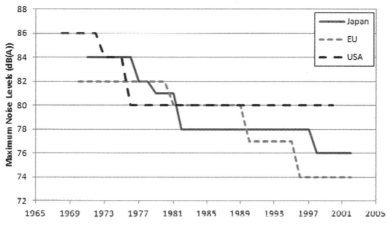

Figure 11.3 The evolution of noise limits (*draft inspired by [11-34]*).

11.3 The Diesel Car as Remedy: Car Cultures and the Perception of Technology

It was this sealed-off capsule which, after the war, evolved into a true cocoon embedded in an increasingly massive flow on the newly constructed freeways, and which started to cause problems for its surroundings.

It began, just like half a century earlier, with problems of visibility: urban smog. Smog as a neologism is a combination of smoke and fog, a term originally used to refer to "London smog," generated by burning fat coal in houses in a damp atmosphere. Then it became coined for the situation in places like Los Angeles, where "photochemical smog" (which did not contain smoke or fog) in the shape of ozone and other chemicals were produced through ultra-violet radiation from the sun on industrial and automobile exhaust products such as hydrocarbons (C_xH_y) and nitrogen oxides (NO_x). Ozone (O_3) is irritating to the eyes, harmful to the lungs, discolors pigments, attacks rubber, and harms vegetation.

It was the Dutch-American biochemist professor Arie Jan Haagen-Smit who isolated in his laboratory the mechanism and connected it to the car. Los Angeles smog is favored by so-called inversion, when static warmer atmospheric layers are on top of colder layers, caught in the wind-free bowl of a valley. Haagen-Smit was subsequently made chair of a Californian committee that helped implement the first ever emission standards in the world. This was the result of the Motor Vehicle Pollution Control Act of 1959 ([11-35], p. 165).

A Crisis in Car Engineering

Apart from causing the smog problem (half of NO_x is traffic-induced), cars also emitted carbon monoxide. That was (and still is) the result of incomplete or imperfect combustion at low engine speeds, cold start, and all those circumstances such as deceleration and acceleration that especially occur in urban traffic (combustion theory calls the generation of only CO_2 and H_2O "perfect" because it does not produce unburned or half-burned by-products) [11-36].

With the public concern about urban pollution slowly rising, car manufacturers had started to remedy some of the problems themselves. For instance, General Motors as early as 1957 cooperated with the Oxy-Catalyst company to develop the means to lower the amount of CO and C_xH_y in the exhaust gas, research that was vehemently opposed internally by an engineering culture heavily dominated by mechanical engineers ([11-35], p. 166–167). In 1961 car manufacturers mounted in cars to be used in California (two years later in all their cars) closed crankcase ventilation systems that led harmful gases in the crankcase back into the air intake to be burnt in the cylinders ([11-37], p. 53). But local authorities, especially on the world's largest and most affected market, California, saw that this was by far not enough to solve their acute air pollution problems, so they started to introduce emission standards, causing a worldwide movement of massive government intervention in matters of motorized mobility.

The result was a fierce struggle between the car industry and governments at all levels, which was aggravated by the reluctance of the former to get orders from outside. Even if, in hindsight, the technology was well developed by the catalyst industry (such as Engelhard and Universal Oil Products OUP in the United States or Degussa and Matthey-Bishop in Europe), and even if deadlines were easily met and extra costs were much less than initially announced, the mood in the higher echelons of the industry was "let's fight." This shows, once again, how technological evolution is much more than a simple linear trajectory of increasing perfection. This happened at the same time that General Motors blundered by intruding in car safety activist Ralph Nader's privacy (as we have seen in the previous chapter), while environmental consciousness among the population grew, so much so that even the not very industry-critical President Richard Nixon had to propose stringent, national abatement measures.

The subsequent legislation was seen by the industry as a slap in the face, which it was. Congress meant to punish the industry for its efforts to delay the introduction of pollution measures, partly by referring to technical constraints, partly by threatening to increase car prices ([11-35], p. 183). Car prices generally are not directly related to costs. According to one calculation, due to regulations, American cars increased in price by one-fifth to one-third of the entire average car price increase ([11-38], p. 16).

From this perspective, the post-war environmental controversy as well as the parallel *scientification* and *electronization* of engineering revealed a remarkably conservative, if not reactionary streak of the mechanical engineering community: since the beginning of the car they had determined what was going on in automotive engineering, and now they not only had to acknowledge the importance of other types of disciplines (acoustics, electronics, mathematics), they also had to rethink their own profession and its truths. In this sense, we can speak of a real crisis of automotive engineering.

Emission Regulation

The Clean Air Act Amendments of 1970 ordered a new government agency, the Environmental Protection Agency (EPA), to make sure that manufacturers would decrease tailpipe emissions of CO, C_xH_y, and NO_x by 90% within five years, a clear example of "technology-forcing." Whereas as late as 1967 the industry did not spend more than 4% of its research and development budget to emission controls, by 1970 it spent more than $147 million a year ([11-35], p. 176). A special Federal Test Procedure (simulating a stylized car trip) was developed by a remarkably activist EPA to assess the efforts expected from the manufacturers in quantitative terms (in grams of harmful substances per test). All this happened at a moment that the first energy crisis (1973) occurred, which prompted the U.S. government to impose fuel economy measures according to the so-called CAFE (corporate average fuel economy) standards in 1975 [11 39], [11 40].

To the surprise of those involved within the EPA itself, the agency, which quickly gained an authoritative amount and quality of expertise in the matter, managed to grow into a powerful stakeholder that would crucially influence automotive technology and the way we think about it ([11-41], p. 188). In short, the environmentally conscious era led to a powerful new player outside the industry who could not be fooled about the intricacies of automotive technology.

As a response, the manufacturers saw two ways to solve the problem, and they pursued both: they could search for alternative propulsion systems that more easily complied with the law or they could try to improve the existing otto (spark ignition) engine. In the latter case, because of the stringent EPA proposals, they had not much more choice than opting for the catalytic converter.

While the first energy crisis of 1973 convinced the American government to give them more time to comply, the installation of the so-called oxidizing catalyst (which made CO and C_xH_y react with oxygen from the air to CO_2 and H_2O, thus reaching complete combustion) at the same time put a process of U.S. fuel economy improvement in motion. This would later be further enhanced by a second generation, the three-way catalytic converter, which also reduced NO_x to less harmful nitrogen compounds. The introduction of the three-way catalyst could be postponed by mounting exhaust gas recirculation systems (EGR), which like the crankcase ventilation, brought the NO_x back into the combustion chamber (Figure 11.4 and Figure 11.5). This happened according to the 1977 Clean Air Act amendments. European car engines were less gas guzzling than their American counterparts, so when the green wave hit Europe and Japan, the catalyst had to compete with other solutions, such as the lean-burn principle, as we will see in the next section.

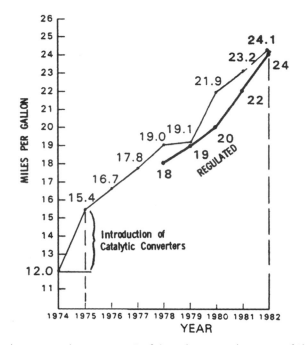

Figure 11.4 Fuel economy improvement of American cars because of the introduction of oxidizing and three-way catalysts. Reprinted from SAE R-226 [11-37], p. 118, Fig. 6-11.

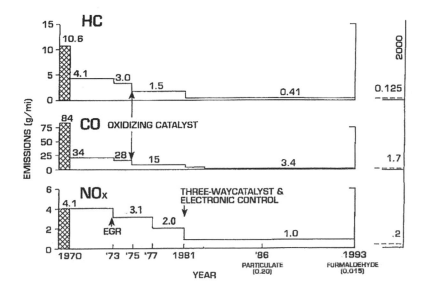

Figure 11.5 Early U.S. emission standards. Reprinted from SAE R-226 [11-37], p. 81, Fig. 6-1.

The Diesel Engine as Alternative

The second option was to go after alternative propulsion systems. This not only led to a revival of the attractiveness of the electric vehicle (as we will see in Chapter 13), but it also brought the diesel engine on the stage as a clean fuel saver. This engine type (which we will study the gestation of by Rudolf Diesel himself in the next chapter) was developed into a useful car propulsion system during the 1920s. In that decade, German Bosch repeated its ignition development feat (as told in Chapter 3) by producing a high-precision fuel injection pump. It was mounted for the first time in the Mercedes 260D, which appeared on the Berlin car show of 1936 ([11-42], p. 62, 81–82).

On the basis of earlier work by inventors such as Prosper L'Orange and Franz Lang (the latter's patents in possession of the American Crude Oil Corporation ACRO; see Chapter 12), Bosch first developed a pump for a Benz truck in 1927 as a replacement for Rudolf Diesel's clumsy air compressor, which kept the diesel engine bulky and slow. Bosch sold its 1 millionth diesel pump after the war, in 1950, and had virtually a world monopoly ([11-43], p. 742).

The next stage in this process was dividing the combustion chamber of the envisaged passenger car diesel engine in two, and injecting the fuel into a separate, lower-pressure high-turbulence chamber. Bosch then designed a fine-mechanical pump and injector, with tolerances of one to two thousands of a millimeter, produced with such precision that a patent was hardly necessary because they were so difficult to copy (Figure 11.6). Early diesel engines used either a pre-chamber or a swirl-chamber, the latter characterized by the generation of turbulence within the chamber, in which combustion is nearly completed before it comes out again in the main combustion chamber. Such passenger car diesel engines produced less diesel knock (because the pressure build-up within the cylinders was less abrupt and less high), but they were less fuel efficient than the directly injected versions. This was caused by the pumping losses in the cylinder, of the gases flowing into and out of the chamber. Direct-injection diesels were installed in trucks [11-44].

Figure 11.6 Precision production of the passenger car diesel fuel pump at Bosch ([11-44], p. 55, 56, 60) (*courtesy of Robert Bosch GmbH*).

Although American engine manufacturers (such as Cummins) also developed passenger car diesel engines with direct injection, after the war, when General Motors became interested in the diesel engine as passenger car propulsion because its emissions promised to comply to the 1977 federal standards, it was the indirectly injected diesel that served as a base for further development. After having considered the 2.1 L diesel engine from its European subsidiary Opel, GM's Oldsmobile division decided to derive its own diesel from an otto engine, opting for the pre-chamber, which emitted less smoke and ran more smoothly than its swirl-chamber counterpart. The result, produced with similar components as the otto version, not only had a new quick-start system (now taking 7 seconds instead of 60), but also promised to comply to the new CAFE fuel efficiency standards. In a short surge of enthusiasm, GM's passenger diesel output increased to 8% in 1981, but fell back to 0.8% in 1984, two years before production was stopped. Blown cylinder head gaskets, broken camshafts and crankshafts made the 5.7 L V8 the worst engine of the last 50 years according to an election of *Automotive News* in 2005 [11-45].

General Motors was not the only manufacturer opting for the cheap solution. At the same time, Volkswagen followed a comparable trajectory by developing a diesel engine from its otto engine production line. But unlike GM, it chose the components very carefully according to classes of tolerances. Opting in favor of the swirl-chamber concept (which was cheaper to develop and had the lowest fuel consumption of the chamber-type diesels), Volkswagen represented the European small-car segment (just as passenger-diesel giant Peugeot did in France), whereas Mercedes, with its larger middle-class cars (and its large export to the United States), stuck to its pre-chamber design. Like Volkswagen, however, Mercedes too handpicked components from a common production line with its otto engines.

Ironically, when Volkswagen decided to produce its Golf Diesel in the United States (called the Rabbit), its sales dropped because of bad production quality, among others. Other factors were the feeling of American customers that the Rabbit was underpowered, and that it belonged to a segment of compact cars that by the early 1980s had gathered a bad reputation as a crisis car. When Mercedes decided to develop a diesel engine (the OM 601) from scratch, it did not experience the same difficulties as Volkswagen, but it too was a victim of a change of mood among American customers. This was caused by what the German historian of technology Christopher Neumaier, who performed an in-depth study in this phenomenon, called a "cancer scare" among the population, which saw the soot emitted by the diesel as carcinogenic. This happened after President Richard Nixon had declared a War on Cancer in 1971 followed a year later by a National Cancer Act. Despite the second energy crisis of 1979 (caused by the war in Iran), diesel sales plummeted from an all-time high of 6.1% of the American market in 1981 to less than 1% in 1985.

The Social Construction of Car Technology

Meanwhile, in Europe, the same energy crisis gave the diesel an eco-friendly reputation. There, the diesel car was seen as a vehicle ready to face the new challenge of climate change, especially after dedicated emission standards promised to reduce the production of both NO_x (the result of high cylinder temperatures) and particulates (Figure 11.7).

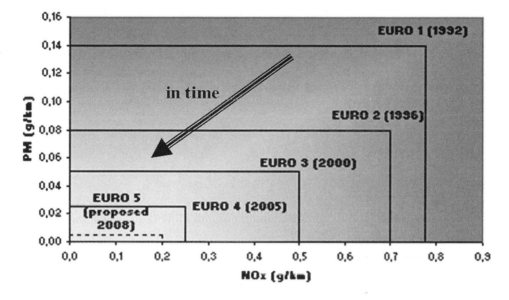

Figure 11.7 European NO_x and particulate emission standards for diesel cars, 1992–2005 (*reprinted from [11-46], p. 1380, Fig. 7, with permission from Elsevier*).

In this redefined environmental setting, local emissions such as CO, C_xH_y, NO_x, and SO_2 (the latter emitted by diesel engines) became less important than the main product of complete combustion itself: CO_2. Thus, while the soot problem seemed to be solved, for the time being by a particulate filter, the diesel engine benefited from an ironical twist of history: its higher efficiency seemed to benefit both the climate and the wish to have more power [11-47].

At first sight, this is different from many other technical innovations, but that is a thinking error: even the turbo compressor, which we have learned to see as a power enhancing device (and thus detrimental to energy conservation), could have been developed to improve fuel efficiency, but the manufacturers (perceiving a customer need) chose not to. Thus, this story illustrates how technical properties should be defined with care: many of these are not intrinsically benign to the environment, but are made as such. A diesel engine emits more soot than an otto engine, but whether this is considered decisive or not depends on the local automotive culture, which is tributary to the local culture in general. Technology, in short, has two, three, or even more faces; which one we see depends on our cultural and social, as well as geographical background. The next section will illustrate this important insight, which we call, as we have already seen on several occasions in the previous chapters, *social construction*.

11.4 Lean-Burn versus Catalyst: The Struggle for a Clean Car

The diesel engine was not the only alternative manufacturers investigated. The *wankel engine* for instance (called after its inventor, the German engineer Felix Wankel), having a rotating piston, was applied by German NSU in its famous Ro80 during the decade after 1967. NSU's peak in sales of 4,203 in the year just before the energy crisis of 1973 also generated enthusiasm in the United States (GM presented a Wankel Corvette in 1973). But it caught on especially in Japan, where Mazda became the rotary engine's major backer and applied it in its sports models (of which it exported 54,000 to the United States in 1973 alone). However, the energy crisis destroyed the chances of this engine with too low fuel economy [11-48].

Another option was the *gas turbine*, also a rotating engine but with continuous, external combustion (like the steam engine), and with a potential of low fuel consumption. Several major American and European car manufacturers (such as Rover, General Motors, Chrysler) invested large sums in this seemingly promising technology. In this case the Pluto Effect did its beneficial work, as the internal-combustion engine inherited the turbo-compressor from this development work, although this was subsequently not used to lower fuel consumption and emissions but to increase specific power (in kW per kg) [11-49], [11-50].

Such engines were a part of a general enthusiasm for automotive innovation during the years of Exuberance of the 1950s and early 1960s, an era in which also the atomic car and even flying cars were proposed (and presented at car shows to an eager public, Figure 11.8). In this chorus of potential substitutes to the internal-combustion engine car, the electric vehicle was revived as well, and we will dedicate an entire Chapter 13 to the question of why this option did not have a chance either.

Figure 11.8 Ford Leva car proposed in the Exuberance phase of the 1950s and 1960s, when enthusiasm around the future ran high. *Source: author's collection.*

Lean Burn

While the choice in the United States soon fell undeniably on the catalyst as an add-on solution for which not much on the engine itself had to be changed (on the contrary, as we have seen: at least a portion of the gas-guzzling properties could be eradicated at the same time), in Europe the situation was quite different. U.S. manufacturers soon found out that the extra $100 of the catalyst could easily be offset by the gains in fuel economy. Especially General Motors products had a "terrible gas mileage" ([11-51], p. 8) as an American adviser to his government said. In Europe, however, some countries were in favor of drastically improving the existing engine itself, instead of adding a catalytic converter to a basically unchanged engine. Such countries produced small cars: Italy in the first place, with 84% of its car park having engines of 1.4 L or less, but also France with 60%. Both manufacturers were afraid that the catalyst would disproportionally increase their prices. They intended to avoid using this add-on technology by introducing special combustion procedures, such as lean-burn and stratified charge.

Both principles were known among specialists for a long time (Nicolaus Otto himself had claimed in his patent that his combustion took place as a layered combustion ([11-52], p. 162). But now they were investigated to drastically lower fuel consumption *and* harmful emissions at the same time. In contrast, catalytic converters theoretically threatened to increase fuel consumption as the price for a cleaner exhaust, for two reasons: placed in the exhaust system they increased the counterpressure, making engine cylinder scavenging more difficult, while its later electronically controlled versions used a richer mixture strength than was necessary for a good combustion, the extra petrol needed to support the catalysis. Also, while the Big Three (General Motors, Ford, Chrysler) pressed the oil companies to get the lead out of their gasolines to prevent it from poisoning the catalytic substrates, in Europe the lead was first and foremost seen as a health issue, poisoning children (playing in urban neighbourhoods next to the roads), rather than catalysts.[1]

The fronts in favor of either improved combustion or the catalyst were not clear-cut [11-53]: while Ford in the United States supported the catalyst, its European subsidiary worked intensely on a lean-burn concept called Proco (programmed combustion). But Chrysler initially also favored lean burn, eager to outbid the other two members of the Big Three. Ford of Europe made a deal with the British government to produce the new engine in England, which may explain why the British negotiators in Brussels found themselves initially in the Southern-European camp of Fiat, Renault, and, to a certain extent, Peugeot [11-54]. Their argument of higher costs for small cars was somewhat hypocritical because European small cars equipped with catalytic converters were meanwhile exported to the United States. It is indicative of the fierceness of the struggle between industry and national governments.

---

1. For American research into childrens' health related to lead pollution and blood-lead levels see ([11-35], p. 181).

Technology Forcing and the Victory of the Catalyst

By contrast, West Germany, with its heavy export of larger cars to the United States, was squarely in favor of the end-of-pipe solution. Germany was supported by some small, non-car-producing (and in terms of environmental policy more radical) countries such as Denmark, Luxemburg, Ireland, and the Netherlands. These smaller countries, especially Denmark and the Netherlands, happened to play a decisive role in the following struggle, as a Dutch analyst later observed. Especially the Dutch representatives in the technical committee in Brussels followed a "strategy [that] directly contributed to the collapse of the qualified majority" that had supported a lenient version of a European emission standard. This prompted the European parliament to propose a more radical version more or less akin to the US83 standards. Japan had meanwhile done so, too. ([11-55], p. 4, 11).

The Dutch strategy to get the harsher regulations accepted by the European authorities in Brussels consisted of a tax incentive for cars with catalysts. This idea threatened to be taken over unilaterally by other countries. It was based upon a $3 million Guilders research project financed by the Dutch Ministry of Spatial Planning and Environment. Senior research engineer Rudolf Rijkeboer, one of these Dutch technical representatives working at the national research laboratory TNO (comparable to the American Argonne Laboratory near Chicago, or the German Fraunhofer Institute), when interviewed in 2010, did not have any doubts in his mind that the catalyst was the only solution to comply with the more stringent standard proposed at that time.

Danish social scientist Kanehira Maruo, on the contrary, basing his conclusion on the responses to a letter sent to the Japanese car manufacturers, is convinced that lean burn did not get its justified chance. He claims that General Motors, teaming up with the oil companies as early as (or even earlier than) the signing of the Clean Air Act Amendments by President Richard Nixon in 1970 had forced the inferior technology upon a world eager to solve an acute local pollution problem (11-56], p. 51–60).

Another study makes convincingly clear that none of the eight alternatives were ready at the moment the Clean Air Act of 1970 was passed and clearly favored the end of pipe solution. In this case, the timing of this act has been more crucial than any of the contemporaries realized, and this timing was the result of a fierce political struggle between President Nixon and his Democratic rival Senator Edmund Muskie. Social scientists call such a strategy *technology forcing* (a strategy often preferred by American authorities), but in this case it looks like the forcing of one technology to the detriment of another ([11-53], p. 103).

One may indeed argue that especially Honda's CVCC (compound vortex controlled combustion) engine looked promising and should have been given a bit more time to be developed into an engine complying with the emission standards. The Honda engine could not meet the 1976 Clean Air Act's NO_x emission limits, while lean mixtures also did not allow for the use of reduction catalysts, so special catalysts would have to be developed ([11-53], p. 107). Ford-Europe revealed its Proco system in 1970, whereas Chrysler-Europe worked on a special carburetor that promised a cleaner and lower fuel consumption. In Europe, the most famous stratified charge proponent was Swiss engineer Michael May, with his Fireball system, who had added an extra inlet valve in a separate chamber where high turbulence was created, enabling the engine to run at a much leaner mixture of an equivalence ratio λ (see Chapter 3) of 2 ([11-57], p. 84–106).

The assessment of the role of lean burn and stratified charge in the catalyst controversy depends on the moment used as point of measurement: originally, when the first emission limits were published, the combustion solution seemed feasible, but when the final goals were set (especially the NO_x standard ([11-51], p. 10)) it was clear that only the catalysts would do. And these final goals were the result of a complex process of negotiations, behind-the-scenes politics, and pure contingency, such as Republican Richard Nixon's wish to get re-elected and his confrontation with a Congress dominated by Democrat Senator Muskie backed by a surge in popular environmental activism, which reached its zenith in the years 1970–1971. Nonetheless, we can observe, with hindsight, that many of the insights during the lean-burn and stratified charge research flowed into subsequent engine research once the catalysts were installed. So, again, the Pluto Effect makes sure that not very much of engineering effort really goes to waste. The irony of the American case is that because of the energy crises of the 1970s, the implementation of the emission standards was delayed on several occasions until the beginning of the 1980s, but by then all engineering effort was already fully focused upon the end-of-pipe solution. Honda, however, managed to stay loyal to its lean-burn principle until 1987, the year that all its cars were equipped with three-way catalysts ([11-53], p. 112, 166).

The Catalyst as Part of a Standardized World System

The switch to the catalyst had enormous consequences for automotive technology: not only was its evolution now crucially influenced by government constraints on outright technical solutions (such as the catalyst instead of lean-burn), but it became truly international. South African mining interests signed an agreement with GM to deliver 420,000 troy ounces of platinum, and GM paid the mining company $100 million dollars to develop a new mine quickly. Every troy ounce needed 12 tons of rock to be processed. Ford closed a comparable deal with another company.

The three-way catalyst added another international dimension, when Swedish Volvo engineer Stephen Wallman got the idea to combine it with a sensor that could measure the relative amount of oxygen in the exhaust gases. It thus could be used to control the mixture formation system such that the equivalence ratio stayed within a very narrow band width around the value of 1. Only around λ = 1, all three gases can be converted, as Figure 11.9 shows ([11-35], p. 185, 190), ([11-51], p. 14), [11-58].

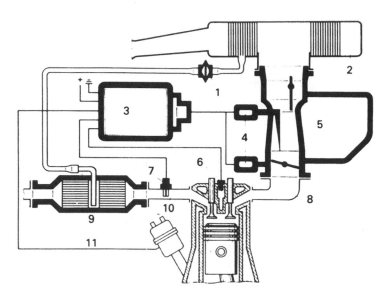

Figure 11.9A

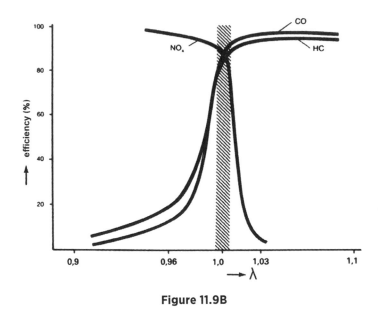

Figure 11.9B

Figure 11.9 Closed-loop control of an electronic carburetor with a lambda sensor (7); B: Conversion graph of exhaust gases as a function of the equivalence ratio λ ([11-59], p. 186, Fig. 2.06, p. 189, Fig. 2.108c).

It was the three-way catalyst that became integrated in a fully electronically controlled engine management system, with the air or lambda sensor as its technical core. Like the ignition and diesel injection pump before, it was again Bosch that developed an existing idea through to a sophisticated technology. In this case it was a sensor consisting of a ceramic probe covered with platinum, which becomes permeable at 300° C for oxygen ions and thus is able to detect a difference in oxygen strength on both sides of the probe. If one side makes contact with the exhaust flow and the other with the ambient air, the difference in oxygen content can be measured as an electric tension. This in turn can be used to control the mixture strength in an electronically controlled carburetor or injection system, a true close-loop control system on the basis of car electronics. The sensor was introduced in California in 1976 in a 1977 Volvo 240, and Bosch sold 10 million of them within a decade. Bosch started series production of a heated sensor in 1982 (for a quicker response at cold start) and a lean-mixture sensor in 1986, the latter used for lambda values between 1 and 1.5 (heated up to temperatures of 800° C) ([11-59], p. 186–191).

When the electronically controlled three-way catalyst was introduced, General Motors also refrained from using its pelleted version (a box with catalytic "pills") and joined the majority of manufacturers who had opted in favor of a "monolith" made of metal or ceramics and covered with precious metals like platinum, rhodium, and palladium. As true catalysts, these noble metals do not take part in the chemical reaction ([11-37], p. 100–107).

By 1980 all American cars had some form of emission control, half of them catalytic converters. Figure 11.10 illustrates this, showing the same mechanism as we encountered earlier in the case of the radial tire, where an *intermediary technology* paves the way for the subsequent real breakthrough of the innovation (see Figure 7.12).

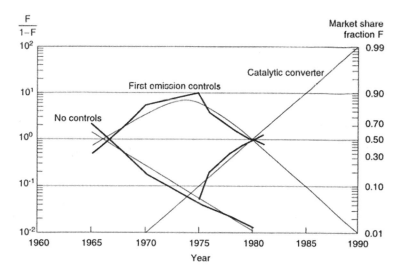

Figure 11.10 Transition to catalytic converters in the United States ([11-60], p. 64) (*courtesy of Arnulf Grübler, IIASA*).

A European Turn-around, and an American

As to the fierce catalyst-or-lean-burn controversy in Europe: in 1989 the European Community turned around unanimously (after consumer representatives had made the switch between 1985 and 1988) and decided to introduce the catalyst in 1993. At that moment, the US83 or similar standards were already in force in Sweden, Switzerland, and Austria, countries that did not form a part of the European Community. First, it was the so-called Luxembourg Compromise of 1985 that allowed lean-burn engines in smaller cars (up to 2 L), but in 1988 a special European Small Car Directive clearly proposed the catalysts as the only viable solution: the window for lean-burn had closed by 1988 ([11-53], p. 119–120).

As a by-product, the unification of the engineering energy invested toward the catalyst also led to a higher stage of European integration, which would allow its car industry and bureaucrats alike to set up an effective defence against cheaper Japanese car imports during the following decade ([11-54], p. 104, 181]). In Europe, too, as in the United States, the emission struggle favored globalization.

But when the EPA, perhaps elated because of its victory over the Big Three, started to address the next subject on the list, the American motorist, a massive countermovement arose. It began in California, where the EPA wanted to force the state to use parking rights to influence its citizens' driving habits. "If we're going to hell, we might as well drive there," was one of the angry responses of a motorist, and he thus eloquently expressed how he and his fellow motorists only had accepted so-called technical fixes (performed by the car and oil companies) to attack car pollution, but that they were not prepared to change their behavior. Apparently, technical fixes do not work in a situation of Southern California, where the population had grown from 3 million in 1940 to 10 and 15 million in 1970 and 2000, respectively. By this demographic effect, EPA risked to lose most of the environmental benefits it had gained through technology ([11-35], p. 204, 206).

The end result was that small cars acquired the smell of crisis: they had peaked in 1961 to nearly 40% of the market, had fallen down in share to 15% in 1967, rebounded during the next half decade to 40% again. Then, from 1976 onwards, American motorists started to show a clear hostility toward the compacts (shortly interrupted by the Iranian crisis in 1979). True, the *compactification* had led to the demise of the V8 in favor of the four- and six-cylinder engine, and had helped to shift toward front-wheel drive and electronic fuel injection, but as soon as they had the chance, the Baby Boomers started to buy pickups, minivans and sport utility vehicles (SUVs) with which the manufacturers managed to circumvent the CAFE standard of 27.5 miles per gallon.

Meanwhile, the car industry had regained much of its power, and used the second energy crisis of 1979 to put Congress (and Senator Muskie) on the defensive, backed by the powerful union UAW, who feared for jobs. Henry Ford II called small cars "little shitboxes" and, indeed, apart from a dip during the crisis of 2008/2009, light truck sales have been higher than cars since 1999 ([11-35], p. 212, 227, 231ff), [11-61], ([11-62], p. 15). Since 1990 (for trucks: since 1996) the CAFE standards have not been changed, and only very recently an initiative by President Barack Obama shook the automotive industry as it intends to improve consumption to 34.1 mpg in 2016, which is only a stepping stone to the ultimate target of 54.5 mpg in 2025 ([11-63], p. 88), [11-64].

Meanwhile, interest for lean-burn and stratified charge revived, for instance through the founding of the Partnership for a New Generation of Vehicles (PNGV) by the federal government in 1993, initiating cooperative research between car companies and national laboratories. New, lean-burn catalysts were developed within this program, based on so-called zeolites. Also, the Revised Clean Air Act of 1990 had reawakened the interest in alternative fuels, such as reformulated gasoline, containing the additive methyl tertiary butyl ether (MTBE), which was in 2003 banned in California because of its carcinogenic qualities, or dimethyl ether (DME). As a result of these and related measures, air quality from the early 1970s began to improve, but several of these improvements were compensated by the increase of the population of cars and their use ([11-37], p. 194, 197, 206, 221–222).

In 2003, a calculation of the average fuel economy (based on 2002 sales) resulted for Europe in 7.55 L/100 km, for Japan in 7.89 L/100 km, and for the United States in 10.13 L/100 km ([11-63], p. 92).

11.5   Conclusions

The cases of car pollution and road safety, as exposed in this and the previous chapter, teach us again that technology and its evolution can only be understood within its historical context of national, regional, and global power play and politics. So much is at stake meanwhile!

There cannot be any doubt about the enormous reductions in casualties and emissions of noise and noxious gases and particles, but these gains are not only easily wiped out at the individual level (by more risky behavior, increasing the amount of injuries; by driving longer trips because it is "clean," both examples of risk compensation behavior), but most of all at the collective level of the population (because of demographic developments). With the Kyoto Protocol signed by European countries in 1997, the expectation to drastically lower the impact of road traffic in the production of greenhouse gases like CO_2 (nearly one-third of the total European CO_2 production stems from transport sector, and about 12% from passenger cars) could become reality. In Europe, car manufacturers volunteered (which was later made into law) to reduce the passenger car's average energy production to 120 g CO_2 per test kilometer.

Will we therefore see a revival of the diesel car in the United States, which did not sign the Kyoto Protocol? According to Dan Sperling, one of the U.S. authorities on alternative transportation, the diesel engine promises to be between 10 and 30% more efficient than the otto engine. The high diesel car shares on the markets of France (more than 50%), Germany, and other European countries versus the virtual nonexistence on the American market can be seen as an expression of the strength of environmental consciousness and the belief in the harmfulness of diesel soot [11-65].

In a way, one can even argue that the seemingly benign effect of the environmental and safety measures of the last half century have turned nasty from the moment the vast populations of Asia, Latin-America, and Africa decided to motorize on a massive scale. It is not clear whether diesel engines and catalysts will be able to meet these future challenges, which are only expected to further increase. One thing is clear, though: without a further scientification of car technology, such a challenge cannot be faced. We will therefore first have to look into the evolution of automotive engineering knowledge (in the next chapter) before we ask ourselves whether we can expect the electric vehicle to solve the problem for us, as a radical shift away from internal combustion on the basis of fossil fuels. After we have done that, we will have to come back to the context and give it the last word: in Chapter 15 we will see how solutions in the traditional configuration of Europe and the United States will not work in the 21st century. The car pollution case makes this abundantly clear: the solutions have to be global, because the problems are global, too.

References

11-1.   Gijs Mom, *Atlantic Automobilism: The Emergence and Persistence of the Car, 1895–1940* (New York and Oxford: Berghahn Books, forthcoming).

11-2.   Gijs Mom, "Compétition et coexistence; La motorisation des transports terrestres et le lent processus de substitution de la traction équine," *Le Mouvement Social* No. 229 (October/December 2009), 13–39.

11-3.   Gijs Mom, *The Electric Vehicle; Technology and Expectations in the Automobile Age* (Baltimore: Johns Hopkins University Press, 2004).

11-4.   P.G. Heinemann, "The Automobile and Public Health," *Popular Science Monthly* 84 (March 1914), 284–289.

11-5.   W.M.M. Pilaar, *De hygiënische beteekenis van automobielgassen* (Amsterdam: H.J. Paris, 1932).

11-6.   Sally H. Clarke, *Trust and Power; Consumers, the Modern Corporation, and the Making of the United States Automobile Market* (Cambridge: Cambridge University Press, 2007).

11-7.   Gijs Mom and Ruud Filarski, *Van transport naar mobiliteit; De mobiliteitsexplosie (1895–2005)* (Zutphen: Walburg Pers, 2008).

11-8.   Karin Bijsterveld, *Mechanical Sound; Technology, Culture, and Public Problems of Noise in the Twentieth Century* (Cambridge; MA/London: The MIT Press, 2008).

11-9.   Mike Goldsmith, *Discord: The Story of Noise* (Oxford: Oxford University Press, 2012).

11-10. Gijs Mom, "Orchestrating Automotive Technology: Comfort, Mobility Culture and the Construction of the 'Family Touring Car', 1917–1940," *Technology and Culture* 55, no. 2 (April 2014) 299–325.

11-11. Karin Bijsterveld, "Acoustic Cocooning; How the Car became a Place to Unwind," *Senses & Society* 5 (2010) No. 2, 189–211.

11-12. J. G. Vincent and W. R. Griswold, "A Cure for Shimmy and Wheel Kick," *Journal of the Society of Automotive Engineers* 24 No. 4 (April 1929) 388–396.

11-13.   *Allgemeine Automobilzeitung* 9 No. 51 (December 1926).

11-14.   Stefan Krebs, "The French Quest for the Silent Car Body; Technology, Comfort, and Distinction in the Interwar Period," *Tranfers* 1 No. 3 (Winter 2011), 64–89.

11-15.   K. L. Herrmann, "Some Causes of Gear-Tooth Errors and Their Detection," *Journal of the Society of Automotive Engineers* 11 No. 5 (November 1922), 391–397.

11-16.   Earle Buckingham, "Transmission Noise and Their Remedies," *Journal of the Society of Automotive Engineers* 17 No. 5 (November 1925), 460–462.

11-17.   F.C. Stanley, "Causes and Prevention of Squeaking Brakes," *Journal of the Society of Automotive Engineers* 18 No. 2 (February 1926), 160–162.

11-18.   Thomas L. Fawick, "Two Desirable Quiet Driving-Ranges for Automobiles," *Journal of the Society of Automotive Engineers* 21 No. 1 (July 1927), 99–106.

11-19.   Theodore M. Prudden, "Noise Treatment in the Automobile," *Journal of the Society of Automotive Engineers (Transactions)* 35 No. 1 (July 1934), 267–270.

11-20.   "Noise Studies Now Important in Design." *Journal of the Society of Automotive Engineers (Transactions)* 34 No. 2 (February 1934), 62.

11-21.   "New Problems in Body Design; Low Cars and High Speed Necessitate Chassis Engineers' Aid, Buffalo Section Is Told," *Journal of the Society of Automotive Engineers* 21 No. 6 (December 1927), 629–630.

11-22.   Alan R. Fenn, "The English Light-Car and Why," *Journal of the Society of Automotive Engineers* 20 No. 4 (April 1927), 483–488.

11-23.   "Recent Progress in Automobile Design," *Journal of the Society of Automotive Engineers* 26 No. 2 (February 1930) 246–249.

11-24.   C.L. Humphrey, "Noise and Heat Control in the Automobile Body," *Journal of the Society of Automotive Engineers* 30 No. 5 (May 1932) 208–210, 214.

11-25.   *La Vie Automobile* (25 May 1929) vi.

11-26.   Hermann A. Brunn, "Trends in Body Design," *Journal of the Society of Automotive Engineers* 22 No. 6 (June 1928), 679–683.

11-27.   Kingston Forbes, "The Body Engineer and the Automotive Industry," *Journal of the Society of Automotive Engineers* 8 No. 5 (May 1921), 436–439.

11-28.   R. E. Chamberlain, "This Body Business," *Journal of the Society of Automotive Engineers* 25 No. 2 (August 1929), 107–109.

11-29.   "10 S.A.E. Activities—Where They Started and How They Grew," *Journal of the Society of Automotive Engineers* 37 No. 1 (July 1935), 18–27.

11-30.   H. Ledyard Towle, "Projecting the Automobile into the Future," *Journal of the Society of Automotive Engineers* 29 No. 1 (July 1931), 33–39.

11-31.   H.M. Crane, "How Versatile Engineering Meets Public Demand," *Journal of the Society of Automotive Engineers* 41 No. 2 (August 1937), 358–392.

11-32. Henry M. Crane, "The Car of the Future," *Journal of the Society of Automotive Engineers (Transactions)* 44 No. 4 (April 1939), 141–144.

11-33. *Autocar* 67 no. 7 (23 October 1931).

11-34. http://www.ncte.ie/environ/noise.htm (consulted on January 17, 2013).

11-35. Tom McCarthy, *Auto Mania; Cars, Consumers, and the Environment* (New Haven/London: Yale University Press, 2007).

11-36. Jan Polman, Course Combustion Engine Technology, HTS-Autotechniek, Apeldoorn, The Netherlands, 1978–1982, personal notes by this author.

11-37. J. Robert Mondt, *Cleaner Cars: The History and Technology of Emission Control Since the 1960s* (Warrendale, PA: SAE International, 2000).

11-38. Daniel Sperling, Ethan Abeles, David Bunch, Andrew Burke, Belinda Chen, Kenneth Kurani, and Thomas Turrentine, "The Price of Regulation," *Access* 25 (Fall 2004), 9–18.

11-39. David Gerard and Lester B. Lave, "Implementing Technology-forcing Policies: The 1970 Clean Air Act Amendments and the Introduction of Advanced Automotive Emissions Controls in the United States," *Technological Forecasting and Social Change* 72 (2005), 761–778.

11-40. George R. Heaton and James Maxwell, "Patterns of Automobile Regulation: An International Comparison," *Zeitschrift für Umweltpolitik & Umwelrecht* 7 (1984) No. 1, 15–40.

11-41. John Quarles, *Cleaning Up America; An Insider's View of the Environmental Protection Agency* (Boston: Houghton Mifflin Company, 1976).

11-42. Eugen Diesel, *Die Geschichte des Diesel-Personenwagens* (Stuttgart: Deutsche Verlags-Anstalt, 1955).

11-43. Holger Bingmann, "Chapter 14: Competence; Case C: Antiblockiersystem und Benzineinspritzung (Anti-Blocking System and Fuel Injection)," in Horst Albach, *Culture and Technical Innovation; A Cross-Cultural Analysis and Policy Recommendations* (Berlin/New York: Walter de Gruyter, 1994), 736–821.

11-44. Friedrich Schildberger, *Bosch und der Dieselmotor* (Stuttgart: Robert Bosch GmbH, 1950) (Bosch-Schriftenreihe, Folge 3).

11-45. Christopher Neumaier, "Design Parallels, Differences and...a Disaster; American and German Diesel Cars in Comparison, 1968–1985," *ICON; Journal of the International Committee for the History of Technology* 16 (2010), 123–142.

11-46. Marc Dijk and Masaru Yarime, "The emergence of hybrid-electric cars: Innovation path creation through co-evolution of supply and demand, " *Technological Forecasting & Social Change* 77 (2010), 1371–1390.

11-47. Christopher Neumaier, "Eco-Friendly versus Cancer-Causing: Perceptions of Diesel Cars in West Germany and the United States, 1970–1990," *Technology and Culture* 55, no. 2 (April 2014) 429–460.

11-48. Diana de Pay, "Chapter 17: Commitment; Case G: Der Wankelmotor (The Rotary Engine)," in Horst Albach, *Culture and Technical Innovation; A Cross-Cultural Analysis and Policy Recommendations* (Berlin/New York: Walter de Gruyter, 1994) (Akademie der Wissenschaften zu Berlin, Research Report 9), 1013 1039.

11-49. C. Besant, K. R. Pullen, M.R.S. Etemad, A. Fenocchi, M. Ristic, N. C. Baines, and W. Dunford, "Hybrid Traction—A Solution for a Not Too Distant Future," in G.P.A. Mom, J. W. Möhlmann, J. C. Vorsterman van Oijen, H. J. Weegenaar, and C. Wiers-Latooij, *Yearbook Autotechnical Trends 1993/Jaarboek Autotechnische Trends 1993* (Apeldoorn: HTS-Autotechniek, 1992), 3.3.1–3.3.13.

11-50. Max Bentele, "Learning from History: Fundamentals," in G.P.A. Mom, J. W. Möhlmann, J. C. Vorsterman van Oijen, H.J. Weegenaar, and C. Wiers-Latooij, *Yearbook Autotechnical Trends 1993/Jaarboek Autotechnische Trends 1993* (Apeldoorn: HTS-Autotechniek, 1992), 4.4.1–4.4.18.

11-51. Daniel Dexter, "Case Study of the Innovation Process Characterizing the Development of the Three-Way Catalytic Converter System" (Final Report, prepared for the U.S. Department of Transportation National Highway Traffic Safety Administration, Office of Research and Development, November 1979) (Report No. DOT-TSC-NHTSA-79-36; HS-804-791).

11-52. C. Lyle Cummins, *Internal Fire* (Lake Oswego, Oregon: Carnot Press, 1976).

11 53. Jan Nill and Jan Tiessen, "Policy, Time and Technological Competition: Lean-burn Engine versus Catalytic Converter in Japan and Europe," in Christian Sartorius and Stefan Zundel (eds.), *Time Strategies, Innovation and Environmental Policy* (Cheltenham/Northampton, MA: Edward Elgar, 2005), 102–132.

11-54. Roland Stephen, *Vehicle of Influence; Building a European Car Market* (Ann Arbor: The University of Michigan Press, 2000).

11-55. Charlotte Kim, *Cats and Mice: The Politics of Setting EC Car Emission Standards* (Brussels: Centre for European Policy Studies, May 1992) (CEPS Working Document no. 64; CEPS Standards Programme: Paper no. 2).

11-56. Kanehire Maruo, "The Three-way 'Catalysis': How the Three-Way Catalyst Became the Ruling Technical Solution to the Automobile Emission Problem," in Mikael Hård (ed.), *Automobile Engineering in a Dead End: Mainstream and Alternative Developments in the 20th Century* (Gothenburg: Gothenburg University, 1992), 45–61.

11-57. Jan Norbye, *The Complete Handbook of Automotive Power Trains* (Blue Ridge Summit: TAB Books, 1981).

11-58. James Scoltock, "Stephen Wallman," *Automotive Engineer* (July/August 2009) 7.

11-59. J. Kasedorf, *Automobielelektronica; principes en toepassingen* (Deventer: Kluwer Technische Boeken, 1987).

11-60. Arnulf Grübler, *Technology and Global Change* (Cambridge/New York/Melbourne: Cambridge University Press, 1998).

11-61. Shane Gunster, "'You Belong Outside'; Advertising, Nature, and the SUV," *Ethics & the Environment* 9 (2004) No. 2, 4–32.

11-62. Caetano C.R. Penna and Frank W. Geels, "The Co-evolution of the Climate Change Problem and Car Industry Strategies (1979–2012): Replicating and Elaborating the Dialectic Issue LifeCycle (DILC) Model" (paper presented at the workshop Electrification of the car: will the momentum last?, Technical University Delft, 29 November 2012).

11-63. John Mikler, *Greening the Car Industry: Varieties of Capitalism and Climate Change* (Cheltenham: Edward Elgar, 2009).

11-64. Kevin Jost, "Mapping the Road to 54.5 mpg," *Automotive Engineering online* (16 October 2012) (www.sae.org/mags/aei/11461, retrieved on 1 April 2013).

11-65. Belinda Chen and Dan Sperling, "Analysis of Auto Industry and Consumer Response to Regulations and Technological Change, and Customization of Consumer Response Models in Support of AB 1493 Rulemaking; Case Study of Light-Duty Vehicles in Europe" (David, CA: Institute of Transportation Studies, June 2004) (report UCD-ITS-RR-04-14) (downloaded from http://www.its.ucdavis/edu on 9 December 2012).

Chapter 12
Scientification: The Co-evolution of Engineering Knowledge

12.1 Introduction: How Do Engineers Know?

In a historical study on the evolution of airplane technology, famous among historians of technology, the former aeronautical engineering professor-turned-historian Walter Vincenti tried to grasp *What Engineers Know and How They Know It* [12-1]. Although a comparable study on automotive engineering does not exist, Vincenti's topic is close enough to the automotive field to be of use to our investigation into the characteristics of automotive engineering knowledge, especially so as the Society of Automotive Engineers considered aeronautical engineers to be as automotive as car engineers: early SAE conferences dealt as much with aviation as with cars.

Vincenti's work is part of a tradition of more general research into engineering knowledge, which meanwhile resulted in a clear rejection of the naïve idea that engineering is applied science, knowledge directly derived from scientific expertise. Using insights from recent innovation studies, another historian of technology, Edwin Layton, Jr., showed how this idea is purely an ideological construct, stemming from the immediate post–WWII period when corporate managers thought that they only had to put some scientists in a building surrounded by some suburban green, far from the factory, and the flow of marketable suggestions started running ([12-2], p. 64).

Others, such as the philosopher Carl Mitcham, have tried to characterize engineering practice as distinct from research in science; they observe that engineering is aimed at "the creating of something new rather than the finding of something already there but hidden" (invention versus discovery), that "opposed to conceiving or even imagining (as in science, GM), [engineering] is the concrete transformation of materials" and that "engineering design is (…) an effort (at first sight, of a mental sort) to save effort (of a physical sort)" ([12-3], p. 93, 97). Another philosopher, Frenchman Jean-François Lyotard, calls engineering "a game pertaining not to the true, the just, or the beautiful, etc., but to efficiency," and Layton even finds that engineering involves "a creative form of cognition, engineering design, akin to that practiced by artists" ([12-3], p. 100), ([12-2], p. 64). And although later research into the practices of scientists and engineers, especially executed within or around the school of the French sociologist Bruno Latour, has questioned this strict distinction between scientists and engineers, he too rejects the idea of engineering as applied science [12-4].

In the previous chapters we have already met several engineers and inventors and their work. We have called Arthur Krebs (carburetors, Chapter 2), Robert Bosch (magneto ignition, Chapter 3), Charles Kettering (battery ignition and starter motor, Chapter 3), and Jean Albert Grégoire (constant-velocity joint, Chapter 4) typical for their time, despite their differences: men (all men, indeed) who influenced the course of early automotive engineering in a decisive way. We have also seen examples of a persistent lack of engineering knowledge, causing new reverse salients. The struggle to get the shimmy phenomenon out of the front wheel suspension (see Chapter 7) was not only long and intense because it was part of one of the largest shifts in the history of automotive technology (the shift from rear-wheel to front-wheel drive), but also because manufacturers, engineers, and users alike were biased toward the propulsion system and only gradually realized that a car is much more than an engine on wheels. The emergence of this insight coincided with the introduction of new engineers and scientists to the automotive world, such as chemists and mathematicians (as we have seen in the case of tire technology in the 1920s; see Chapter 7), acoustic engineers and biochemists (see the previous chapter). In other words, to come to this insight, automotive engineering had to evolve into a scientifically inspired field, as we will see in this chapter, and it did so only after impulses emanating from aeronautical engineering.

In this chapter, we can only touch upon some salient issues of the co-evolution of engineering knowledge and technology. We will start with Rudolf Diesel as a living example of the failure to directly transform scientific insights into engine technology (Section 12.2). We will then acknowledge that engineering knowledge increasingly became the result of team work, and we will visit some laboratories spread over the automotive world (Section 12.3). This will be followed by some examples of knowledge dissemination, through conferences, education, and handbooks (Section 12.4), and we close this chapter with conclusions (Section 12.5).

12.2   Rudolf Diesel: Failure or Success?

The story of Rudolf Diesel's efforts to develop what he called a "rational engine" is an interesting case to study the complex and certainly not linear and one-directional transfer of scientific knowledge into an engineering design.

Diesel knew at age 14 that he wanted to become an engineer. Once at the Stuttgart technical university he wrote during a course on Thermodynamics in his notes: "can one construct steam engines that perform the perfect working cycle without being too complicated?" ([12-5], p. 114). The perfect cycle was the Carnot cycle, represented by a pV diagram (see Chapter 2) that showed two isothermal and two adiabatic processes, together resulting in the production of the highest possible amount of work taken from a given quantity of heat. Historian of technology Lynwood Bryant has done detailed research in how Diesel conceived this idea and how he failed to accomplish it [12-6], [12-7], [12-8].

The ideal cycle was called after the French thermodynamics pioneer Sadi Carnot, who had published a famous book in 1824 on the theoretical limits of what we now would call the thermal efficiency of an engine. He showed that the addition of heat and the subsequent removal of heat to and from the working medium should happen without any change in the medium's temperature (isothermal). Also, the expansion of the medium (generating mechanical work on the piston) and its subsequent compression (necessitating work loss by the piston) should take place without any exchange of heat with the surroundings (adiabatic). During the remainder of the 19th century, and before the electric motor came out of this struggle as victor, at least a hundred inventors tried to replace steam by other working media, mostly air, with the aim to decentralize power and to support the small manufacturer against the violence of the emerging large companies. Diesel stands in this tradition of what he later would call *Solidarismus* (Solidarism): he emphatically saw his engineering creativity as the technical equivalent of a kind of middle class socialism ([12-5], p. 132–133, 367), [12-9], [12-10].

Mimicking Carnot

Bryant insists that Diesel did not succeed in realizing his ideal, which would have meant an engine with an efficiency of about 73%, working at a maximum compression ratio (the relation between the volume above the piston in the lowest and in the highest position in the cylinder) of 60. As the pV diagram in Diesel's patent already showed for the knowledgeable observer (see Figure 12.1), the compression pressure had to be extremely high, and the resulting surface of the diagram was too small to generate enough work to do more than moving the piston and overcoming internal friction losses of the other moving parts within the engine. Diesel had to compromise and therefore could not bring about isothermal combustion. He had intended to realize this through the compression of the air in the cylinder and injecting, at very small amounts and very gradually, fuel into the air which was previously heated by the compression. The mixture should then ignite without increasing the temperature, because of the resulting expansion of the air.

Rudolf Diesel's son, who wrote his father's biography, mentioned that he was inspired to use this principle by the demonstration of a pneumatic igniter at the university, which worked through the ignition of a piece of tinder by compressing (and through this: heating) an amount of air ([12-5], p. 86–87, 164).

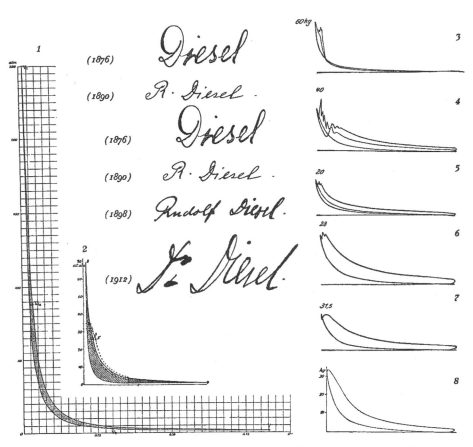

Figure 12.1 The ideal Carnot cycle and what Rudolf Diesel made of it. His son Eugen, from whose book these figures are borrowed, suggests that his father was so involved in his "black mistress" that he even adjusted his signature along the lines of its indicator diagram; 1: the Carnot cycle; 2: Diesel's idealized compromise; the eight pV diagrams to the right are indicator diagrams taken from Diesel's engine and intended to mimic Carnot as much as possible ([12-5], p. 478–479) (*courtesy Historical Archive of MAN Augsburg*).

When Diesel found an industrial backer (*Maschinenfabrik Augsburg*, the later engine manufacturer MAN) he spent about a decade working on a commercially feasible engine (which he called his "black mistress," *schwarze Geliebte*) by trying to solve myriad practical problems. Such problems were injecting the tiny amounts of fuel through a separate air pump (the propulsion of the pump, which had to create a pressure higher than that within the cylinder, caused additional work loss) and realizing something in between the two extremes of an isothermal combustion à la Carnot and a constant volume combustion as invented somewhat earlier by Nicolaus Otto.

His engine, Diesel knew quite well, realized a constant-pressure combustion, resulting in an efficiency of only 40% (the same as the best steam engine of his days), but much higher than the other internal combustion engine, the otto engine. And while we now see the diesel engine as a compression-ignition engine, Diesel himself insisted that his real insight was the answer to the question how one could realize isothermal combustion. As proof of this, one can refer to his decision to approach Robert Bosch in 1894 to inquire about the possibility of using Bosch's famous magneto ignition to solve his fuel injection problem.

Trying to Develop an Automotive Diesel Engine

At a moment that his engine was applied, shortly before the First World War, in hundreds of ships (and started to be mounted in the submarines of the German Navy), Diesel declared: "I am still convinced that also the car engine will come and then I consider my life work completed" ([12-5], p. 226, 395), [12-11], [12-12]. As we saw in the previous chapter, it took another two decades before Robert Bosch would develop a fine-mechanical diesel fuel injector, which would enable to miniaturize the diesel engine into a machine for propelling trucks and passenger cars.

Bosch's strategy was similar to the one developed for its magneto ignition: high-quality precision engineering. Typical for Bosch's reputation, the German company bought the rights on an existing system (developed by engineer Prosper L'Orange at American Crude Oil Company ACRO), and developed it through to fine-mechanical near-perfection. In this case, Bosch engineers combined two separate pump elements (one for transporting the fuel, the other for metering it) into one adorned with the famous control ridge (*Steuerkante*) taken over from ACRO as well (Figure 12.2) ([12-13], p. 271–277), ([12-14], p. 740–743), ([12-15], p. 62).

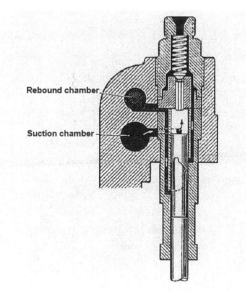

Figure 12.2 Control ridge (*Steuerkante*) in the first diesel injection pump from Bosch ([12-13], p. 275) (*courtesy of Robert Bosch GmbH*).

Although during later development work a second type of diesel combustion could be realized (a mixed form of constant-volume and constant-pressure), isothermal combustion remained unattainable, because the reality of constructing an engine that could withstand the high pressures and other real-world conditions prevented the theory from being realized. No wonder that some professors within the university community in Germany attacked Diesel's patent, and the resulting controversy may well have caused Diesel to commit suicide by jumping off the ferry to England, on his way to negotiate an industrial deal. His son, at least, opts for this explanation, and adds Diesel's financial problems, but he argues against the conspiracy theories that want to have Diesel murdered by German intelligence who wished to prevent valuable knowledge to be given to the future enemy ([12-5], p. 213–217, 394–401, 421).

Whether killed by Intelligence or academic intelligence, the question of whether Diesel's work can be considered a failure or a success depends on the question one asks. From a personal point of view, he did not live up to his own ambitions (and those of several of his scientific critics). But from an evolutionary point of view, there cannot be any doubt that the diesel engine (as compression-ignition engine developed for cars and trucks) decisively shifted the course of automotive technology. Diesel himself rightly claimed, in the middle of the controversy with the professors, that his work clarified the position of the Carnot process ([12-5], p. 200). It was therefore all the more ironic that it was the scientific world that was his greatest opponent. Also, it is certainly not unique that an invention turns out to be important because of other reasons than intended by the inventor, a phenomenon called "unintended consequence" in the history of technology.

12.3   Team Work in Laboratories: Scientification of Car Dynamics

A comparable tension between science and engineering played a role in the emergence of scientifically inspired engineering laboratories for automotive research. This process can be described as the (successful) effort, during the inter-war years, to emulate aeronautical research.

Aeronautics from an early stage was not only adorned with a more scientific engineering reputation (dominated by aerodynamic wind-tunnel research and by research into light materials), but also benefited from the radiance of the glamorous world of the flying aces as civilian as well as war-time pilots. This was especially the case in Germany, forbidden by the Allied powers to continue its aeronautical work after the WWI defeat, and, as compensation, eager to develop a scientific approach to automotive technology ([12-16], p. 4–75).

Although during the first phase a laboratory of automotive research was founded in Berlin close to the fledgling car industry led by professor Alois Riedler, laboratory research on the car as a whole only broke through during the inter-war years.[1] The FKFS (*Forschungsinstitut für Kraftfahrwesen und Fahrzeugmotoren Stuttgart,* Research Institute for Automobiles and Engines Stuttgart) founded in 1930 became one of the largest in the country because it aligned its research program along with German national (and National Socialist) civilian and military interests. Its research received an extra boost when the Nazis came to power and automotive research was put in the service of developing cars to be used on the newly built *autobahnen* (freeways) within the general Nazi desire of *volk motorization.* This set the scene for a major shift to high-speed automotive technology which in fact would result in a new conception of what an automobile was all about: a high-speed, reliable, long-range, family machine. This shift could not have been accomplished without a scientific approach of the design process, especially of the car as a complex body in movement.

The central figure who emancipated German car engineering from its tinkering roots was Wunibald Kamm, a car enthusiast educated in aeronautical engineering [12-19]. Like Arthur Krebs (see Chapter 2), he had started his career in a military context, working on fixed balloons, whose instability, he found out, was caused by their aerodynamic shape. After the war, he worked for a while on Mercedes racing engines, developed and built his own light-weight car with independent wheel suspension, became the head of a test institute for aviation engines, and then was called, in 1930, to the Technical University of Stuttgart (close to the German car industry) to become a professor in automotive technology. There he founded the FKFS. By that time, the application of theory by engineers was mainly limited to strength calculations of components and the testing of prototypes, and the measurement instruments were mainly engine and roller test benches, where stationary loads were applied in duration tests and propulsion performances were measured. Some fuel test engines and tire test benches were available, but there were no automotive wind tunnels. The contribution of Kamm's team (which grew to 400 by 1940) to this fully undeveloped field was a comprehensive theory of vehicle dynamics ([12-19], p. 45–4).

Personally, Kamm introduced aerodynamics into car research, claiming that the aerodynamic drag coefficient of current cars (about 0.6 to 0.7) could be halved. But he also realized (reminded of his fixed balloon experiences during the war) that streamline and driving stability were opponents that had to be reconciled by careful redesign of suspension and steering. To make German cars fit for the *autobahnen,* basic research was needed to increase engine longevity as well as its efficiency and specific energy and power production (in Nm/L or Nm/kg, respectively kW/L or kW/kg) (Figure 12.3). In short, Kamm's work can be seen as instrumental in a full redesign and rethinking of the car in a system dominated by high-speed freeways.

---

1. Two reports testify of this early scientific approach to the testing of cars in Germany; see: [12-17] and [12-18].

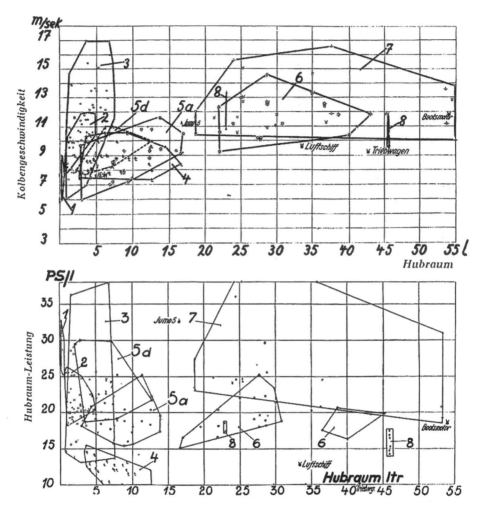

Figure 12.3 Wunibald Kamm's FKFS research institute in Stuttgart, Germany, started its scientific research on engines by investigating the state of the art in terms of piston speed (in m/s) and specific power (in hp/L) as a function of cylinder capacity (in L). 1 = small cars; 2 = German cars; 3: foreign cars; 4 = trucks; 5-8: aviation engines ([12-19], p. 49) (*courtesy of FKFS*).

Wunibald Kamm and Car Dynamics

From the nearly 500 reports that Kamm's research institute produced until the end of the war, it becomes clear which types of engineering knowledge and skills were needed for such a major transition. The reports show a clear focus on car dynamics (including aerodynamics) and engine research ([12-19], p. 97–136).

The FKFS contributed nearly one-fifth of all German research on the Volkswagen project, in which "three adults and one child" had to be seated, but which also should have enough ground clearance to afford war applications ([12-16], p. 83), [12-20]. Tire research revealed the progressive relationship between rolling resistance and speed, and resulted in a construction of a diagram called *Kammsche Reibungskreis* (Kamm's friction circle) showing the relationship between traction forces and sideslip. Kamm's employees P. Riekert and T. Schunck developed a theory of car behavior (their book, published 1940, was a post-war classic in automotive theory), whereas the air resistance measurements in wind tunnels of several sizes led to the development of the so-called *Kamm-Heck* or *K-Heck* (Kamm stern or backside).

The *K-Heck* is a nice example of scientifically guided engineering research (Figure 12.4). Reminiscent of Diesel's struggle with "nature," the *Kamm-Heck* was part of a car body that complied to the basic law according to which at high speeds the reversed droplet was the ideal shape. For practical reasons, however, the droplet's tail had to be cut off in a special way in order to keep as many as possible of the theoretical advantages of the droplet shape. Reminiscent of Diesel, too, was the subsequent patent struggle with other contenders, such as the racer and aerodynamicist Reinhard Freiherr Koenig-Fachsenfeld, who claimed that the *K-Heck* should in fact be called the *K&KF-Heck*, but also with Paul Jaray, whose parallel work on streamlined body shapes represented "Jewish science" and thus may not have had its fair chance in the anti-semitic culture of Nazi Germany ([12-16], p. 182).

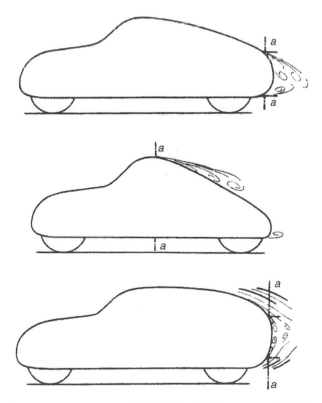

Figure 12.4 Kamm-Heck ([12-19], p.117) (*courtesy of FKFS*).

Just like science, both the Diesel and the Kamm case show how modern engineering should be seen as a political practice as much as a scientific one. Kamm managed to maintain his priority claim, even after he was kidnapped at the end of the war by the American army and brought to the United States as war booty, together with 13 other professors of his university. Although he was removed from his professorship, and his (and his institute's) role has been criticized for supporting "the building of the German war machine" and "the construction of [the] National Socialist violence regime" in the near-official history of the FKFS ([12-16], p. 101), in 2009 he became one of the 16 German personalities within the total of nearly 250 "great men" in the history of the automobile who have been included in the Automotive Hall of Fame in Detroit. His handbook from 1936 (*Das Kraftfahrzeug*; The Automobile) was the first comprehensive scientific handbook on car technology after Arnold Heller's handbook of 1911 (see Chapter 8), while his work can be seen as crucial preparation for the post-war boom in German automotive engineering. Wunibald Kamm is remembered as one of the first who introduced systems thinking into automotive technology, but his biographers do not hide the fact that, in the historical circumstances of his time, he could do so only by becoming a rabid technocrat and opportunist ([12-19], p. 331–335, 187).

Translating Aeronautics to Automotive Technology

Kamm was not the only one who developed a theory of dynamic car behavior between the wars. Historian of technology Ann Johnson even claims that the real center of this theoretical work was the UK, where scientists and engineers at the Road Research Laboratory "translated" mostly aeronautical knowledge to automotive knowledge, by "transforming mathematical models of airplane stability into mathematical models of a car's handling characteristics."

The parallels between airplane and car stability may have inspired Maurice Olley (from the British General Motors branch, see Chapter 6) to develop his theory of oversteer (instable) and understeer (stable). Although Olley's career did not start in aeronautics, he must have been familiar with this discipline's characteristic as reflecting "the ongoing struggle between pilots and engineers over ownership of (…) design decision" related to their craft, a struggle that is comparable to what happened at the same time in the automotive field when it came to the deskilling of the driver, as we have seen in Chapter 9 [12-21].

But despite the work of Kamm and Olley, the American post–WWII pioneers of the mathematical modeling of vehicle dynamics, William Milliken and David Whitcomb of the Cornell Aeronautical Laboratory in Buffalo, New York, were "surprised at the lack of sophisticated theoretical tools used by automobile designers," and amazed about the general "(r)esistance to mathematization" of the automotive engineering community. Johnson mentions the French engineer Jean Odier (working at brake specialist Ferodo) and German engineer Manfred Mitschke (working at Bosch, but from 1966 director of the *Institut für Fahrzeugtechnik* [Institute of Automotive Technology] at the University of Braunschweig) as European post–WWII examples of men who pushed car dynamics in the direction of braking behavior. At that moment, because of its complexity, but also because safety was in the air, so to speak, braking analysis was considered more prestigious than work on steering and handling ([12-22], p. 27, 73, 86, 88, 94–97). On top of that, it happened in a phase when top managers of research believed in the linearity of knowledge development, from science to technology (as we have seen in Section 12.1).

Charles Kettering and General Motors Research

Another example of laboratory research is the hunt for a knock-free fuel in General Motor's central laboratory, led by Charles Kettering. After his successful development of the starting-lighting-ignition system (see Chapter 3), Kettering was invited by Vice President Alfred Sloan to become a vice president himself as head of a new central laboratory at General Motors. Like in Germany, during the first phase "General Motors, along with its competitors, had done little in the way of research except for modest efforts at quality control, including a staff of about twenty chemists and metallurgists which ran a testing laboratory in Detroit," as Kettering's biographer asserts ([12-23], p. 95). Corporate research on a large scale (in teams, working according to a program, at a distance of production) had been initiated before the First World War in other, more science-based industries, such as the chemical industry (DuPont, Kodak) and the electro-technical industry (General Electric, AT&T), also in Europe (Siemens, the German dye-stuff producers, the predecessors of Alsthom in France, Philips in the Netherlands). Certainly in the United States with its increasing public and political pressure against large industrial conglomerates, such laboratories were set up from a defensive point of view: to protect the monopolistic positions of the big industry. Famous individual inventors such as Thomas Edison, a study of the history of corporate research concludes, were "not interested in the kind of work that industrial research would stress—ongoing process and product improvement," but he, too, depended on a team of researchers ([12-24], p. 125), ([12-25], p. 193). By 1921, only 15 U.S. firms each had more than 50 researchers employed ([12-26], p. 141).

During the "golden era of American industrial research during the interwar years" ([12-24], p. 126), however, university-style research was introduced in industry (leading to science-based inventions such as nylon in 1934 and the transistor in 1947), but it seems that in the automotive sector a mixture of the new and the old approach was dominant. When Kettering agreed to become GM's vice president of research, President Pierre du Pont advised him "to focus on the question of reliability" ([12-26], p. 142). Although not the first within the car industry (Ford had started some modest research in 1913, but this was largely limited to inspections and component testing), Kettering was the first to realize that such research had to be conducted close to production and marketing ([12-23], p. 148).

As a matter of fact, the scientifically inspired approach was not only introduced in technological research. Also, production (through the introduction of statistical methods, and in general scientific Taylorist time and motion studies, see Chapter 14), and consumer research was set on a scientific footing ([12-26], p. 132). But the General Motors Research Corporation, founded 1920, deviated considerably from a university institute: Kettering combined the eccentricity of the famous inventors before him (such as Thomas Edison) with well-planned team research, when he organized his laboratory along the following lines: Combustion Fundamentals, Engine Fundamentals, Engine Materials, Carburetion and Distribution, and Car and Dynamometer Test section. By 1927 the laboratory counted 260 personnel, which was more than doubled by 1938 (Ford by then had 250; Chrysler 500). Meanwhile, 120 firms each had more than 50 researchers employed ([12-26], p. 142).

Engine Knock

In contrast to the German case dealt with previously, the organization of Kettering's laboratory showed that research was clearly engine biased. This was Kettering's choice: upon his nomination his venture was a laboratory looking for a problem, so he opted for topics closely related to his own vision of the salient problems of the industry. His first project was an outright failure: driven by the (in hindsight: false) idea that the market was waiting for an engine with low fuel consumption he developed the so-called copper-cooled engine, air cooled (and light) with cooling fins made of copper for better heat conduction, a radical innovation taken from aircraft technology and meant to compete head-on with Ford's Model T. Even if it has never been established with certainty that his engine was to blame, it was quite clear that during the four years of this project he had done a bad job in convincing the powerful engineers who were leading the car divisions. What did not help either was his anti-theoretical can-do approach (Kettering was no specialist in engine design), which clearly was not enough to get all bugs out of such a complex project in time before production. But what killed the project in the first place was his (and his superiors') misjudgement of the market: American motorists had entered the gay twenties and were not at all interested in a light, cheap, and fuel efficient car. Sloan, who within a decade would formulate his well-known philosophy of annual model change, convinced Kettering not to resign. Later, Kettering, famous for his wisecracks, would comment on this phase by saying "You must learn to fail intelligently, for failing is one of the greatest arts in the world" ([12-23], p. 148).

In his second big project Kettering's can-do approach worked. First of all, the new project, finding a solution for the devastating engine knock, fitted seamlessly in the laboratory's engine bias during the early 1920s, a time when officials at the car and oil industries had begun to worry about the limited oil reserves [12-27], [12-28], [12-29]. They knew that increasing gasoline sales had forced suppliers to distill more gasoline from the same amount of crude oil, resulting in a less volatile fuel that worsened both the starting of the engine and the distribution of fuel to the cylinders, causing rough engine running. At the same time, the problem threatened to block another trend toward higher compression in the cylinders, necessary for increasing the performance of the engine.

The reason was that these heavier gasolines contained more hydrocarbon types that were sensitive to knock, a phenomenon of uncontrolled heat release leading to the destruction of the pistons, but this causal relationship was as yet unknown to the actors involved. The strategy that was chosen to attack this problem was typical for the United States: backed by the government's Bureau of Standards, the Society of Automotive Engineers took the initiative of a Coordinated Fuel Research program. This program *de facto* suspended intercompany competition and led oil companies to develop the thermal cracking process, which not only produced more gasoline from the same amount of oil, but—as an *unintended consequence*—generated types of hydrocarbon that were less prone to knock ([12-30], p. 146–147).

They could do so, however, because of the path-breaking work of Kettering and his laboratory. At that moment the oil industry did not do much laboratory research. Military research in plane engines during the war, in which Kettering had been involved, had led to the discovery of certain gasoline types (containing many cyclic compounds such as aromatics and cycloparaffins) that burned better in airplane engines than types containing long chains of carbon atoms (such as paraffins). Reminiscent of the struggle between electric and combustion-engine cars at the last turn of the century, a test pilot during the war had remarked that with Kettering's selection of gasoline types his engine "gave one the impression of flying behind a steam engine or an electric motor… [the pilot at the time was sitting behind the engine, GM]" ([12-23], p. 80).

In his new hunt for a knock-free fuel in the early 1920s, Kettering seemed to have remembered this lesson from the war. For him, the problem somehow was of a chemical nature. Typical for his anti-theoretical approach, however, he made a mechanical (and not a chemical) engineer, Thomas Midgley, head of the project team that started working with a single cylinder engine equipped with an indicator (to register pressure as function of volume in the cylinder). In the light of the upcoming struggle between mechanically inclined chassis engineers and new types of engineers (chemists, acoustic engineers, physicists) we can also see Kettering's mistrust of theory as an expression of a mechanical engineer's bias. However this may be, thus started what Kettering later would call a "scientific foxhunt" reminiscent of what his famous colleague Thomas Edison had done when he searched for the right material combination of his nickel-iron battery ([12-23], p. 153), [12-31].

First, they thought physical characteristics (such as specific gravity and boiling point) would do the trick. Assuming that the volatility of the fuel had to do with knock, they dyed the fuel red (assuming that it would evaporate faster because the red color helps to absorb heat better) and indeed found that the engine knocked less. The reason, however, was not the color but the type of compound they had used (jodine), and they saw this as a confirmation that the problem was chemical indeed. This is remarkable as at the same time, in England, another head of a research lab, Harry Ricardo, assumed a mechanical basis of the knocking problem, and he and his team, too, found a solution: the so-called turbulent head, a shape of the combustion chamber forcing the mixture to take on a turbulent state, thus eliminating pockets in the chamber where knocking could occur. A better example of the *social construction of technology* can hardly be found: the two teams on both sides of the ocean found what they were looking for. Science, certainly looked at from an engineering angle, is not as hard as is often suggested.

Kettering's team then set out to search systematically (according to the periodic system of the elements) for a cheap and safe compound. One of the promising materials was selenium oxychloride, but this was rejected because it made all laboratory workers smell of garlic. As we will see in the next chapter, for a brief period they considered to go for alcohols (especially ethanol) as a full replacement of oil. Indeed, during the 1920s many governments started to mix their national gasoline stock with 5 to 10% of alcohol, mostly with protectionist and autarkic motives in mind, such as independence from oil and protection of national agriculture [12-23], [12-33]. Why Kettering then opted for gasoline is not known, but in the end he stumbled upon tetra-ethyl lead (TEL) as the compound that, added in tiny quantities to the fuel, would reduce knock-proneness. Later, the material of the valves had to be adjusted to resist the presence of the new compound in the fuel, while chlorides and bromides had to keep the compound afloat in the cylinder and avoid deposits on the cylinder walls and the spark plugs [12-34]. This shows how an innovation never comes alone: the receiving system has to be adjusted at the same time.

The Marketing of TEL

The subsequent marketing of Ethyl Gasoline was very controversial, as TEL appeared to be poisonous (Figure 12.5). When ten workers in the Ethyl factory died after attacks of mental illness, production of the "Looney Gas" was interrupted until the U.S. surgeon general declared the fuel safe. In 1985 the *American Journal of Public Health* apologized to its readers that it had printed this message without a critical note ([12-53], p. 4).

Only later, Kettering decided "to put a little more science in their trial and error" (as his biographer remarks) by developing a special indicator with which he could measure the anti-knock quality in relation to a calibration fluid. In 1930, the Society of Automotive Engineers accepted a proposal of one of Ethyl's employees to use octane as a calibration fluid, and since then gasolines are classified according to their octane number, which indicates the knock phenomenon (measured as pressure peaks) in comparison to iso-octane (officially called 2, 2, 4 tri-methyl pentane) ([12-23], p. 149–180), ([12-36], p. 1).

Figure 12.5 This cartoon appeared in the mid-1930s, when the American public already had accepted Ethyl as a knock-free dope in gasoline ([12-34], p.149).

However, a history of the invention of TEL that would stop here (as is often done) would miss an important insight into the intricacies of engineering research. When one reads the *Journal of the Society of Automotive Engineers*, it appears that the full potential of the anti-knock fuel was by far not used by the industry during the following decades (contrary to the aviation industry that struggled with the same problem). The reason for this was that higher combustion pressures increased heat transmission and mechanical forces in the engine and would have necessitated a total redesign. Instead, the engine's compression ratio was only incrementally adjusted to the new fuel and knock was avoided through designing more compact combustion chambers in smaller cylinders (making use of Ricardo's suggestion to increase turbulence), which on its turn mounted the pressure to increase the number of cylinders to 6, 8, and even 12. For two decades, TEL and cracked (and later: reformed) un-doped gasoline competed on the market. As late as 1939, 80% of American car engines were still tuned down in order not to disturb the incremental annual changes allowed by the mass-production process (Figure 12.6) [12-37].

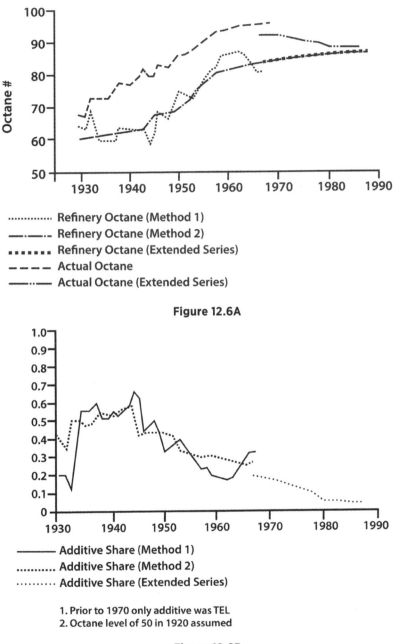

Figure 12.6A

1. Prior to 1970 only additive was TEL
2. Octane level of 50 in 1920 assumed

Figure 12.6B

Figure 12.6 Competition between refining and TEL. A: Refining (represented in this graph by the curves of Refinery Octane) added more to the emergence of knock-free fuels than the addition of special dopes (mostly TEL), which bridges the gap between Refinery Octane and the curves of Actual Octane. B: The amount of dopes (especially TEL) decreased after a steep rise in the 1930s, until it was forbidden by law in the 1970s (*reprinted from [12-30], p. 150, Fig. 3 with permission from Elsevier*).

Kettering and Science

Kettering's TEL case should not convey the impression that his engine bias was absolute, however. During his subsequent career, his laboratory was instrumental in getting General Motors into four-wheel brakes (universal within GM in 1927, see Chapter 8), the synchromesh transmission in 1930 (Chapter 5), and during the 1930s also hydraulic brakes, the all-steel body, and independent front-wheel suspension, topics reminiscent of Kamm's more comprehensive approach in Germany. At the same time, Kettering had heeded President DuPont's warning to focus on reliability: between 1926 and 1939 the car's operating costs per mile were halved, although it is not clear how much of this can be put on the account of Kettering's laboratory ([12-26], p. 144).

Nonetheless, Kettering's influence went far beyond his research leadership. His fame as a public speaker was legendary, as was his conservatism and his reputation of a rabid technocrat. Like Rudolf Diesel, his career was haunted by the tension between engineering and science, suffering, as his biographer concluded, from "an inferiority complex" toward scientists. When Kettering on a visit to Princeton University was to be introduced to Albert Einstein, the latter exclaimed: "Oh, yes, the auto mechanic" ([12-23], p. 327–328). Although this may be seen as typical for the condescending way (inspired, one may assume, by ignorance of the knowledge claims of engineering) of scientists, for Kettering, people had to bow for the "progress" of technology, even for the atom bomb: "you've got to go on with it even if we're all blown to hell with it" ([12-26], p. 341). While Einstein, after the war, opposed the use of the bomb, Kettering and Sloan founded a cancer research institute, and as we will see in the next section, his name adorns the former engineering school General Motors Institute (GMI), now called Kettering University, where also his personal papers are conserved.

Research on Comfort

The third case in this overview of automotive laboratory research is the problem of vibration, which became acute once the industry started to shift toward the closed body (and told more extensively in [12-38]). As we have seen in Chapter 10, this was one of the major shifts in the history of automotive technology, and its repercussions were felt over the entire field.

One of the problems was how to define comfort. Comfort came up as a result of consumer research, together with other general functions, such as economy, performance, and speed, in a period (the second phase of Persistence, between the wars) that engineering had emancipated from a myopic stare at the level of the single component and the component assembly to the holistic approach of the (sub)system, in short: the structure of the car. Car comfort, many American engineers had witnessed when in Europe during WWI, came from Europe: "Our cars come in for a great deal of criticism. They say we sit on our cars while they sit in theirs, and when you ride in their cars you agree with them. We spent 10 days in different makes of European cars. They ride remarkably 'easy.' (..) Their cars are most comfortable and they are very low" ([12-39],

p. 523). Soon, American engineers were showing much more self-confidence, especially when they started to criticize the European trend to small cars with their high-revving engines and their stiff springs. Thus the stereotype of the American "lazy" and comfort-minded motorist was constructed: "The American car is the most convenient vehicle in the world to start and drive," Maurice Olley opined as early as 1921. "It is best able to get [to] any place whether there is a road or not and is the paradise of the lazy driver. It is year-round transportation with the minimum of grief, but it is not a speed monster. It is not pleasant to steer, with certain exceptions, and its brakes are deplorable" ([12-40], p. 111).

Comfort was a crucial element in this myth formation, but it was clear that engineers did not give in to every wish from the customer. "One of the greatest difficulties encountered is to produce a body which will please both Mr. Short and Mr. Long. It is equally difficult to construct a seat cushion so that Mr. Fat will be comfortable as well as Mr. Lean" ([12-41], p. 273, 271).

Such reading of the major American automotive engineering journal is instructive because it shows that not only the engineering community but also that of the users were not a monolithic entity. In the case of engineers, it consisted of factions, schools (often related to car makes), and leading individuals who managed to convince their peers of the necessity and the direction of problem solving. As we have already seen in the case of gasoline, the role of the engineering association in this *translation process* is often one of smoothing opposing interests and initiating research in fields that either are too difficult to manage by one single manufacturer or are too far from the production process. Realizing that the American car industry had "no satisfactory yardstick with which to measure riding-comfort," but that at the same time "vibrations (..) are by far the greatest annoyance in a car," the Society of Automotive Engineers itself took the initiative, in 1925, to approach psychology professor Fred A. Moss at George Washington University in Washington, D.C., and set up a Riding-Comfort Research Subcommittee. They developed a "wabble-meter" that could measure "fatigue" as an indicator of the much more elusive concept of comfort. By placing living humans on a specially prepared "vibrating chair," the human body was used as a seismograph (Figure 12.7).

Later, this research was expanded to Purdue University, where professor Ammon Swope subjected 135 men and women (mostly students) to vibration tests in order to measure groups of "sensory qualities." These included motion, sound, sight, smell, spatial relations, and aesthetics, each the result of a myriad of measurements, well distinguished between men and women, such as speed, noise from brakes, desirable amount of visibility, leg room, and kind of floor covering. Comfort, it appeared soon, was a very complex topic which needed the scrutiny of scientifically educated specialists, acoustical engineers among them [12-43].

Figure 12.7 Vibrating chair ([12-42], p. 482, Fig. 1) (*courtesy of SAE International*).

Unfortunately, there is not much further historical research done on this early form of university research into automotive comfort. This is a pity, because we would probably be able to find here a major difference with the situation in inter-war Europe. Whereas in Germany the emphasis was on car dynamics of the body and its suspension system, American research also included the interior enabled by a buffer function of the Society of Automotive Engineers. This special attention for the passenger may also partially explain why in car safety research the United States was the first to shift its interest from the pedestrian as accident victim to the driver and the passenger, as we have seen in Chapter 10.

12.4 Constructing the State of the Art: Conferences, Education, and Books

Ann Johnson, in her recent analysis of the development of ABS (see Chapter 10), has coined the term *knowledge community*, encompassing the "producers" and "keepers" of such knowledge. ([12-22], p. 1-22), [12-44. In our case, perhaps the term engineering community is more appropriate, as such a community not only produces knowledge, it also deploys practices, aimed at performing an intended task. We call this amalgam of knowledge and skills, attitudes and practices the *state of the art*.

The state of the art is much more than the collection of existing artifacts, although their hands-on character of what one could call "frozen knowledge" makes them especially suitable for disseminating the current and previous states of the art to colleagues and users alike. The state of the art is a collective phenomenon without much structure, different for every member of the community, but consisting of a common core of proven (physical) types and (mental) truths surrounded by myriad loose ends of possibilities, fantasies, and practices not further pursued. As we will see next (as well as in the next chapter), the state of the art also consists of *expectations*, which are forceful incentives for engineering practices. In short, the state of the art is the collective memory and the reservoir of expectations of an engineering community.

The Importance of Expectations

This community, or members thereof, is constantly busy trying to get a grip on (and trying to improve upon) the state of the art, and they do so along two trajectories, by categorizing artifacts into taxonomies and by constructing records, written cross-sections of the state of the art typical for a certain period of time, or a geographical unit (a country) or for a certain application (the engine; trucks). *Taxonomies* of artifacts are built on the basis of types (archetypes, ideal types, and average types), as we have seen in Chapter 1.

Records are the second way of getting a grip on the state of the art ([12-45], p. 45). They come, grossly speaking, in three forms: conference proceedings, journals, and handbooks, forming what one could call an active and a passive (or background) part of the state of the art, respectively. The same is true, by the way, for the population of artifacts: there, too, an active and a passive subpopulation exist, as the entire record of all cars produced and proposed is often not included in the construction of the specific "type." Whereas journals testify of the weekly or monthly development of the state of the art (including reports of important wrap-ups such as annual engineering conferences and exhibitions of artifacts), handbooks need a longer gestation time, and they are often expressly aimed at formulating a normative picture meant for educational purposes: this is the collection you, as reader, should know when entering and working in our field, these handbooks seem to convey. They form the substrate, the proven technologies of the weekly and monthly developments that have been presented and debated in the journals, especially those contributions that intend to give an overview of the current state of the art, often meant as an effort to reach consensus within the community. As such, the development of handbooks, their growth in size, their gradually changing content, provides a vista on the "deep changes" of the technology, as the "noise" of the daily fashion has to a certain extent been weeded out.

For studying the latter, periodical conferences are the ideal scene. There, the (often tacit) knowledge about the state of the art was and is mobilized at every business meeting and conference session, adjusted time and again, tested against the knowledge of colleagues, confirmed and corrected by senior engineers in their keynote speeches, functioning as the very backbone of automotive engineering culture. This happened and still happens because of a fundamental uncertainty about "what the public wants." Proceedings are the record form of such conferences.

"Engineering has no hint of the absolute, the deterministic, the guaranteed, the true," one analysis of engineering practice concludes, "Instead, it fairly reeks of the uncertain, the provisional, and the doubtful" ([12-45], p. 41). Modern research by social scientists into engineering therefore emphasizes the crucial importance of expectations, "resources (…) that help to interlock activities and to build up agendas." The purpose of formulating and sharing common expectations is to decrease uncertainty: "by sketching a future, others will find reasons to participate." Social scientists call these verbal exercises performative, which means that they are able to trigger action. "When you apologize, or when you promise something, you do not just give a factual description of some reality 'out there', but you alter social reality." The same applies to expectations and "futures" ([12-46], p. 177, 187, 191), [12-47]. The ideal type (see Chapter 1) is such a future, constructed in the language and imagery of the automotive engineer. We should be aware, however, that such futures are fantasies or, as one student of the phenomenon concludes, "forceful fictions," all the more so if it concerns a special type of expectations, so-called credible expectations ([12-46], p. 239), ([12-48], p. 158).

Given the fundamental prospective character of much engineering work, engineers are eager to decrease this uncertainty. During the inter-war phase for which we studied the *Journal of the Society of American Engineers* and some French, German, and English trade journals, one can observe that one method to decrease this uncertainty was to formulate a shortlist of rather abstract functions that more or less covered the typical user, at the same time allowing enough freedom to engineers for alternative designs. How to concretely fill in these concepts, and in how far developing *properties* were expected to allow these *functions* was a subject of continuous debate, necessary to reach at least a minimum of consensus. Whoever has wondered sometimes why automotive engineers are often so open about their work (for instance during conferences), and why they put so much effort in contributing to the state of the art, finds the answer here: Without this minimum of consensus the uncertainties are simply too numerous, in which case much more alternatives would have to be developed, the majority of which would be rejected by the users, anyway.

In her exemplary analysis of the "construction" of ABS, Ann Johnson reveals that patents are very important in this process. Although they perhaps cover only 1 to 2% of the total knowledge, their proprietary character allows engineers to freely roam in the remaining 98 to 99% of the state of the art. Engineering communities, Johnson claims, are about sharing information ([12-22], p. 150). Nowadays, Motor Vehicles patents are still among the largest patented bodies of knowledge in the world, after patents on Electronic Components & Accessories ([12-49], p. 11).

Handbooks

Different from the journals and the (often periodical) conference proceedings, *handbook* authors have to organize their material in a more or less taxonomic way, using one or more of the types mentioned in the preceding section. They use the conventions of book writing (such as the partitioning in chapters of reasonable length), but also the professional division of labor within the field (such as the subfields dealing with engines, gearboxes, electrical components, sales) to achieve a first structure of the material presented. For the inter-war years of this study, for instance, we have used a set of French, German, English, American, and Dutch technical handbooks to reconstruct the passive state of the art and its evolution.[2]

The earliest handbooks were geared toward manufacturers and users alike as a tinkering community. This is for instance apparent in the handbook published by W. Poynter Adams in 1907, who dedicated 89 of the 199 pages of his volume on the petrol car (45%) to what he called management of the car, which included maintenance, general working of components, fault finding, and so forth [12-51]. The sections on use tended to disappear in later handbooks. This is not to say that the users were no longer among the intended readership; on the contrary, as the extensive titles testify. Most handbooks announce in their titles that they were practical and illustrated. Handbooks explicitly aimed at the calculating engineer (such as the one by Heller and by Kamm; see Chapter 8 and this chapter, respectively) were few and far between in this phase.

In terms of topics covered, the earliest handbooks mostly contain three parts (or even separate volumes), on steam, electric, and gasoline propulsion. Often, the gasoline car received the largest share in terms of the relative number of pages, probably because there were more different models on the market and the authors refused to typify too much: they still tried to cover everything.

But even in the most practical treatises some form of taxonomic intervention was always necessary. Most authors used a combination of average types and archetypes to make their material manageable, whereby the archetypes, the use of existing models and components that stood for (part of) the population, clearly dominated.

Deep Trends

Some deep trends can be observed if one goes through these handbooks in the order of their appearance. The first trend is a *bias toward propulsion*, in most cases the internal combustion engine that fascinated the handbook writers as much as it did the users. This can be explained not only because the engine was the component with by far the

---

2. The collection consists of the handbooks present at the Netherlands Center for Automotive Documentation NCAD in Helmond, the Netherlands, and this author's private collection. See for an overview of American handbooks ([12-50], p. 106).

highest number of (constantly changing) parts, but its properties also generated many defects and maladjustments. Also, the desire to master the car's acceleration and speed had to be satisfied by constantly adjusting the engine controls (ignition, mixture formation), so a thorough knowledge of the component assembly was necessary to inform the tinkering community. And yet, the other function that fascinated users, the magic of braking in such a short distance of such a heavy car [12-52], was much less covered in the early handbooks.

The second observed trend is a *gradual increase in dealing with theory* and general working principles, although their share remained quite modest: the handbooks of this period continued to be practical treatises for a technically educated readership, but this education definitely started to include some basic engineering knowledge.

Related to this was a third trend: a shift from a pure description of the artifacts and its components to overviews where the *control* of the technology was gradually included. Instead of a machine on wheels, the car was more and more presented as a tool to be handled well, a tendency that also had to do with the owner-driver: instead of a producer perspective, a user perspective started to dominate, and the technology was explained in function of this envisaged practice.

FISITA

Nobody has as yet charted the post–WWII global engineering community from an evolutionary point of view, but it is clear that the globalization of the automotive industry has brought new actors on the scene, such as the international engineering association FISITA (*Fédération Internationale des Sociétés d'Ingénieurs des Techniques de l'Automobile*; International Federation of Automotive Engineering Societies), organizing regular world conferences since its foundation in 1947. Figure 12.8 illustrates the diversity of topics presented at the conferences held between 1960 and 1995, with a remarkable bias towards propulsion systems, especially engines. Every eight years or so, special sessions about the future of the car are set up, as a periodical survey of the state of the art [12-53].

Perhaps the most significant change in terms of knowledge production in the post-war phase is the expansion of the engineering community into a more loosely organized knowledge community, in which new types of knowledge are added to the state of the art from without the traditional engineering circles, such as the environmental movement (Chapter 11) and the safety movement (Chapter 10), as well as social science research (Chapter 13).

This expansion of the state of the art in terms of knowledge coincided with a deepening of the same: the scientification of engineering went into a new phase as it came to be applied to vehicle dynamics, thus strengthening the importance of the vehicle versus the engine, and the new engineering of acoustics, vibrations, and computer-aided simulations versus the classical chassis engineers of an earlier phase.

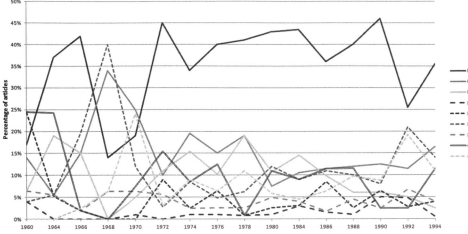

Figure 12.8 Variation of topics presented at the biennial FISITA conferences 1960–1995 ([12-53],p. 6).

12.5 Conclusions

This chapter showed the long road toward scientification of the automotive sector, in terms of both production and consumption. As we have seen, engineering research leads to two results: an artifact (a car, or a part of it) and knowledge, to be disseminated through conferences, handbooks, journals, patents, and education. The resulting state of the art is no monolith, certainly not in a globalized world: there is a clear distinction between an American and a European engineering culture, but their commonalities are perhaps more important as they are both guided by a form of knowledge of a specific character, aimed at innovation.

In terms of education, the differences between an American and a European system remain apparent, despite the existence of the Kettering University in Flint, Michigan. In most other car-producing countries in the Atlantic sphere, dedicated schools do not exist, probably because the industry is confident that they can "adjust" generally educated engineers toward the intricacies of automotive technology when newly graduated engineers get employed in their factories. In continental Europe, only in two small countries without much car industry do such specialized schools exist: the HTS-Autotechniek (Polytechnic for Automotive Technology) in Arnhem, the Netherlands, and a comparable school in Biel, Switzerland. Perhaps this state of affairs confirms the idea that automotive technology is not a scientific discipline, but rather a constantly changing amalgam of subdisciplines, where lately a shift toward electrical engineering can be observed. At Eindhoven University of Technology (the Netherlands), for instance, an Automotive Bachelor started in 2012 at the Electrical Engineering Department, where the automotive world seems to be put upside down: Mechanical

Engineering is treated as supportive, but control engineering and programming form the very core of the new curriculum. This book is part of a separate course called Cars in Context I (in the first year) and II (in the second), emphasizing the insight that the automotive industry of the 21st century needs a new type of engineer: she (hopefully more "she's" than previously) should also be able to reflect on her own knowledge acquisition and be able to put technology in its societal context.

What these students learn, is not science, nor a derivative of science. Ann Johnson, in her research on the history of ABS (see Chapter 10) relates how "even scientists working in the theoretical field of vehicle dynamics admitted that the theories they were building were not science. There are no *laws* of antilock braking systems; the laws governing the operation of the system are contingent on how the system is designed. Furthermore, the goal of these models is predictive rather than explanatory" ([12-22], p. 152).

However, scientification is not the same as becoming a science. Scientification means using procedures (such as statistics, literature searches, systematic research policies, and mathematical simulations) and principles (such as basic laws, the periodic system of the elements), as well a similar reporting procedures (testified by the production of texts with literature references and notes, such as this book) developed typically within a scientific community, to solve technical problems. Most of all, scientification is about mathematization of engineering theory, including modeling, which, as we have seen in this chapter, automotive engineers borrowed especially from aeronautics.

Instead of trying to solve the question whether engineering is scientific or not, it is perhaps better to look at the *practices* of research (Table 12.1), especially after the peak of the linear model (which saw science at the beginning and engineering at the end) around 1965 [12-54]. By that date company labs often developed close links to university labs and cut back on what is called generic research drastically. Engineering, and automotive engineering especially, is a world on its own, with its own knowledge production.

Table 12.1 Three types of knowledge production, and their associated products and capabilities. ([12-54], p. 431 (Table 1)).

Type of knowledge producing activity	Aim of this activity	Products	Capabilities
Research	Create understanding (answers "why" questions)	Ideas	Do research
Development	Develop technology (answers "how" questions)	Designs	Create economic value
Testing	Collect data (answers "what" and "how many" questions)	Data	Enhance control

References

12-1.   Walter G. Vincenti, *What Engineers Know and How They Know It; Analytical Studies of Aeronautical History* (Baltimore/London: Johns Hopkins University Press, 1990).

12-2.   Edwin T. Layton, Jr., "A Historical Definition of Engineering," in Paul T. Durbin (ed.), *Critical Perspectives on Nonacademic Science and Engineering* (Bethlehem - Lehigh University Press/London and Toronto: Associated University Press, 1991) (Research in Technology Studies, Volume 4), 60–79.

12-3.   Carl Mitcham, "Engineering as Productive Activity: Philosophical Remarks," in Paul T. Durbin (ed.), *Critical Perspectives on Nonacademic Science and Engineering* (Bethlehem - Lehigh University Press/London and Toronto: Associated University Press, 1991), 80–117.

12-4.   Jim Johnson (= Bruno Latour), "Mixing Humans and Nonhumans Together: The Sociology of a Door-Closer," *Social Problems* 35 No. 3 (June 1988), 298–310.

12-5.   Eugen Diesel, *Diesel; Der Mensch, das Werk, das Schicksal* (Hamburg: Hanseatische Verlagsanstalt, n.y. [1939?]).

12-6.   Lynwood Bryant, "Rudolf Diesel and His Rational Engine," *Scientific American* 221 (August 1969) 108–117.

12-7.   Lynwood Bryant, "The Role of Thermodynamics in the Evolution of Heat Engines," *Technology and Culture* 14 No. 2 (April 1973), 152–165.

12-8.   Lynwood Bryant, "The Development of the Diesel Engine," *Technology and Culture* 17 No. 3 (July 1976), 432–446.

12-9.   C. Lyle Cummins, *Internal Fire* (Lake Oswego, Oregon: Carnot Press, 1976).

12-10.   Horst O. Hardenberg, *The Middle Ages of the Internal-Combustion Engine 1794–1886* (Warrendale, PA: Society of Automotive Engineers, 1999).

12-11.   C. Lyle Cummins, *Diesel's Engine; Volume One: From Conception to 1918* (Wilsonville, OR: Carnot Press, 1993).

12-12.   M. Hård and A. Jamison, "Alternative Cars: The Contrasting Stories of Steam and Diesel Automotive Engines," *Technology in Society*, 19 no. 2 (1997), 145–160.

12-13.   Olaf von Fersen (ed.), *Ein Jahrhundert Automobiltechnik; Personenwagen* (Düsseldorf: VDI Verlag, 1986).

12-14.   Holger Bingmann, "Chapter 14: Competence; Case C: Antiblockiersystem und Benzineinspritzung (Anti-Blocking System and Fuel Injection)," in Horst Albach, *Culture and Technical Innovation; A Cross-Cultural Analysis and Policy Recommendations* (Berlin/New York: Walter de Gruyter, 1994) (Akademie der Wissenschaften zu Berlin, Research Report 9), 736–821.

12-15.   Eugen Diesel, *Die Geschichte des Diesel-Personenwagens* (Stuttgart: Deutsche Verlags-Anstalt, 1955).

12-16. Jürgen Potthoff and Ulf Essers (with Helmut Maier and Barbara Guttmann), *75 Jahre FKFS - Ein Rückblick; Eine Chronik des Forschungsinstitut für Kraftfahrwesen und Fahrzeugmotoren Stuttgart - FKFS - aus Anlass seines 75-jährigen Bestehens, 1930–2005* (Stuttgart: Forschungsinstitut für Kraftfahrwesen und Fahrzeugmotoren Stuttgart FKFS, 2005).

12-17. A. Riedler, *Wissenschaftliche Automobil-Wertung; Berichte I–V des Laboratoriums für Kraftfahrzeuge an den Königlich Technischen Hochschule zu Berlin* (Berlin/Munich, 1911).

12-18. Alois Riedler, *Wissenschaftliche Automobil-Wertung; Berichte VI–X des Laboratorium für Kraftfahrzeuge an der Königlich Technischen Hochschule, Teil II* (Berlin, 1912).

12-19. Jürgen Potthoff and Ingobert C. Schmid, *Wunibald I.E. Kamm - Wegbereiter der modernen Kraftfahrtechnik* (Heidelberg/Dordrecht/London/New York: Springer, 2012).

12-20. Wolfgang König, "Adolf Hitler vs. Henry Ford: The *Volkswagen*, the Role of America as a Model, and the Failure of a Nazi Consumer Society," *German Studies Review* 27 No. 2 (May 2004), 249–268.

12-21. William F. Milliken and Douglas L. Milliken, *Chassis Design: Principles and Analysis; Based on Previously Unpublished Technical Notes by Maurice Olley* (Warrendale, PA: Society of Automotive Engineers, 2002).

12-22. Ann Johnson, *Hitting the Brakes; Engineering Design and the Production of Knowledge* (Durham/London: Duke University Press, 2009).

12-23. Stuart W. Leslie, *Boss Kettering* (New York: Columbia University Press, 1983).

12-24. John Kenly Smith, Jr., "The Scientific Tradition in American Industrial Research," *Technology and Culture* 31 No. 1 (January 1990), 121–131.

12-25. Michael Lind, *Land of Promise; An Economic History of the United States* (New York: Harper-Collins, 2012).

12-26. Sally H. Clarke, *Trust and Power; Consumers, the Modern Corporation, and the Making of the United States Automobile Market* (Cambridge: Cambridge University Press, 2007).

12-27. C. F. Kettering, "Cooperation of the Automotive and Oil Industries," *Journal of the Society of Automotive Engineers* (January 1921) 43–45.

12-28. J. B. Hill and T. G. Delbridge, "Seek Cracking Process for Production of Non-Detonating Fuel," *Journal of the Society of Automotive Engineers* (March 1926), 271–274.

12-29. "Fuel Research Summarized," *Journal of the Society of Automotive Engineers* (February 1927), 193–195.

12-30. Robert U. Ayres and Ike Ezekoye, "Competition and Complementarity in Diffusion: The Case of Octane," *Technological Forecasting and Social Change* 39 (1991), 145–158.

12-31. Gijs Mom, "Fox Hunt: Materials Selection and Production Problems of the Edison Battery (1900–1910)," in Hans-Joachim Braun and Alexandre Herlea (eds.), *Materials: Research, Development and Applications (Proceedings of the XXth International Congress of History of Science (Liège, 20 - 26 July 1997)* (Turnhout: Brepols, 2002) 147–154.

12-32. Gijs Mom and Diekus van der Wey, "Ethanolverdamping en de dieselmotor; Literatuuronderzoek naar en ontwerp van een ethanolverdamper t.b.v. het gecombineerd gebruik van dieselbrandstof en ethanolgas in een DAF-dieselmotor" (Thesis, HTS-Autotechniek, Apeldoorn, The Netherlands, June 7, 1982).

12-33. Hal Bernton, William Kovarik, and Scott Sklar, *The Forbidden Fuel; Power Alcohol in the Twentieth Century* (New York: Boyd Griffin, 1982).

12-34. Joseph C. Robert, *Ethyl; A History of the Corporation and the People Who Made It* (Charlottesville: University Press of Virginia, 1983).

12-35. Gijs Mom, "Ongelode benzine: Een anti-klopjacht op wereldschaal," *Auto Service* 2 no. 11 (25 March 1988), 1, 3–4.

12-36. Gijs Mom, "Ongelode benzine: Een anti-klopjacht op wereldschaal (2)," *Auto Service* 2 no. 12 (25 April 1988), 1, 4–5.

12-37. "80% of Cars Reported Below Peak Efficiency," *Journal of the Society of Automotive Engineers* 44 No. 1 (January 1939), 21.

12-38. Gijs Mom, "Orchestrating Automotive Technology: Comfort, Mobility Culture and the Construction of the 'Family Touring Car', 1917–1940," *Technology and Culture* 55, no. 2 (April 2014) 299–325.

12-39. David Beecroft, "Conditions in the Automotive Industry Abroad," *Journal of the Society of Automotive Engineers*, 4 No. 6 (June 1919), 521–525.

12-40. M. Olley, "European and American Automobile Practice Compared," *Journal of the Society of Automotive Engineers*, 9 No. 2 (August 1921), 109–117.

12-41. E.W. Goodwin, "Automobile Body Design and Construction," *Journal of the Society of Automotive Engineers*, 2 No. 4 (April 1918), 271–278.

12-42. W. E. Lay and L. C. Fisher, "Riding Comfort and Cushions," *Journal of the Society of Automotive Engineers (Transactions)* 47 No. 5 (November 1940), 482–496.

12-43. Gijs Mom, *Atlantic Automobilism: The Emergence and Persistence of the Car, 1895–1940* (New York and Oxford: Berghahn Books, forthcoming).

12-44. Gijs Mom, "Constructing the State of the Art: Innovation and the Evolution of Automotive Technology (1898 - 1940)," in: Rolf-Jürgen Gleitsmann and Jürgen E. Wittmann (eds.), *Innovationskulturen um das Automobil; Von gestern bis morgen; Stuttgarter Tage zur Automobil- und Unternehmensgeschichte 2011* (Stuttgart: Mercedes-Benz Classic Archive, 2012) (Wissenschaftliche Schriftenreihe der Mercedes-Benz Classic Archive, Band 16), 51–75.

12-45. Billy Vaughn Koen, "The Engineering Method," in Paul T. Durbin (ed.), *Critical Perspectives on Nonacademic Science and Engineering* (Bethlehem: Lehigh University Press/London and Toronto: Associated University Press, 1991), 33–59.

12-46. Harro van Lente, *Promising Technology; The Dynamics of Expectations in Technological Developments* (Delft: Eburon, 1993).

12-47. Gijs Mom, "'The Future Is a Shifting Panorama': The Role of Expectations in the History of Mobility," in Weert Canzler and Gert Schmidt (eds.), *Zukünfte des Automobils; Aussichten und Grenzen der autotechnischen Globalisierung* (Berlin: edition sigma, 2008), 31–58.

12-48. Sjoerd Bakker, *Competing Expectations; The Case of the Hydrogen Car* (Oisterwijk: BoxPress, 2011).

12-49. Gaston Heimeriks, Floortje Alkemade, Antoine Schoen, Lionel Villard and Patricia Laurens, "The Evolution of Technological Knowledge: A co-evolutionary analysis of patterns of Corporate Invention," (paper presented at the ECIS seminar, Eindhoven University of Technology, November 8, 2012).

12-50. Michael L. Berger, *The Automobile in American History and Culture; A Reference Guide* (Westport, Connecticut/London: Greenwood Press, 2001) (American Popular Culture series, ed. M. Thomas Inge).

12-51. W. Poynter Adams, *Motor-car Mechanism and Management; Part I: The Petrol Car* (London: Charles Griffin, 1907).

12-52. J. Edward Schipper, "Passenger-Car Brakes," *Journal of the Society of Automotive Engineers* 10 No. 4 (April 1922), 273.

12-53. Emilio Reyes, Alejandro Medina, Alexandru Dinu, Coen de Winter, and Frank Rams, "World automotive congress FISITA" (Student report of the course Cars in Context, Eindhoven University of Technology, 2012).

12-54. Arjan van Rooij, "Knowledge, Money and Data: An Integrated Account of the Evolution of Eight Types of Laboratory," *British Journal for the History of Science* 44 No. 3 (September 2011), 427–448.

Chapter 13
Decarbonization: Searching for Radical Alternatives

13.1 Introduction: The Importance of Expectations

Many popular histories of the automobile start with the observation that once, in the pre-history of the car, the steam and electric cars reigned supreme. When the gasoline car then appeared on the scene, it quickly took over the lead, because steam cars were too difficult to handle while driving and starting, and electrics lacked enough energy density in their batteries. In this chapter we will investigate this "myth of the electric" (as in other histories, we will neglect the steam car, although its study might generate interesting additional insights, also in the environmental history of the car [13-1], [13-2], [13-3]). In order to get a grip on this elusive topic, we will have to come back to the distinction between two sets of artifact characteristics we made in Chapter 1 and which we called the *dual nature of technology*.

Ever since the internal-combustion engine car was locked in as the quintessential personal mobility tool at the beginning of the 20th century, engineers and users alike have continued to search for alternatives, out of dissatisfaction of the current choice or simply because they liked to try something else. From an evolutionary point of view, this is nothing new: at all levels of the car's structure as well as in all corners of the automobile system's hierarchy, *variation* is continuously emerging, choices are made, and innovations are added only to start the cycle all over again. But sometimes, alternatives were sought because people started to have doubts about the future of the car as such, as artifact or even as system. This desire for a radical change was especially the case when there was a fear that oil would soon be depleted, or, more recently, that climate change was imminent (if not already happening).

In this chapter we will investigate such radical alternatives and the role *expectations* play in this respect. In Section 13.2 we will focus on the first perceived energy crisis of the 1920s, leading to a frantic search for oil substitutes, especially alternative fuels like alcohols. In Section 13.3, we focus upon the most famous, if not iconic, alternative of all, the electric vehicle, and ask ourselves why this alternative looked and looks so attractive as a substitute and why it did not work, up to now, to get the combustion car replaced. Here, too, expectations seem to be the predominant by-product. We will conclude this chapter in Section 13.4.

13.2   The End of Oil! In the Early 1920s!

The search for alternative fuels is as old as the car itself. Initially, there was a solid engineering reason for leaving oil-based fuels behind: the gasoline types distilled from crude oils found in the United States appeared to be very transparent (which was seen as a token of quality), in contrast to the "dirty" gasolines harvested in Malaysia, for instance. It took some time for engine specialists to find out that the clean types were causing problems of what soon became known as knock. Thus, clear petrol not only was very dangerous because it was extremely volatile, it also became known as "pinging water" ([13-4], p. 1).

One of the earliest researchers into the knock phenomenon was Harry Ricardo, who already in 1913 discovered the anti-knock characteristics of aromatic compounds, unsaturated cyclic carbohydrates like benzene, toluene, and xylene. Up to then, Shell, with which Ricardo's laboratory cooperated, had burned its aromatic fractions because it considered density as the fluid's only quality measure.

A second way to combat knock was less well known in the west, because it was developed by a Russian chemical physicist called Semenov, director of the Institute for Chemical Physics in Moscow, who was awarded the Nobel Prize for his theory of layered combustion, the same procedure used in the Honda CVCC engine (see Chapter 11).

The third road to anti-knock measures was the chemical one, followed, as we have seen in the previous chapter, by Charles Kettering, who had to struggle with the highly knock-prone oil types from Pennsylvania with their long alkane chains. Remarkably, Charles Kettering's search for a knock-resistant fuel during and after the First World War, can be seen in the context of the struggle between the magneto ignition and his invention of the SLI (starting - lighting - ignition) system (see Chapter 3), as the proponents of the former accused the latter of causing knock. Some engineers used the term *spark knock* to characterize the phenomenon (although magneto ignition, as we saw, also works with a spark). Thus, Kettering's "scientific fox hunt" was much more than a search of an alternative at the sub-artifact level: it concerned the reputation of the entire (American) car. Both Ricardo and Kettering also investigated alcohol as a remedy against knock [13-5], [13-6].

Alcohol as Alternative Fuel

After the First World War, the search for alternatives was intensified because of two reasons. First of all, American car and oil companies started to warn for an impending oil depletion. For Ricardo, it was "a matter of absolute necessity to find an alternative fuel. (..) By the use of a fuel derived from vegetation, mankind is adapting the sun's heat to the development of motive power, as it becomes available from day to day; by using mineral fuels, he is using a legacy—and a limited legacy at that—of heat stored away many thousands of years ago." And a French engineer opined: "The time will arrive when the industry of Europe will cease to find those natural sources, so necessary for it. Petroleum springs and coal mines are not inexhaustible but are rapidly diminishing in many places" ([13-5], p. 10). And although by the end of the 1920s this fear had subsided because of the find of large oil reservoirs, especially in the United States, national governments remained interested in alcohol as a motor fuel, because agricultural surpluses could thus be used to fuel the nation's cars, a strategy which became acute during the financial crisis of 1929 and the subsequent Depression.

In the course of the 1930s most motorizing countries introduced policies of obligatory addition of between 20 and as much as 50% ethanol to gasoline, in Brazil even up to 90% ([13-7], p. 225–231). France, which had a long tradition of adding alcohols and other substances (such as benzene) to petrol, opted for 50%. Sweden, impressed by the coming oil shortage in the United States (Figure 13.1) started to convert its wood pulp to ethanol. Its production peaked for 30 years, from the early 1940s to the late 1960s. Here, too, expectations were high: "As long as the sun shines over our fields, they will produce forever and we have never to fear that this will disappear. Furthermore, why use potatoes and corn for alcohol production? Foodstuffs serve the best as human food" ([13-8], p. 16).

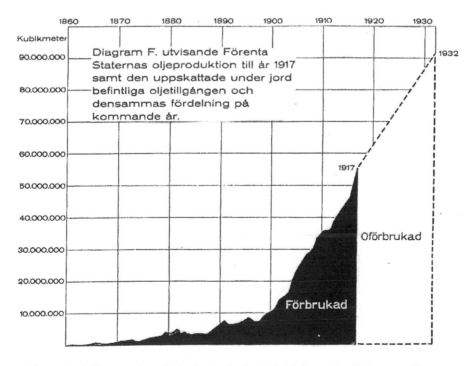

Figure 13.1 Threatening oil depletion in the United States by 1932, according to a Swedish estimate. The black surface under the curve represents the consumed oil, the area under the dotted line is the expected consumption ([13-8], p. 15).

The only European countries who continued to use pure gasoline (apart from some additions of benzene) were Greece, Portugal, Norway, Romania, and the Netherlands, the latter no doubt because of the dominating presence of Shell (one of the prime ministers during the Depression, Dr. Colijn, was a former Shell employee).

A second wave of alcohol application in car engine fuel started in the second half of the 1960s, with a shift from ethanol to methanol, especially in Europe. At the same time, the approach of the problem shifted from a national to an international one, parallel to the same phenomenon of globalization we have observed in Chapter 11 about environmental issues. Apart from *environmental* arguments, the *energy* crisis of 1973 gave an extra impulse: diminishing oil dependence became of national (if not: Western) strategic importance. In Germany, government-sponsored research into methanol and hydrogen started in 1974, whereas the Swedish Methanol Development Co. two years later organized the first international conference on this topic under the title Methanol as a Fuel.

The American revival of interest for fossil fuel alternatives started ten years earlier, but here, too, the energy crisis formed the catalyst: General Motors, NATO, and the American Department of Energy all started research or initiated conferences, the most well known being the ISAFT meetings (International Symposia on Alcohol Fuel Technology), held since 1976. Substituting lead in gasoline now became an additional argument.

Developing Countries

The post-war wave led to a number of shifts within the automotive fuel field and in the search for alternatives. One of these shifts was the inclusion of developing countries such as Brazil and India as leading nations in the alcohol fuel area. Also, the diesel engine now came in the orbit of the search for alternatives. The problem here is that alcohols have a rather high octane number and thus a low cetane number (a measure for the ignition proneness of diesel fuels), so for diesel engines other alternatives had to be found, such as vegetable oils. In the background, another shift played a role: the increasing international dominance of the oil companies, who started research into the possibilities of converting natural gas into gasoline and diesel fuel. From GTL (gas-to-liquid, for instance: diesel fuel made from natural gas), through oxygenates (oxygen containing fuels like esters and ethers), to reformulated gasoline, bio fuels and alcohols, a plethora of alternatives meanwhile was being investigated. But fossil fuel remained dominant, despite the warning calls for peak oil, the threat of a definitive depletion of oil reservoirs in the 21st century. Others point at the existence of large reservoirs of shale oils and wells to be opened deep under the sea or in remote places like Alaska or the Poles [13-9]. The paradox of this history is, that if the price of conventional oil starts to rise, the alternative harvesting methods which now are still prohibitively expensive or are too risky from an environmental point of view, become attractive.

With the threat of climate change in the air, some of these alternatives suddenly got another meaning, as they would fit in a *sustainable mobility concept*. But from the point of view of sustainability, the electric vehicle seemed (and seems) to have the highest promise.

13.3   The Promise of the Electric Vehicle: A Perpetual Car of Tomorrow?

In order to analyze the emergence and subsequent development of electric traction, we use a distinction in *generations*, defined in function of the changes in technology of the electric vehicle. Thus, we distinguish between a first generation of electrified horse-drawn carriages, a second generation with the same vehicle types but with drastically improved tires and batteries, and a third generation, after the victory of the competition, of petrol car look-alikes. The common-sense conviction that the electric car has a serious technical flaw (its short radius) has been derived from the first generation: electrified horse-drawn carriages mostly applied in urban taxicab fleets.[1]

As a matter of fact, the often-quoted dominance of the electric vehicle among the three propulsion alternatives in the United States shortly before 1900 is nearly fully based on this businesslike application: if we only count cars in private use the steam car, not its electric counterpart, was no doubt in the majority.[2] If, however, one would have asked the experts who, shortly before the turn of the century, started to motorize some taxicab fleets in London, Paris, and several American cities like New York and Boston, the answer would have been clear: the electric vehicle offered by far the most reliable technology. This was also testified by the fact that around 1900 not a single large-scale taxicab experiment was set up based on combustion engine technology, whether functioning on the basis of external (steam) or internal combustion (petrol).

The Failure of the First Generation

All these early initiatives failed, through a combination of technical and managerial problems, in the United States aggravated by destructive financial speculation. The core of the trouble was clearly technical: active mass falling from the grid plates in the lead batteries causing short circuiting and resulting in limited longevity, and fast wear and tear of the pneumatic tires incapable of carrying the extra weight of the batteries. In other words, the businesslike atmosphere of these early experiments was detrimental, because cost increases soon led to abandoning the initiatives. On the other hand, petrol cars, much less reliable, were in use by an elite who wouldn't care less about costs: many of them had several cars. Engineers (and many others in their wake) who are still convinced that the electric vehicle is no option because of its technical flaws, base their opinion on this first generation's failures.

One can ask why not many more would-be motorists opted in favor of the much more reliable technology. Apart from a reluctance to engage in maintenance of the battery based on chemistry (early motorists likened electric car maintenance to what happened in the laboratory), private gasoline car users meanwhile developed a culture dominated

---

1. The following is based upon [13-10], but also see [13-11], and for the post–WWII period, especially [13-12].
2. This misconception is still very much alive, especially in those studies that do not conduct historical research themselves, such as ([13-13], p. 23).

by automotive adventure, of racing and touring, but also of tinkering and a general preparedness for "trouble," as they called it, much like the early adopters of personal computers would later behave: for them, the imperfections of the gasoline car were a virtue rather than a problem, a cause of pleasure and relaxation. Such masculine users were not interested in a vehicle that, as the Briton T. Chambers remarked in his diary in 1907, "always functions." There is "not much sport in driving an electric carriage," he wrote. "It is far too unexciting to be attractive. The fascination of the petrol engine to the man who is born with an engineering instinct is largely due to its imperfections and its eccentricities" ([13-10], p. 41).

And yet, although the failed early experiments caused a lot of hesitation about fleet motorization among entrepreneurs during the following years (and in their wake, as we currently can observe, among car engineers and manufacturers), the experiments provided interesting lessons that, unfortunately, for the most part have been forgotten. One of these lessons is the charging of mass instead of energy practiced in the United States: shortly before the turn of the century, the Electric Vehicle Company installed a charging station on Broadway, New York, where battery sets could be exchanged semi-automatically in about 75 seconds (faster than filling a gas tank) (Figure 13.2). Similarly, in France a proposal was launched to enable an electric Tour de France with battery exchange stations, a proposal which even had a following, also aborted, in the Netherlands. This lesson proves that the radius argument, so often heard nowadays, is not valid: if we had wanted, we could have set up exchanging stations every, say, 50 km, just as this recently has been proposed again by a company called Better Place ([13-14], p. 214–215).

Figure 13.2 An electric "hansom cab" at the Electric Vehicle Company's charging station on Broadway, New York 1902, where the battery set was exchanged semiautomatically in 75 seconds ([13-10], p. 81).

A second lesson from this early period is that we should be careful to compare electric and combustion cars along the same yardstick. Taxicabs and privately used cars thrive (or fail) in quite different contexts, just like race cars and cars for everyday use. The electric *La Jamais Contente* with which the Belgian race car driver Camille Jenatzy broke the speed record of 100 km/h in 1899 (Figure 13.3), not only did so over a distance of not more than one (flying) kilometer, the subsequent Grands Prix and other speed contests were fully geared to the petrol car. The electric did not fit in the racing infrastructure, but not because it did not function properly. *La Jamais Contente* shows that the racing culture was tailored to the adventure machine: speed, in other words, is contingent upon its context, the period and locale of its application. Technology, we have to conclude again, is *socially constructed*.

Figure 13.3 *La Jamais Contente*, electrically propelled record car driven by the Belgian race car driver Camille Jenatzy. *Source: collection of author.*

The Successful Second Generation

At first sight, the second generation of electric road vehicle propulsion did not differ much from the first: similar old-fashioned vehicles from a bygone age (also in the eyes of the contemporaries) were used in taxicab fleets in Germany and the Netherlands, as well as in several urban fleets of fire engines in Germany, starting around the middle of the first decade of the new century. But the battery and tire technologies had meanwhile considerably improved. This, however, was not the crucial difference. Around this improved technology a business organization was set up that compensated for every technical flaw these components could produce.

All European electric cab initiatives were done with the same French-German (Kriéger-Namag) taxicab design solidly backed by a central maintenance station where specialists exchanged defective plates and worn tires whenever necessary. For a modest sum of some cents per driven kilometer the fleet owner had a guaranteed, up-to-date set of vehicles which in the case of Amsterdam (the German fleets were abandoned in 1914 because of the war) resulted in a performance clearly better than petrol car alternatives that meanwhile also had been set up.

It was, however, at the fringes of these experiments, in the very tiny taxicab fleets (down to one cab driven by father and son), that gas engine propulsion had a chance: for these applications the investment in a central maintenance infrastructure was not realistic, although, *if* by then a general scheme of changing stations had been set up, one-cab enterprises also could have made use of it. In fact, this cooperative use of central charging happened in Berlin before the First World War for a short while.

Likewise, the German fire fighters were very pleased with their electric (and even hybrid, some of them steam-electric) fleets of six-ton vehicles, leading to the third lesson we can draw from this history: the idea that electric traction cannot cope with heavy loads should be abandoned since Berlin fire corps chief Maximilian Reichel, a steam engine enthusiast, decided in favor of electric traction for his motorization project, soon followed by his counterparts in other German cities (Figure 13.4) 13-15]. Here, too, the petrol alternative crept alongside in those small towns that only could use one or two of such vehicles.

Figure 13.4 A fully electric fire engine "train" in Berlin before WWI: each car weighed more than 6,000 kg. The electric motors are of the Lohner-Porsche type (Porsche started his career as an electrical engineer) built in the wheel hubs [13-16].

A fourth lesson is even more important: the second generation cab and fire engine initiatives clearly show that technology can never be a convincing argument for failure: assuming a minimum reliability, such failures can always be compensated by organization.[3] This, however, is only true when we consider fleet applications. An individual car owner had to use the public urban garage systems then in the process of being built up: there, mechanics could take care of this side of car culture. Again, however, this culture was soon abandoned in favor of curb parking, where the (petrol) car was always near at hand, and could be used without much planning or foresight.

The First World War played a decisive role in this general development, because of the clear preference of the military for the gasoline alternative. During this war, heavily expanded production facilities, but also thousands of young men trained in driving gasoline trucks, laid the basis for a post-war boom, an explosion of lower-middle class motorization which in one massive blow made the electric alternative obsolete. This was especially the case in the United States, where during the war a wave of electric vehicle use seemed to unfold, resulting in about 30,000 electric passenger cars which, together with the about 10,000 electric trucks, formed the most extensive occurrence of an electric car culture during the first half of the previous century (Figure 13.5). In principle, "electric tourism" (as it was called then) was possible, as a map showing the network of charging stations around New York testifies, but as the route had to be planned carefully, electric touring differed quite substantially from "anarchic" mainstream car tourism.

Figure 13.5 "Electric tourism" was possible in the United States, with 30,000 EVs around 1915; here, a Detroit Electric, the most sold EV make in the United States, is being charged.

---

3. And yet, engineers still claim that "God does not want us to have full-function electric vehicles" ([13-10], p. 275). This opinion, vented by a representative of Ford at a U.S. Congress hearing in 2000, is sometimes echoed by social scientists who have to confide in the engineers' judgment. For instance, ([13-14], 209) point at "limited technological progress" as the main failure factor of the Californian electric vehicle experiment of the 1990s. For a critique of this "better battery bugaboo" see ([13-17], p. 183).

The Third Generation

Meanwhile, the competing technology had taken over so many of the electric's advantages that it became increasingly difficult to deviate from the pattern of mainstream automotive culture. Electrics that in this phase were put on the market, hid the electric character of their propulsion system by mimicking the general layout and body shape of the mainstream, petrol car. When such cars are the result of a simple swap of the drive train, they are called conversions. The petrol car culture, as we have seen, was still to a large part pleasure-driven, and started to develop into a culture of long-range touring.

Meanwhile, the Pluto Effect reigned supreme: the gasoline proponents not only took over the closed body, the automatic starting and the silent running of the propulsion unit from the electric, they also integrated the sturdy cord tire (originally developed for the heavier electric) into their paradigm, and one can even argue that the insight that the (unreliable) vehicle has to be backed by a reliable maintenance infrastructure, and the general idea of service, stems from the early experiments of electric vehicle fleets.

A Fourth Generation?

But would the Pluto Effect work until the bitter end, or will it turn into a Sailing Ship Effect (see Chapter 2)? Now that we are entering the transitional phase between the past and the future, are the same arguments still valid when it comes to judging the developments of the current electric vehicle? Let's try to answer these questions by, again, taking one step back and start in the immediate post–WWII phase and first ask ourselves whether there is any reason to distinguish a new, fourth generation, which, perhaps, does not follow the pattern we have observed for the previous generations.

One of the striking observations about the attractiveness of the electric vehicle is that their manufacturers have never invested much time and money in carving out their own characteristic body shape. But that does not mean that we cannot see a new type of car emerging here. Apart from the pseudo-aerodynamic *La Jamais Contente*, the Kriéger taxicab and especially the products of the American market leader Detroit Electric suggest a *city car* idiom that for half a century would lay dormant before it was rediscovered by post–WWII petrol car manufacturers.

Electric drive literally disappeared from public space, and found a niche as industrial truck, of which thousands were built during the inter-war years. Especially Germany (led by battery giant AFA, later Varta) had an extensive utilitarian electric truck culture, but in the UK during the 1930s the biggest fleet ever of tens of thousands of electric milk delivery vans was built up, which would make this country into the most important electric vehicle country in the world during two decades or so after the Second World War, with a peak, during the 1960s, of 60,000 vehicles [13-12].

Whatever one thinks of the automotive character of these contraptions (Figure 13.6), it cannot be denied that they were carving out a special utilitarian urban niche. This niche was enriched with environmental concerns during the post–WWII revival of electric vehicle experiments, which started during the 1960s, well before the energy crises of the 1970s that are often seen (erroneously so) as the trigger for the new EV wave. Most of these cars (and vans and light trucks), developed out of an environmental concern about urban livability, were still conversions. Such reconfigurations of existing vehicle designs were often seen as rolling laboratories, of new electric motor types but especially new battery types, preferably of the high-energy variants such as zinc-air or sodium-sulfur [13-18].

Figure 13.6 Electric Maserati (*source: author's collection*).

There is, however, at least one argument why one can speak of a new generation. This wave resulted also in the first of a new "species" of cars, tiny city vehicles reminiscent of the *voiturettes* and cyclecars of half a century ago (see Chapter 6) and built by small companies who wanted to show the world that the big manufacturers had it all wrong. Perhaps as a response to the dozens of these initiatives around the motorized world, the established carmakers also started to propose prototypes and fully new developments of their own. One of the most controversial of these was the Smart, a mini-urban car, originally designed as a pure electric city car by a Swiss watch maker, but when taken over by Daimler-Benz is was equipped with a petrol engine. This was a token of the fact that electric drive now depended on the constantly changing preferences of manufacturers, governments and fleet owners, as well as the consumers, and the constant insecurity among the large car manufacturers about the size of the future electric vehicle market ([13-19], p. 318).

Another reason why it seems justified to distinguish the post–WWII electric car evolution from pre-war practices is the specific shape of the experiments, especially of the second post-war wave. The first post–WWII wave took place during the 1970s, driven by the search (in vain) for the "wonder battery" (see below). The second wave started after a period of lull during the 1980s when, in 1990 the California Air Resources Board (CARB) was founded that started to push for a massive introduction of electric driving among the citizens of the state ([13-20], p. 27), [13-21]. By 2003, the CARB ordained, one-tenth of the big companies' vehicles sold in California should be Zero Emission Vehicles (ZEV) (Figure 13.7). Inspired by this initiative, smaller-scale experiments took place in Europe, such as on the German island of Rügen (still driven by the hunt for the high-energy battery by the major manufacturers) or the cities of La Rochelle and Mendrisio in France and Switzerland, where the end user was targeted by big (PSA) or small companies, respectively.

(vehicles < 3750 pounds, 50,000 miles durability)

Note: In this table, non-methane organic gases (NMOG) replaces total hydrocarbons (THC)

	NMOG (g/mi)	CO (g/mi)	NO$_X$ (g/mi)
TLEV	0.125	3.4	0.4
LEV	0.075	3.4	0.2
ULEV	0.04	1.7	0.2
ZEV	0.0	0.0	0.0

Figure 13.7 California emission standards for zero emission vehicles (ZEV) and several types of low emission vehicles (LEV) (Reprinted from SAE R-226 [13-22] p. 146).

The Hunt for the "Miracle Battery"

However, there are also many similarities with the pre-war electric vehicle developments. The most remarkable is the continued, and even much intensified, hunt for the "miracle battery," which started already way back in the early 1900s, when Thomas Edison worked on his nickel-iron steel battery, promising both a higher energy density (more kilometers of driving for a given kilogram of battery) and a quicker charging without the need to recourse to battery exchange. Since then, we have seen nickel-cadmium appear (and vanish again), followed by the fuel cell (since the early 1960s as a derivative of its application in outer space) and at least a dozen less-exotic variants, such as the nickel–metal hydride, the sodium-sulpher version (tested at Rügen, and functioning at a temperature of 300°C), the maintenance-free lead-gel and, quite recently, the lithium ion combinations. They all testify to the uncertainty of the major stakeholders about the real reasons why electric drive never was enthusiastically embraced by the motoring public.

Like the technical conversions, this public needed to be converted too, as several projects initiated by social scientists attempted: once they had experienced the new style of electric driving, such was the expectation, users would develop a more environmentally conscious and a more planned attitude toward car use [13-23]. This idea, that the public was deliberately kept from a benign form of transport and only had to be re-educated to enjoy it, also spread to other sectors of the fledgling electric vehicle movement. This was for instance shown most eloquently by the film *Who Killed the Electric Car?* and its sequel *Revenge of the Electric Car* by the same director [13-24], [13-25].

And although one does not need a conspiracy theory to explain the chain of events that together constitute the history of the electric car in the 20th century, the panic-stricken behavior of General Motors, killing off the relative success of its EV1, and the covert actions and overt campaigns of the oil companies to get the California ZEV initiative off the agenda, suggest that here a real turning point was almost reached. It was followed by deep frustration and disillusionment among many stakeholders, who now are often working on hybrids, especially after the successful introduction of the Toyota Prius shortly before the turn of the century, of which more than 2 million have been sold [13-26].

The Hybrid as Transition Vehicle

This is a second proof of the electric car's continuity during the past century. Just like the miracle battery, the hybrid is as old as the plain electric. It functioned and functions as a transition vehicle in between two paradigms: internal-combustion versus electric. Against the price of added complexity (which meanwhile seems to be manageable with modern electronics) the hybrid allows to drive electrically in town and at full speed on the freeway using the internal combustion engine. Ironically, the fully fledged hybrid questions the very last solid argument why pure electric propulsion should be introduced at all: to avoid local pollution in a smog-ridden inner city. While this target seems to be better served by electrifying the many delivery vans with their frequent stop-and-start behavior (as has already been tried on several occasions in the past hundred years), the efforts to electrify second cars in the household by introducing funnily-styled city cars do not seem to lead to less local pollution, but to more urban space taken up by the automobile.

Recent analyses in the social sciences of the past twenty or so years of electric vehicle activity unearth a history of hypes, cycles of enthusiasm and disappointment, which can be substantiated by counting the number of articles on the different alternatives in major American newspapers, of Congressional Records and of patents (Figure 13.8). Such studies now observe a turning point in 2006, a real turnaround in a qualitative and a quantitative sense toward an electric vehicle society. This is confirmed by the fact that since 1996 more patenting activity among OEMs (original equipment manufacturers) has been registered on hybrids than on pure electrics [13-27].

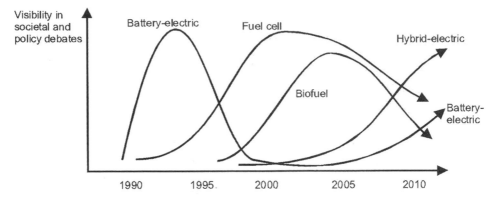

Figure 13.8 Hypes of enthusiasm for and disappointment in alternatives to the mainstream automobile (*reprinted from [13-27], p. 2, Fig. 1 with permission from Elsevier*).

According to such studies, which nowadays include the management of *expectations* (because media hypes are bad for an alternative's reputation[4]), shifts can be observed to other types of vehicles (as we have seen: the city car), but also other countries and even other cultures of use. As to the latter: electric vehicle proponents point to shifts in preferences among the digitized youth, who seem to be less dependent on the cyborg experience with the car to enjoy a happy life. They see the car as an iPad on wheels rather than a transport contraption. Hopeful statistics are presented showing that car using is stabilizing in the western world, and observers try to discern what the result of the current economic depression is and what is really new. Yet, it remains unclear whether this would bring them on the path toward electric drive.

In terms of countries, Norway now seems to have the largest market share (5% in September 2012) of electric vehicles, followed by Switzerland. While in the former country financial incentives may explain the shift, in Switzerland it must be something else (an in-born environmental consciousness?) because the country did not introduce many incentives [13-31], ([13-32], p. 11).

13.4 Conclusion

One-third of man-made CO_2 addition to the atmosphere is caused by the burning of oil, the largest share among all CO_2 producers. Transport's energy demand is similar: about 28% of the total amount of 550 EJ (2008).[5]

The question, now, is how (and how fast) transport is going to prepare for a shift away from fossil fuels and toward sustainability. Modern back-casting studies (calculating

---

4. The best explanation of the role of expectations in the current phase is [13-28]. For an earlier, more fundamental treatment see [13-29]. Applied to car history: [13-30].
5. ([13-33], p. 34–35, 49–51); 1 EJ = 1 exajoule = 10^{18} joule.

back from the goals agreed during the Kyoto conference and later conferences) indicate that drastic measures are asked for [13 34]. The question, then, is why we want to introduce electric driving, when about everyone knows now that it is the ignition key, not an engine swap, that seems to provide the real key to the problem how to diminish the car's contribution to climate change and global warming. This question is difficult to answer, all the more so because the Pluto Effect has meanwhile made sure that a "mimicking" of electric vehicle advantages continuously decreases the gap between the two competing technologies: well-to-wheel calculations are ambiguous at best and do not seem to point in the direction of a clear advantage for the electric alternative, unless one lives in countries with a lot of nuclear energy (like France) or "white coal" (like Norway).

There is no doubt that when one is interested in an absolute freedom of local pollution (to combat smog and fine particles), electric drive has no equal (although we saw in 13.3 that even that is now questioned by a well-designed hybrid). In fact, one should say: nearly no equal, because walking and especially cycling could substitute for many motorized trips. The reluctance, also among social scientists, to seriously analyze why the most recent and seemingly very promising experiments of the 1990s went foul, is a case in point: one wonders whether we really are willing to learn from mistakes in the past, given the eagerness with which stakeholders move on to the next experiment as if they fear that looking back into yesterday will never bring tomorrow any nearer. Historians know that the opposite might well be nearer to the truth.

Whatever the outcome of such analyses, still to be performed, it is quite clear that the major historical role of the electric vehicle, along with several other serious alternatives to the internal combustion petrol engine (such as the gas turbine in the 1960s; or internal combustion with alternative fuel such as natural gas or biodiesel), has been to invite, incite, and even force the hegemonic technology to accelerate its technical and organizational evolution, such that in the end the incentive to change to the other technology is weakened considerably. Our myopic stare at the artifact (that the one should be replaced by another, for the benefit of us all) hides the fact that it is under the hood, in an evolutionary rather than revolutionary perspective, that most changes have taken place and there is no reason to assume that this will be different in the near future. It also hides the fact that perhaps we are not at all interested in this change, but would like to keep the Car of Tomorrow as an alibi for what we secretly like to do: running around in a speedy, vibrating, but "electrified" combustion car, changing to natural gas, biofuels, or even hydrogen when we run out of oil. It is, indeed, very seductive to portray the electric vehicle as the car of tomorrow. But whether the electric will come or not, one thing is certain: we have already entered a phase of electrified (and as we saw in Chapter 10: electronified) automobility.

As expectations (and disappointed ones especially) seem to be a crucial driver of future practices and policies, it is perhaps good to realize that such a future is not constituted by "reality," but by how we narrate this reality. The possible worlds of the future need to be told, to ourselves, to others. Such stories, we conclude, should start in the past.

References

13-1. M. Hård and A. Jamison, "Alternative Cars: The Contrasting Stories of Steam and Diesel Automotive Engines," *Technology in Society*, 19 no. 2 (1997), 145–160.

13-2. Andrew Jamison, *The Steam-Powered Automobile: An Answer to Air Pollution* (Bloomington/London: Indiana University Press, 1970).

13-3. David Beasley, *Who* Really *Invented the Automobile; Skulduggery at the Crossroads* (Simcoe, ON: Davus Publishing, 1997) (rev. ed.; first published as *The suppression of the automobile* (Westport, CT: Greenwood Press, 1988)).

13-4. Gijs Mom, "Ongelode benzine: Een anti-klopjacht op wereldschaal," *Auto Service* 2 no. 11 (25 March 1988) 1, 3–4.

13-5. William Kovarik, "Special Motives: Automotive Inventors and Alternative Fuels in the 1920s" (paper to the Society for the History of Technology, October 18, 2007).

13-6. Gijs Mom and Diekus van der Wey, "Ethanolverdamping en de dieselmotor; Literatuuronderzoek naar en ontwerp van een ethanolverdamper t.b.v. het gecombineerd gebruik van dieselbrandstof en ethanolgas in een DAF-dieselmotor" (BSc thesis, HTS-Autotechniek, Apeldoorn, The Netherlands, 7 June 1982).

13-7. Hal Bernton, William Kovarik, and Scott Sklar, *The Forbidden Fuel: A History of Power Alcohol; new edition* (Lincoln/London: University of Nebraska Press: Boyd Griffin, 2010).

13-8. Bo Sundin, "From Waste to Opportunity: Ethanol in Sweden during the First Half of the 20th Century" (paper presented at the Society for the History of Technology conference, October 19, 2007).

13-9. "Schalieolie zet Dakota op zijn kop," *De Ingenieur* (March 22, 2013) 61.

13-10. Gijs Mom, *The Electric Vehicle; Technology and Expectations in the Automobile Age* (Baltimore: Johns Hopkins University Press, 2004).

13-11. David A. Kirsch, *The Electric Vehicle and the Burden of History* (New Brunswick,New Jersey/London: Rutgers University Press, 2000).

13-12. Dietmar Abt, *Die Erklärung der Technikgenese des Elektroautomobils* (Frankfurt am Main/Berlin/Bern/New York/Paris/Vienna, 1998).

13-13. René Kemp, Frank W. Geels, and Geoff Dudley, "Introduction: Sustainable Transitions in the Automobility Regime and the Need for a New Perspective," in Frank W. Geels, René Kemp, Geoff Dudley, and Glenn Lyons (eds.), *Automobility in Transition? A Socio-Technical Analysis of Sustainable Transport* (New York/London: Routledge, 2012), 3–28.

13-14. Renato J. Orsato, Marc Dijk, René Kemp, and Masaru Yarime, "The Electrification of Automobility; The Bumpy Ride of Electric Vehicles Toward Regime Transition" in Frank W. Geels, René Kemp, Geoff Dudley, and Glenn Lyons (eds.), *Automobility in Transition? A Socio-Technical Analysis of Sustainable Transport* (New York/London: Routledge, 2012), 205–228.

13-15. Gijs Mom, "Wie Feuer und Wasser: Der Kampf um den Fahrzeugantrieb bei der deutschen Feuerwehr (1900 – 1940)," in Harry Niemann and Armin Hermann (eds.), *100 Jahre LKW; Geschichte und Zukunft des Nutzfahrzeuges* (Stuttgart: Franz Steiner Verlag, 1997) (Stuttgarter Tage zur Automobil- und Unternehmensgeschichte; Eine Veranstaltung des Daimler-Benz Archivs, Stuttgart, Bd. 3), 263–320.

13-16. Feuerwehrmuseum, Berlin.

13-17. Michael Brian Schiffer (with Tamara C. Butts and Kimberly K. Grimm), *Taking Charge: The Electric Automobile in America* (Washington/London, 1994).

13-18. Remigius Johannes Franciscus Hoogma, *Exploiting Technological Niches; Strategies for Experimental Introduction of Electric Vehicles* (Enschede: Twente University Press, 2000).

13-19. Roland Wolf, *Le véhicule électrique gagne le coeur de la ville* (n.p., n.d. [Paris, 1995]).

13-20. Daniel Sperling (with contributions from Mark A. Delucchi, Patricia M. Davis, and A.F. Burke), *Future Drive; Electric Vehicles and Sustainable Transportation* (Washington, D.C./Covelo, CA: Island Press, 1995).

13-21. Mark B. Brown, "The Civic Shaping of Technology: California's Electric Vehicle Program," *Science, Technology, & Human Values* 26 No. 1 (Winter 2001), 56–81.

13-22. J. Robert Mondt, *Cleaner Cars: The History and Technology of Emission Control Since the 1960s* (Warrendale, PA: SAE International, 2000).

13-23. Heide Gjøen and Michael Hård, "Cultural Politics in Action: Developing User Scripts in Relation to the Electric Vehicle," *Science, Technology & Human Values* 27 No. 2 (Spring 2002), 262–281.

13-24. Chris Paine (dir.), *Who Killed the Electric Car?* (Sony Pictures Classics, 2006 [DVD]).

13-25. Chris Paine (dir.), *Revenge of the Electric Car* (WestMidWest Productions and Area 23A, 2011) [DVD]).

13-26. Bruce Pietrykowski, "The Curious Popularity of the Toyota Prius in the United States," in Weert Canzler and Gert Schmidt (eds.), *Zukünfte des Automobils; Aussichten und Grenzen der autotechnischen Globalisierung* (Berlin: edition sigma, 2008), 199–211.

13-27. Caetano C.R. Penna and Frank W. Geels, "The co-evolution of the climate change problem and car industry strategies (1979 - 2012): Replicating and elaborating the Dialectic Issue LifeCycle (DILC) model" (paper presented at the workshop "Electrification of the car: will the momentum last? " Technical University Delft, 29 November 2012).

13-28. Sjoerd Bakker, *Competing Expectations; The Case of the Hydrogen Car* (Oisterwijk: BoxPress, 2011) (diss.).

13-29. Harro van Lente, *Promising Technology; The Dynamics of Expectations in Technological Developments* (Delft: Eburon, 1993).

13-30. Gijs Mom, "'The future is a shifting panorama': The role of expectations in the history of mobility," in Weert Canzler and Gert Schmidt (eds.), *Zukünfte des Automobils; Aussichten und Grenzen der autotechnischen Globalisierung* (Berlin: edition sigma, 2008), 31–58.

13-31. Sander Heijne, "Autorijden, echt iets voor 40-plussers," *De Volkskrant* (24 November 2012), 24–25.

13-32. William Sierzchula, Sjoerd Bakker, and Kees Maat, "The influence of financial incentives on the adoption of electric vehicles" (paper presented at the workshop "Electrification of the car: Will the momentum last? " Technical University Delft, 29 November 2012).

13-33. Thomas B. Johansson, Nebojsa Nakicenovic, Anand Patwardhan, and Luis Gomes-Echeverri (eds.), *Global Energy Assessment (GEA)* (Vienna: International Institute for Applied Systems Analysys, 2012) (downloaded from www.iiasa.ac.at on 26 October 2012).

13-34. *Naar een schone economie in 2050: routes verkend; Hoe Nederland klimaatneutraal kan worden* (The Hague: Planbureu voor de Leefomgeving [PBL]/ Energieonderzoekscentrum Nederland [ECN], 2011).

Chapter 14
Innovation: Production, Diffusion, Use

14.1 Introduction: How to Construct Trends?

Now that we have dealt with trends of the artifacts themselves in their context, it is time to zoom out to the (artifact) population, as it is there, in Darwinian terms, that evolution really takes place and can be observed. We see evolution happen as *speciation*, the phenomenon that, after *bifurcation*, groups of similar artifacts, different from the rest, emerge resulting in a new species, such as the four-wheel brake car in the 1920s in a population with only rear-wheel brakes (see Chapter 8), or the car with automatic transmission versus the car with manual transmission (see Chapter 5). In economic terms, we speak of new market niches, which in the former case expands to encompass, in the end, all cars, and in the latter case develops into a new, coexisting culture, substitution versus coexistence.

Although we may assume that many of these bifurcations are a response to a (perceived) change in user preference (as we have assumed when we explained the *dual nature of technology* in Chapter 1), it is the manufacturer (the producer, the engineer, the designer) who produces the properties enabling the new functions. If we therefore wish to identify trends in the history of car technology, we not only need more information about these properties' quantitative and qualitative importance, but also about the circumstances of their creation, in the car factory.

In other words, apart from trends in the product, we also have to identify trends in the process. In fact, the evolution of car technology presented in the previous chapters was based on a least an intuitive idea about this statistical background. In the following two chapters, we will try to understand the *co-construction* (by producers and users alike) of these trends, and we begin, in this chapter, with the question how such trends can be identified and detected. We will first explain the way a product penetrates the market (Section 14.2), and will then focus upon the major trends in car production that fed the diffusion of the products over the market (Section 14.3). Next, we will zoom out to the user culture and ask ourselves what drove the changes in this culture during the last 125 years or so (Section 14.4). We will close this chapter with conclusions (Section 14.5).

14.2 Diffusion

Diffusion is the technical term to designate a crucial element in the innovation process. A naïve conception of innovation sees it as the last in a linear sequence of separate practices: invention, production, and diffusion. Meanwhile, innovation theories have not only revealed that the inventive process is an incrementally ongoing and nonlinear practice, often executed by a team rather than an isolated genius, with feedback from the users, whereas production, as we will see in Section 14.3, has its own tradition of process innovation, more or less separated from (but related to) product innovation. Often, invention is less important than assumed in the romantic image of the lonesome Ford, Diesel, or Edison, which became clear when we analyzed the emergence of petrol injection systems (see Chapter 3 and Chapter 9) and dedicated some space to teams as subjects of innovation (Chapter 12). Nonetheless, innovation theories agree on the fact that without diffusion, there is no innovation.

In the previous chapters we have given several examples of diffusion of subsystems and component assemblies in a quantitative sense (see, for instance, Figure 1.6, Figure 1.7, and Figure 7.12). Insight in these detailed innovation processes, which—as we have seen—often did not result in substitution but rather in coexistence of several alternatives, help explain the overall diffusion curves of the entire artifact, the car. At the level of the latter, diffusion curves are the result of myriad purchasing decisions by customers: through the diffusion of the entire car, new component assemblies and subsystems diffuse also, often triggering the purchase decision in the first place. By taking purchase decisions by customers in certain car cultures together one can construct diffusion curves per car culture, revealing differences between the average type (see Chapter 1) of the car in Europe versus the United States (Figure 14.1).

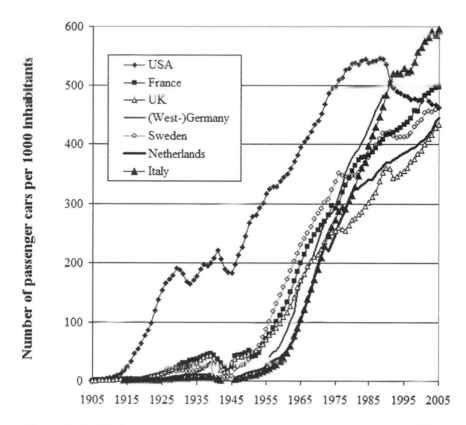

Figure 14.1 Diffusion curve (expressed in car density, passenger cars per 1,000 inhabitants) of the average type of the car in the United States and selected European countries. *Sources:* Number of vehicles (European countries): B.R. Mitchell, *International Historical Statistics Europe 1750–1993* (Vol. 4) (London/ New York: MacMillan Reference LTD/ Stockton Press, 1998), complemented with other sources such as *World motor vehicle data*, American Automobile Manufacturers Association, 1997 (Germany, for 1956/7), *World Road Statistics* 1969, 1973, 1977, 1981, 1985, 1986, 1989, 1993, 1998, 2002 (for Germany, Italy, and UK, in some of the years of the 1962–1999 period), and other sources. For the last years (2000–2005): ACEA, *European Motor Vehicle Parc 2005*, for all the European countries except Italy and UK. Italy: data from Automobile Club d'Italia, Direzione Studi e Ricerche—Ufficio Statistica (1993–2005); UK: Department for Transport (1995–2005); Number of vehicles (United States): U.S. Department of Transportation, Federal Highway Administration, Office of Highway Policy Information, Highway Statistics Series; Population: Mitchell (1998) and U.S. Census Bureau, International Data Base. I thank Luísa Sousa (Lisbon) and Hanna Wolf (Eindhoven) for constructing this graph.

Such curves can also be drawn, if the data exist, to help explain partial diffusion processes, such as the spread of the car per town (Figure 14.2). In fact, such curves illustrate growth processes (of the vehicle population), and hence can also be used to document similar phenomena, such as the growth of networks (of roads, for instance; Figure 14.3) [14-1].

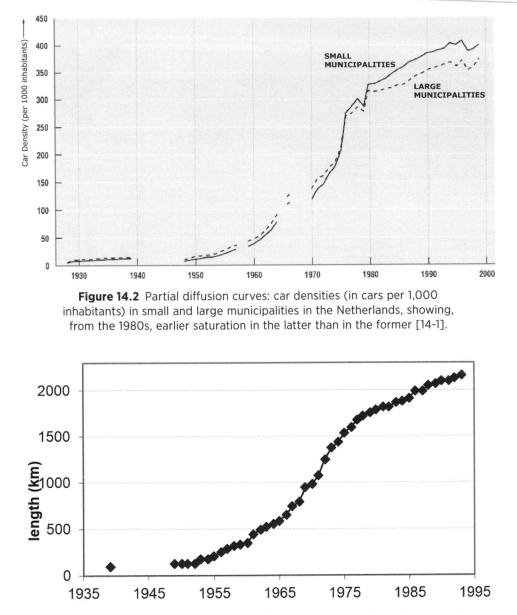

Figure 14.2 Partial diffusion curves: car densities (in cars per 1,000 inhabitants) in small and large municipalities in the Netherlands, showing, from the 1980s, earlier saturation in the latter than in the former [14-1].

Figure 14.3 The growth of a road network represented by a "logistic curve": the freeway network in the Netherlands per December 31 of the years 1939 to 1993 [14-2].

The Logistic Curve

Indeed, quantitative diffusion theory is based on growth phenomena in bacteria, human populations and the spread of diseases, hence the term *contagious diffusion*, indicating that the curves assume that people, when they buy a product, convince neighbors to do the same.

The mathematical formula representing such behavior of the population (of bacteria, viruses, or people) is an exponential expression and the curve is called logistic. Such curves all have in common that they near, when time passes, a growth limit. At first growth accelerates exponentially, then passes an inflection point, and diminishes further when nearing the maximum without ever reaching it (Figure 14.4 is stylized, because it does not show this principally asymptotic character). Applied to social phenomena the S curve rests on the assumption that adoption breeds adoption: the more users decide to adopt the bigger the chance that the rest of the population turns into a user, too [14-3].

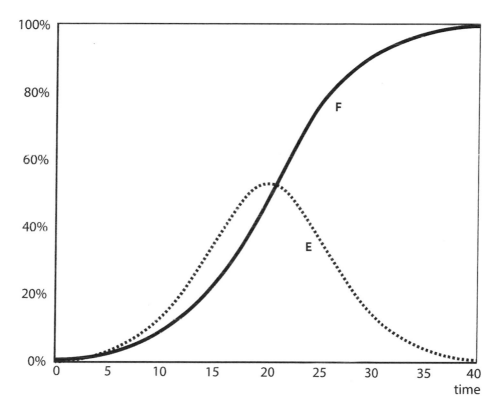

Figure 14.4 General form of the logistic growth curve (F). Curve E can be considered as the first derivative of F, indicating diffusion speed, which is highest at the inflection point (*sketch inspired by [14-4]*).

The attractiveness of such exponential curves is that their logarithm produces a straight line. Drawn in a semi-logarithmic format (with the time axis kept linear) a quick look enables to check whether a certain time series follows the law of growth or not. This is especially apparent at the beginning and end of the curve, as there not much data exists, because the phenomenon is still too new, or the statistician's attention has meanwhile shifted to another phenomenon, respectively. However, the analysis of the diffusion of some commodities such as cars revealed that their spread did not easily fit into the logistic picture. During their diffusion new departures seemed to take place which appeared to be better covered if the overall diffusion would be represented by two (or more) than just one S curve (Figure 14.5).

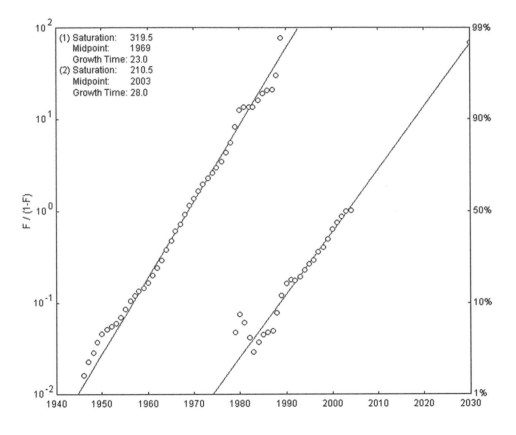

Figure 14.5 Semi-logarithmic representation (Fisher-Pry transformation) of the diffusion of car density in the Netherlands, assuming two consecutive logistic curves. The second curve is drawn under the assumption that car density saturates in 2030 [14-5].

However attractive to mathematically inclined students of car evolution, the growth curve's sophistication cannot hide its finalist and *deterministic* character. A precondition for the construction of a growth curve is the establishment of a saturation level. And unless it regards a phenomenon which has been closed already (such as the substitution of the sailing ship by the steam and motor ship) the established saturation level is always a perceived level. This can have very questionable effects on the logistic's use for policy purposes. The classical example is the sequence of predictions of automobile saturation in Germany by Deutsche Shell A.G., which, in hindsight, all show a structural and massive under-estimation of the real trend.

Another example is the way in which such curves (for instance of the expected growth of the car population) were used by road building departments in most western countries in the 1960s and 1970s in order to give their policy of building large freeway networks a scientific justification. Growth curves then become self-fulfilling prophecies.

Another drawback of using these curves as exclusive explanatory tool is that every curve rests on the assumption that growth is determined by a set of pure constants: a constant infrastructure, a constant artifact, and a constant adopting population. The curves seem to say: once a product appears on the market and proves successful, the nature of this product and of the socio-technical system in which it is embedded, as well as the character and culture of the consumers, does not matter anymore. Remarkably, there are some exceptional cases where consumers seem to behave in such a perfectly logistic way, for instance in post–WWII Italy, where the growth of the car park followed the growth curve with utmost precision. But even there users are no bacteria: they have to make a decision in order to buy, and producers have to select between marketable options.

Through the diffusion curve, we enter a world of black and-white where only one road seems to lead to progress: that of innovation. No attention is being paid to nondiffusion, diffusion rejection, (organized) resistance, consumer interference and nonuse, activities that may change the course of automotive history considerably, as we will see in the next chapter. Additionally, by focusing upon the artifact as the unit of analysis, these approaches tend to overemphasize substitution (injection versus carburetor, prop shaft versus chain) over coexistence of competing artifacts (drum and disc brakes, manual and automatic transmissions).

Even in the limited case of a seemingly clear substitution process, the question remains what is exactly substituted. Often, when substitution graphs are presented, one gets the impression that apples and oranges are compared. The reason for this is the focus upon the artifacts, instead of upon the societal functions these artifacts afford. Take, for instance, the substitution of the American agricultural horse by the tractor (Figure 14.6). Historical evidence shows that tractors were originally introduced for other purposes on the farm than horses were used for. Also, horses and tractors may be better compared on the basis of the mechanical work they provided; on this level, the aggregated horsepower curve (the solid line in Figure 14.6) may be better suited to describe what happened on farms, suggesting continuity rather than a breach [14-6].

Another point of criticism is the starting phase of the innovation process. Because of a lack of data this starting phase is often badly covered, which can be very misleading. It is exactly during this initial phase that the choice by so-called early adopters between alternatives is crucial [14-8]. Thus, the resulting diffusion curve is the curve of the winner and, as such, the curve neglects the Pluto effect.

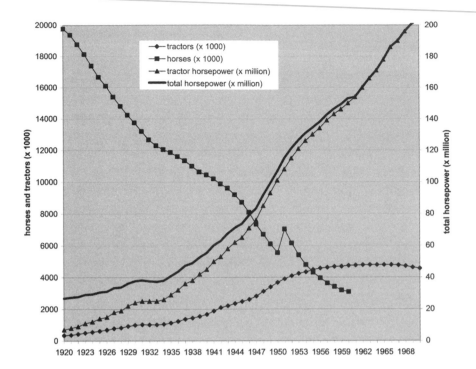

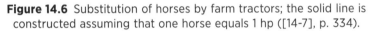

Figure 14.6 Substitution of horses by farm tractors; the solid line is constructed assuming that one horse equals 1 hp ([14-7], p. 334).

All this implies that diffusion cannot be explained on the basis of (economic) efficiency alone. Diffusion curves hide, but nonetheless represent active behavior of people: in order to start and maintain the process of diffusion, to slow down, to accelerate, and certainly to redirect their course, activities have to be deployed, by suppliers, by users, and their intermediary institutional representations such as consumer clubs, consultancy bureaus, and governmental institutions.

Thus, consumers cannot be treated as a population of passive individuals devoid of any structure. Instead, they form institutions that consciously try to influence the diffusion process by proposing changes in the product to be diffused (co-construction) and by collectively resisting adoption (nonuse) or preventing others from adopting (resistance movements).

This means that the entity to be diffused has to be constantly adjusted to keep the diffusion process going, and while the technology changes, its use (and its users) have to be "(re-)invented" as well. In other words, whereas the history of technology has shown that the process of technological change can be analyzed as a form of *co-evolution* (of technology and its use), diffusion studies should recognize the phenomenon of *co-diffusion*, of the artifacts and at the same time of the way they can be used. It is the use (not described by diffusion curves, as they cover only the aggregated act of buying) which constantly inspires producers to adapt their products.

However, it would be a misunderstanding of the previous litany of criticisms to deny that diffusion curves can be a powerful tool to identify anomalies and irregularities in the spread of artifacts, to compare regional and national developments or to help with periodization and related problems.

Spatial Diffusion

A second way to analyze diffusion in a quantitative way is to chart the artifact's spatial spread. It appears on the basis of such studies that spatial diffusion did not follow a path of ever wider concentric circles around a large urban innovation center (as the contagious diffusion idea would suggest), but that certain groups of early adopters were responsible for the emergence of more or less independent smaller innovation kernels: rural U.S. medical doctors, for instance, were among the first in spreading the automobile in and around small cities, and only after this had happened did the general geographic spread follow its regular course around these multiple innovation centers (Figure 14.7).

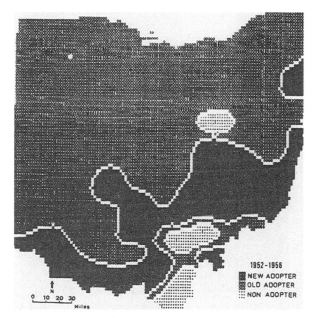

Figure 14.7 The spatial diffusion of cars in Ohio; situation of 1952–1956 (*reprinted from [14-9], with permission from Taylor & Francis*).

14.3 Production

Turning now to the production sphere, the first question to be asked is: what is produced? Several economic historians who studied the history of car production in detail have argued that what is purchased (and hence produced) is not so much an aggregate of components, but a set of characteristics. The economist Pier Paolo Saviotti, for instance, distinguishes between technical and service characteristics, more or less parallel to the distinction between technical properties and relational functions we use in this book [14-10].

Other students of the industry have not so much focused on the *character*, but more on the *frequency* of innovation. William J. Abernathy, for instance, claims that "radical innovation has given way to standardization and to the efficiency of highly complex mass-production methods" ([14-11], p. 3). This happened through productivity increase, and hence the innovation activity shifted from the product to the process and product innovation became more difficult to achieve.

This is especially the case in the capital-intensive automobile industry (Abernathy performed a detailed study of the Ford Motor Company), where innovation tends to be incremental, whereas in the science-based industries such as electronics, pharmaceuticals, and chemicals product innovation may remain radical. Whether Abernathy's conclusions can be upheld for the non-American car industry remains to be seen, and even for the United States itself one can wonder whether his focus on the Ford production of the Model T (of which the design, Abernathy claims, "was not significantly changed" during the two decades it was the world's best-sold car ([14-11], p. 13)) does not overemphasize the idiosyncrasies of one case.

Fordism and Taylorism versus Sloanism

On the other hand, it cannot be denied that the first phase of automobile production was governed by what later has been called Fordism, mass production based on a standardized design to be produced on an assembly line, combined with Taylorism, the disciplining of the workers on the basis of time-and-motion studies of their production activities.

As a matter of fact, Abernathy concluded that it was not so much the production method which made the decisive difference (conveyer belt production was also already used in the slaughter houses of Chicago, and the Olds Motor Works, as early as 1902, produced the first cars in series), but the application of vanadium steel. In fact, it was (again) the *expectations* spun around this miracle metal (namely that it allowed speeding up production of a much more robust and sturdy product) rather than its actual properties that triggered the shift to mass production. Henry Ford saw it applied in a French car and only later found out that it "proved too brittle" and soon was replaced by other special steels. By that time, vanadium steel had already fulfilled its trigger role ([14-11], p. 12, 31–32)

However this may be, mass production led to a first shake-out of the American industry, a process which gained extra momentum when, during the 1920s, Ford was overtaken, in numbers of car output, by General Motors on the basis of a new production paradigm, Sloanism. Named after GM's president, Sloanism was based upon a conscious strategy to avoid major product innovations. This was replaced by an annual model change suggesting a continuous "improvement" of the car, a phase in which the car's body shape became as important as the technology hidden underneath, if not more so ([14-12], p. 349–350). In other words, our belief in continuous progress, which we have so often questioned in the previous chapters when we signalled nonlinearities, has been fed by the car industry itself.

While in the United States these developments gave rise to the Big Three (General Motors, Ford, and Chrysler), in Europe the market was more fragmented. There, production developed along national lines, but the number of manufacturers also diminished drastically, in parallel with the American market (Figure 14.8) [14-13]. It confirms that, despite the differences in car density and size of national markets, the commonalities between the countries of the North-Atlantic car culture are perhaps more important, from an evolutionary point of view.

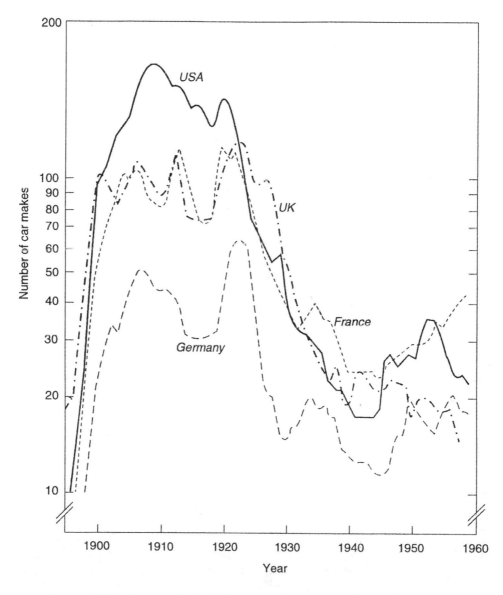

Figure 14.8 A history of industrial ecology: the number of car makes in some industrialized countries ([14-14], p. 53) (*courtesy of Arnulf Grübler, IIASA, Vienna*).

When this North-Atlantic car culture entered what we have called a phase of Exuberance during the 1950s, it happened, quite as a surprise to the American industry, that European manufacturers managed to conquer one-tenth of this market, despite the high protectionist wall built around it. The Big Three felt they had to answer with so-called compact cars, in 1960. It was a first sign that the seeming monopoly of the American car industry started to crumble under the competition from the emerging continents of car production (Figure 14.9).

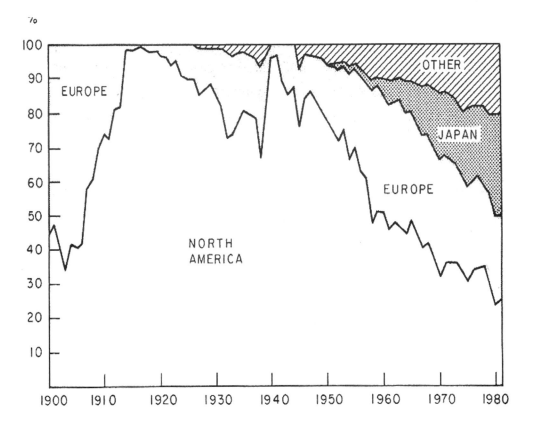

Figure 14.9 Shares of countries and continents in world automobile production (*reprinted from [14-15], p. 13, Fig. 2.1, with permission from Taylor & Francis*).

Toyotism

Soon thereafter, the American automobile industry came under fierce attack, not only for producing vehicles that were unhealthy for people and the environment, an attack that gained extra momentum during the two energy crises of the 1970s, but also because it made dangerous vehicles as Ralph Nader's actions showed (and as we retold in Chapter 10). Similar developments could be observed in Europe. At the same time, a new competitor appeared on the global market and it was again a team from MIT

(Massachusetts Institute of Technology; MIT professor Abernathy was the first) that diagnosed the threat for the American car industry in a famous book: *The Future of the Automobile* [14-15], [14-16].

It was Japan, and especially its lean production methods with just-in-time delivery by the supplier industry known as Toyotism, which according to the MIT team's analysis had beaten Fordism and Sloanism as paradigmatic principles of modern car production. Now, Japanese manufacturers invaded the American market (Toyota alone conquered about 13%) and a new worldwide shake out of the industry took place, indicating that the industry entered a truly global phase. By the mid-1970s, the Japanese car industry had overtaken the American industry in the number of patent applications ([14-15], p. 102). Robotization and ICT enabled to produce personalized vehicles on the assembly line, while manufacturing strategy emphasized the importance of platforms, common technological substrates on which different bodies and models could be built.

By the middle 1980s imports had conquered one-third of the American market (Figure 14.10). For a brief period of time, they had on average a higher power than their American counterparts, while their specific power (in hp/L) was about one-third higher, a token of higher engineering quality in Europe ([14-17], p. 89). Europe also was in the vanguard of production technology innovation, when in the mid-1970s Italian Fiat introduced its Robogate production system based upon robots developed by Comau, a technology even exported to the United States, and "one of the few occasions on which Europe had a superior technique to that which had been developed in the US" ([14-18], p. 12). Soon, Japan would take the lead in production technology.

At the same time the variation in models to choose from increased precipitously: in Europe alone from about 180 models in around 70 versions in 1984 to about 270 models in about 3,200 versions in 2004. A French study of Renault even claims that the number of versions increased from 120,000 for its model R25 in 1989, to more than a million for its model Mégane. Cars could now be ordered à la carte, in a fully personalized way (the Renault Laguna has 800 different door panels on offer alone) ([14-19], p. 5), ([14-20], p. 27–28). In this extremely competitive atmosphere, in which every deviation from the zero-error gospel was punished by massive recall actions, product innovation seems to suffer and seems only to be expected from the smaller companies and the suppliers, as several students of the industry conclude. The trouble the major manufacturers have with the development and marketing of the electric vehicle (as we have seen in the previous chapter) is a case in point. On the other hand, when it comes to innovation in production technology the major manufacturers are still in the lead ([14-18], p. 34).

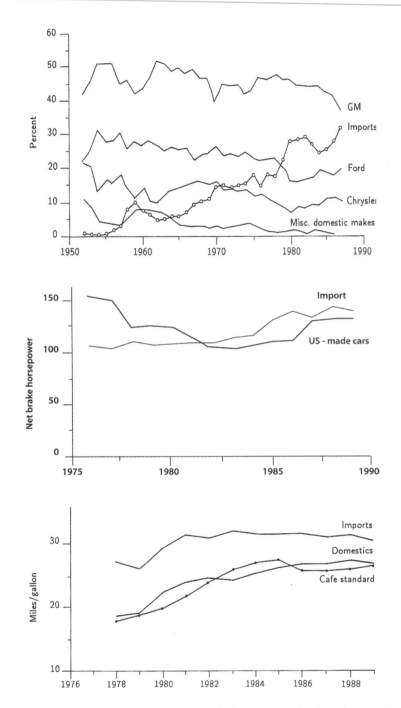

Figure 14.10 Invasion of (higher-powered!) imports on the American market; as a result, fuel consumption of the American cars decreased in parallel with the increased CAFE standards. (*reprinted from [14-17], p. 84, Fig. 2b, 89, Fig. 7a, 90 Fig. 8 with permission from Elsevier*).

14.4 Use

Now that we have seen how, over the course of one century, car manufacturers had to adjust their production process in order to reach ever higher economies of scales and how the spread on the market of the product developed both in time and place, it is time to look at the actor that put this whole process in motion in the first place: the user. It was the behavior of the user who invited (if not forced) the manufacturer to constantly adjust the car's technology in order to *afford* (in the words of James Gibson, see Chapter 9) the practices he and she developed around the car. It is to these practices that we now turn, following the periodization we proposed in Chapter 1, with the aim to understand the major changes in user behavior (see Table 14.1).

Table 14.1 Periodization of car use in relation to the types of use, the types of users and the types of the car			
Period	**User group/market**	**Technology**	**User function/ activity**
Until WWI: Emergence	aristocracy and upper middle class	open car: (clam) shell	touring in "nature"
Inter-war years: Persistence	(new) middle class	closed car: capsule	family outings and some business
Post-WWII until 1975: Exuberance	mass motorization	"affordable family car": cocoon	holidays and commuting
1975–now: Doom & Confusion	fragmentation	smart, personalized car: flow as cocoon	urban and suburban car-based mobility culture

During the first period (the phase of *Emergence*) elite motorists used their open cars to tour in nature [14-21]. They used the car as an adventure machine in three respects: in time (they liked racing and other forms of high-speed use), in space (they liked to roam the countryside without knowing where the trip would end), and in function (they liked, to a certain extent, the unreliability of the technology, and if they did not, they always could delegate this part of the automotive adventure to their chauffeur). The history of early automobility has meanwhile gone way beyond the romantic image of rich fathers of large families toying around in the country repairing their tires after a careful study of the automotive handbooks. They existed, too, but the movement also had a grim side, of arrogance to nonusers (who threw stones and tried to prevent motorists from monopolizing the street), violence to playing children and animals, and, in general, an overwhelmingly masculine (if not macho) attitude toward nature and all "others" (those who did not belong to the inner circle of motorists, including people living in the "periphery," such as the southern parts of Europe and the United States, or in the colonies). When, by the end of this phase, a new type of user (the business man, the medical doctor) started to look for a new, smaller and cheaper, type of car, the die-hards moved on to next macho challenges, mostly in aviation. In other words, the close connection we observed in engineering knowledge between aeronautics and automotive also exists in user culture: both were adventure machines.

During the second period (the phase of *Persistence*) the new users took over, the white collar middle class who used the now closed car for family outings: not only the car's technology, but also the adventures spun around it, became domesticated, tamed as it were, into Fordist (mass-produced, in the myriad autocamps, later in the thousands of motels) recreation for the "nuclear family," consisting of husband, wife, and, say, two kids. The idea that this phase was characterized by the substitution of the earlier adventure machine by a utilitarian car is a myth, carefully constructed by car proponents, first of all the automobile and touring clubs, supported by the car industry. There exists some historical evidence that the business user came up once the improved highway network offered an attractive alternative to the first-class train compartment, but the user profile remained dominated by pleasure and fun. The idea that the car became "serious" was an element in the struggle with a revived resistance against the car in the big cities, especially in the USA, because of its devastating effect on road safety, and also against those governments that still saw the car as a luxury consumption item that needed to be taxed [14-22]. In this sense, the violent and aggressive character of the car remained, but it was largely hidden under alarming statistics of road fatalities, as we have seen in Chapter 10.

During the third period, then (the phase of *Exuberance*), motoring expanded to the entire population of the West (and, as we will see in the next chapter, some related parts of the non-Western world) and a typical pattern of use got universalized: about one-third of the trips was dedicated to commuting and business use, one-third to recreational use, and the rest for shopping, social visits, bringing children to school and other practices. The car holiday became a central element in this culture, despite the fact that most of the trips were short-range.

The fourth period (the phase of *Doom*) brought to the fore, for the first time on such a massive scale, doubts that the car in its current shape was perhaps not the right solution for a global population, and especially its middle class, which slowly but undeniably entered a condition of seemingly endless prosperity. Local pollution, the threat of climate change (as we have seen in Chapter 11), and an ever increasing amount of road casualties (globally, despite the fact that road casualties in the West started to diminish) inspired a broad resistance movement against more asphalt and more congestion, but the diffusion curves of global car ownership continued to rise, on its turn increasing the feelings of doom.

It is during this fourth and in the fifth period (the current phase of *Confusion*) that the use profile of the car started to shift from a Fordist to a "postmodern" pattern, in which not the nuclear family was the core of the car culture anymore. This culture now started to become fragmented into multiple profiles of use not delineated along clear class lines. Figure 14.11 gives an overview of such a culture in the Netherlands, based on a survey of

12,000 respondents who are distinguished in eight groups, from "traditional bour-geoisie" to "postmodern hedonists." What is remarkable is not so much this multiplicity of values (loosely related to lifestyle), but the core of the car culture which appears to be hardly changed: the car and its culture is still carried to a large extent by the middle class, so much so that it is not too far fetched to see the car as a middleclass technology per se [14-23].

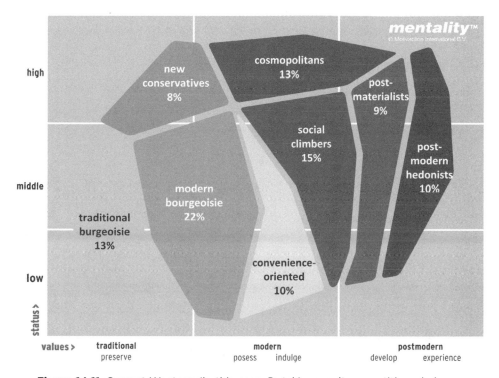

Figure 14.11 Current Western (in this case: Dutch) car culture partitioned along value groups. Motivaction, quoted in ([14-23], p. 37) (*courtesy of Motivaction*).

This sequence of the four or five phases, which we proposed in Chapter 1 as based mainly upon a cultural analysis of "Atlantic automobilism," coincides quite neatly with a technical periodization based upon three respective levels within the automotive structure and system: if the first period focused on the component and the component assembly (a level accessible to the fledgling engineering community as well as that of the tinkering users), the second phase went up to the next level of the subsystem and the car's hierarchical structure, and became the heydays of the classic chassis engineer. The third phase opens up the wider car system as the playing field of engineering (including road building, but also traffic engineering and its related safety concerns), whereas the fourth phase extends the control of both (internal) structure and (external) system through the introduction of electronics, a crucial extension which in the fifth, current phase, results in a possible "smartification" of car and road.

14.5 Conclusions

The brief overview of car production, diffusion and use in this chapter reveals another trend, next to the ones identified in Chapter 9 for the artifact's evolution. Clearly one of the major trends in process evolution was *increased scale*, enabled by the same trend that also governed the product: *automation*. The last phase of this trend, *electronization*, may well have opened the car market to new global players, just as it has also revolutionized car technology, as we saw in Chapter 9. The emergence of new Indian and Chinese makes, which inspired other countries (for instance Venezuela) to take a license and start their own production, indicates that we seem to enter a new phase in the evolution of the car and its production.

Who could have guessed that during the recent credit crisis in the West, General Motors went broke and had to be rescued by the American taxpayer? Who could have predicted that by the beginning of the second decade of the new century the Chinese car market would outgrow that of the United States and that the fate of global warming and the condition of our climate depends on the choice of the hundreds of millions of potential motorists in the BRIC (Brazil, Russia, India, China) countries? And who could have foreseen a decade ago that India would try to revolutionize car technology by developing the Tata Nano, a combination of technical sophistication and sobriety in order to keep the price down for the upcoming millions of the middle classes in these countries? And who, finally, could have realized that the decision of the Chinese authorities whether to invest in the production of electric vehicles or improved combustion-engine cars, would determine the shape and nature of the global car fleet?

Suddenly, in less than a quarter century, the car and its industry had stopped to be an exclusively Western phenomenon to be exported to the "periphery." How could manufacturers, policy makers and users alike have ignored that apart from the West, billions of people decide on a daily basis how to go to their work, to the shops, to visit family and friends and on holidays? To answer these questions, it is time to investigate the changes in global mobility and its impact on the way we live on this planet.

This investigation should first and foremost focus upon the question of the role of the middleclass in the emergence and persistence of the Western (and perhaps also: the non-Western?) car culture. Mass motorization, as it emerged after the Second World War, can be explained to a large extent by the preparatory work done before that war: the building of a road infrastructure, the seductive combination of pleasure and power, fun and violence of car use as a promise of future possession to vast amounts of non-users and, most of all, the dominance of a middle-class mentality based on the nuclear family and the affordable family car, prepared an eager post-war population to the hedonistic aspects of consumer culture which especially could be experienced and practiced in the car. In the next chapter we will confront this explanation of Western mass motorization with mobility experiences in other parts of the world, which will enable us to draw a picture of world mobility.

References

14-1. Gijs Mom, "Mobility for Pleasure; A Look at the Underside of Dutch Diffusion Curves (1920 – 1940)," *TST Revista de Historia; Transportes, Servicios y Telecomunicaciones* No. 12 (June 2007) 30–68.

14-2. *1899–1994; vijfennegentig jaren statistiek in tijdreeksen* (Den Haag: sdu, 1994).

14-3. Gijs Mom, "Frozen History: Limitations and Possibilities of Quantitative Diffusion Studies," in Ruth Oldenziel and Adri de la Bruhèze (eds.), *Manufacturing Technology: Manufacturing Consumers; The Making of Dutch Consumer Society* (Amsterdam; aksant, 2008), 73–94.

14-4. D.S. Ironmonger, *New Commodities and Consumer Behaviour* (Cambridge: Cambridge University Press, 1972) (University of Cambridge, Department of Economics, Monographs series, No. 20).

14-5. Gill Reyes e.a., "Car density analysis involving the Netherlands" (student report, course Autotrends; Long-term developments in automotive technology, Eindhoven University of Technology, Mechanical Engineering, 2005, appendix on CD-ROM).

14-6. Gijs Mom, "Compétition et coexistence; La motorisation des transports terrestres et le lent processus de substitution de la traction equine," *Le Mouvement Social* No. 229 (October/December 2009), 13–39.

14-7. Devendra Sahal, *Patterns of Technological Innovation* (London/Amsterdam/Don Mills, Ontario/Sydney/Tokyo: Addison-Wesley Publishing Company, Inc., 1981).

14-8. Everett M. Rogers, *Diffusion of Innovations* (New York/London/Toronto/Sydney/Tokyo/ Singapore: The Free Press, 1995[4]).

14-9. Avinoam Meir, "Innovation Diffusion and Regional Economic Development: The Spatial Diffusion of Automobiles in Ohio," *Regional Studies* 15 No. 2 (1981) 111-122.

14-10. Pier Paolo Saviotti, *Technological Evolution, Variety and the Economy* (Cheltenham/ Brookfield: Edward Elgar, 1996).

14-11. William J. Abernathy, *The Productivity Dilemma; Roadblock to Innovation in the Automobile Industry* (Baltimore/London: The Johns Hopkins University Press, 1978).

14-12. Jeffrey Robert Yost, "Components of the Past and Vehicles of Change: Parts Manufacturers and Supplier Relations in the U.S. Automobile Industry" (unpubl. diss. Case Western Reserve University, May, 1998).

14-13. Michael T. Hannan, Glenn R. Carroll, Elizabeth A. Dundon, and John Charles Torres, "Organizational Evolution in a Multinational Context: Entries of Automobile Manufacturers in Belgium, Britain, France, Germany, and Italy," *American Sociological Review* 60 (1995) August, 509–528.

14-14. Arnulf Grübler, *Technology and Global Change* (Cambridge/New York/Melbourne: Cambridge University Press, 1998).

14-15. Alan Alsthuler, Martin Anderson, Daniel Jones, Daniel Roos, James Womack, *The Future of the Automobile; The Report of MIT's International Automobile Program* (London/Sydney: George Allen & Unwin Ltd., 1984).

14-16. Jean-Pierre Bardou, Jean-Jacques Chanaron, Patrick Fridenson, and James M. Laux, *The automobile revolution; the impact of an industry* (Chapel Hill, 1982).

14-17. Gerhard Rosegger, "Diffusion Through Interfirm Cooperation; A Case Study," *Technological Forecasting and Social Change* 39 (1991), 81–101.

14-18. Krish Bhaskar, *Innovation in the EC Automotive Industry—An Analysis from the Perspective of State Aid Policy* (Brussels/Luxembourg: Commission of the European Communities, April 1988).

14-19. Giuseppe Volpato and Andrea Stocchetti, "Product-line variety and innovation along product life-cycle in carmarket: Are carmakers' policies really effective?" (paper 14th Gerpisa International Colloquium "Are automobile firms markets-oriented organizations? Myths and realities," 12–13 June 2006, Paris, France).

14-20. Noémie Behr, "Vers la voiture sur mesure; Document de travail–Cerna," *Problèmes économiques* No. 2891 (January 18, 2006), 24–33.

14-21. Gijs Mom, *Atlantic Automobilism: The Emergence and Persistence of the Car, 1895–1940* (New York and Oxford: Berghahn Books, forthcoming).

14-22. Peter David Norton, *Fighting Traffic; The Dawn of the Motor Age in the American City* (Cambridge, Massachusetts/London: The MIT Press, 2008).

14-23. Hans Jeekel, *The Car Dependent Society: A European Perspective* (Aldershot: Ashgate Publishers, 2013).

Chapter 15
World Mobility: Shifting the Focus

15.1   Introduction: The West and the Rest

It is hardly ever acknowledged that the story told so far is not the whole story. On the contrary: a small part of the world's mobility, developed over more than a century in the North-Atlantic world, Australia, New Zealand, and, for the post–WWII period, Japan, has been generalized as the world history of mobility. This easy assumption, that the rest of the world would, now or later, mimic western motorization, is a topic worthy of its own study.

If we wish to describe the mobility history of the global majority, however, it appears that the automobile is not a good vehicle to start, even if many people in the West (and increasingly also in the East) tend to equate mobility with the car. The hegemonic status of this vehicle type tends to blind us for the fact that modernization can happen without a focus on the car. In other words: there are different paths to a modern society (there are "multiple modernities" and thus there are "multiple mobilities"), but as we will see, this will always in the end involve the car and its technology somehow.

Studying the "Rest," as it is so often condescendingly called, not only helps us to relativize the car, it also helps to discover trends that in the previous story were overlooked. It helps us rethinking mobility. Expanding the mobility story to the entire world not only brings in the neglected mobile majority's different mobility cultures, it also forces us to reconsider the West and tell the story differently, with the wisdom gained by our visit to the Rest.

This chapter first tells an expanded story which in one way or another applies to all countries in the world: the struggle between road and rail, which appears to be a struggle between two types of societies. We will tell this story as a Western story, as research so far has been mostly limited to this area (Section 15.2). In a following section (Section 15.3) we will visit the mobile majority through one specific vehicle type, not known in the west: the rickshaw. We will show that not so much its technical *properties*, but the *functions* these properties afforded and still afford, invite us to revisit the

Western car story, and help us discover trends that we had up to now overlooked, such as the collective use of vehicles, or the relation between poverty (however defined, in Bangladesh, or in the United States) and mobility. We close this chapter with conclusions (Section 15.4).

15.2 Road versus Rail: Clashing Mobility Cultures

In the previous chapters we have seen that car technology made major strides during the early 1920s, just at the moment that the elite was superseded by a growing motoring middle class who could speed up on a rapidly improved and paved road network and have its cars maintain in garages. This was also the moment that other vehicle types we have not dealt with up to now, appeared on the roads. First of all, in those countries where cars seemed unattractive (because they were too expensive, for instance), the (lighter) moped and (heavier) motorcycle appeared. What "car countries" accomplished by cars, motorized two-wheeler countries (such as Germany, but also many countries beyond the West) realized with other means. Second, now that roads became reliable, the truck and the bus appeared on the streets, taking over freight and passengers from tram and train, but mostly creating new functions such as the transport of small packages by truck and the group holiday by bus, and they all strengthened the case of the road against the rail. For contemporaries, this phase was experienced as an explosion of mobility, a proliferation of road travel experiences, whether in the car, on the moped, as truck driver or bus passenger [15-1].

Initially, leading railway officers saw the truck and the bus as feeders to their extensive system built up during the 19th century. This system was drastically expanded during the last decades of that century and the first of the 20th by local and regional railroad networks. Railway interests observed an injustice, because the railroads were meanwhile acknowledged as a public good (paying decent salaries to its employees, forced by law to charge minimum tariffs and serve remote and thinly populated parts of the country), whereas the new vehicles functioned under another regime of individualized transport and wildcat entrepreneurship. Whereas the railroads were planned from above, and cross-subsidized bulk freight like coal and sand by the profits on more lucrative freight, the cars, buses and trucks were mostly driven by single owners who anarchistically could roam the mobility domain for lucrative trips.

On top of that, the railroads came out of WWI in disarray, despite their reputation as rescuers of the country in times of war: bad management and relatively high salaries for its increasingly powerful and unionized workers made the railroad companies vulnerable at a moment that trucks, buses, and cars started to grab the most lucrative sides of its business. At the exact moment this was most needed, railway companies lacked the capital to modernize (through electrification, for instance). Many national governments, who had invested large sums in the building of the network, decided to protect the railways by nationalizing them, at the same time leaving the lower-order networks such as the tram and other light rail systems to dwindle.

However, these same governments started to build road networks, showing that governments are not monolithic entities: they consist of factions, reflecting interest groups in society. In this case these factions represented two different types of societies: one geared toward the heavy industry with its needs of bulk material such as steel, sand, coal, and the other geared to the light industry and the retail sector, interested in flexible service of small units of freight and passengers (such as factory workers from nearby villages).

Against this basic background of two clashing mobility cultures, countries tried to (re)shape their own modal split (the distribution of the transport tasks over the different modes or vehicle types): Western Europe managed to rescue its national railroad network for massive passenger transport, while the United States kept its network mostly for freight. They did so by regulation, forcing bus and truck entrepreneurs to acquire a licence, which often was not given when special commissions judged that it was detrimental to the railroads, for instance because a bus line was planned to run parallel to the railway line.

Non-Western countries found other combinations of railways and roads use, often consisting of a coarse-meshed network of railroads and a (mostly urban) deployment of other vehicles such as the bicycle, or the moped, or horse-drawn or human-drawn transport (as we will see in the next section). But also some Western countries deviated considerably from the general picture, such as the Netherlands (Figure 15.1), which up to the end of the inter-war years remained a country dominated by the bicycle, also in passenger-kilometers (the normal unit to express modal split).

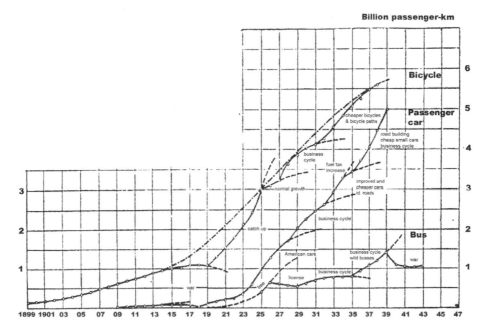

Figure 15.1 Modal split in the Netherlands before and during the inter-war years: the bicycle dominates [15-2].

In Germany, like in the Netherlands, the first type of (railway) entrepreneurs were dominant, so the support for the car, bus and truck culture was modest, until Adolf Hitler tried to mimic his rivals (the United States and England) by a crash-program of *volk* motorization, an effort that failed. Most motorizing countries in the West, however, followed the pattern such as can be seen in France, where the railroads reigned supreme in freight, but first the bus, then the car started to take passengers from the train and create new needs on their own.

"Serious" versus Pleasurable Mobility

Such needs were often not what the planners expected. Transport science in those days was not only mostly economically inspired (and hence often neglected the car because it did not represent a factor in the production of national income, except in the large "car countries"), it also could not cope with modernizing trends such as the pleasurable character of the automobile (for the family to go on holidays, or the weekend spin). Even historians have some trouble getting rid of this economic bias which privileged all "serious" forms of mobility. Although the figures are ambiguous, it is safe to say that in the United States during the inter-war years the most "serious" side of automobility, commuting, certainly was not the dominating function of the car [15-3]. Commuting was gradually added to the spectrum of potential practices, but the base of these practices remained pleasurable.

National transport policy also was biased toward the large companies and hence the regulation of the modal split often went against the single-person enterprise or the small company. Such actors were seen as wild-catters and often were either made illegal, or were subjected to regulation. Figure 15.2 for instance shows how the UK government, known for its liberal policy and often weary of state intervention, more or less decimated the motor bus field to protect the railways. In some countries (such as Belgium), it was the railways that got the freedom from the government to start bus lines and trucking companies themselves, strengthening the railroad monopolies.

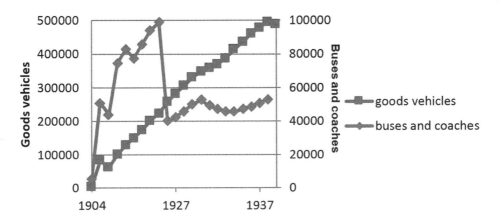

Figure 15.2 Buses, coaches, and goods vehicles in the UK 1904–1939 (attention: the x axis is not linear) ([15-4], p. 144–145, Tables V and VI).

After WWII, the West became a true car paradise. This period of Exuberance saw mass motorization emerge as the threshold to purchase a car became increasingly lower. At the same time the road network received an extra layer of freeways for a new generation of high-speed cars, motorcycles, buses, and trucks.

A City Crisis

Only during the last two or three decades of the century a countermovement came up which tried to find an answer to the car crisis, which we have described in Chapter 11, by reviving public transport, and especially the railways. It was France where a new generation high-speed railroad network started to compete with short-distance aviation, a move so successful that now even the United States started considering implementing this new system.

At the same time, as we also have seen in Chapter 11, the car crisis became a city crisis, so new urban public transport systems were created such as light rail, tramways and bicycle lanes to combat local congestion and deterioration of the atmosphere. Even the major car companies cannot deny anymore that if the car and city crises are not solved somehow, their future looks really grim, so they started founding study departments where their products are embedded, conceptually and actually, in broader systems of (urban) mobility. This not only is expressed by some car companies selling bicycles with their cars, but also by others promoting car sharing projects, sometimes on the basis of electric traction. At the same time these companies are very busy trying to bring about the next generation of car technology: the "connected car" that communicates not only with the infrastructure outside the cocoon, but also with other cocoons so as to create intelligent flows, promising less pollution, higher efficiency, and more fun (see Chapter 9).

Are we on the brink of a new automotive era? Before we can try answering this question we first have to zoom in on our cities, and here we truly will have to leave the West to find out what options are left as global solutions.

15.3 Global Urban Mobility: The Case of the Rickshaw

The rickshaw was invented in Japan in 1869, its name derived from Japanese *jinrikisha*, meaning: human-powered carriage; there were 200,000 in 1890. It was a hybrid cyborg: drawn by human labor but equipped with Western-styled wheels, it was first exported to Hong Kong (1874) and Singapore (1880), arrived in Chinese Beijing in 1886 (and appeared there on the streets for public hire in 1898), while in Indian Calcutta it was used for freight carriage from around 1900, and 14 years later for passenger service. It also appeared in Durban, South Africa ([15-5], p. 140), ([15-6], p. 27), ([15-7], p. xiii).

In all these cities they were an immediate success because of their "feeder" character (filling the gap of the last kilometers to the train, or the ship) or very short-range commuting and shopping. It replaced the *palanquin* (India) or sedan chair (China), which meant decreasing human labor by half (and when improved wheels with ball bearings were introduced, also making redundant a pusher at the back) and became popular among the growing middle class. But the pullers belonged to the major victims of the emerging automobile, in more than the common sense of traffic safety. In Singapore, of which the history has been researched the best, pullers were recruited from massive migration flows of poor peasants from two Southern Chinese provinces. The same applied to the vehicle owners, who owned between 1 and 20 rickshaws and who often were not seen as capitalists but as mediators in a complex culture of migratory flows. The official number of registered rickshaws reached its peak in Singapore in 1923 (with nearly 30,000 vehicles), but as so often in this domain, the informal economy was much more important (Figure 15.3 and Figure 15.4). By then, decline had already set in because of the same phenomenon as in the West: the emergence of bus and truck.

The connection between extreme poverty and the use of wheeled mobility may have influenced the reluctance in several Western countries to keep to the bicycle. In the UK or France, for instance, cycling has, since the emergence of the car, been associated with poverty. Figure 15.5 shows a cartoon in which the function of the rickshaw as a social and economic safety net has been depicted.

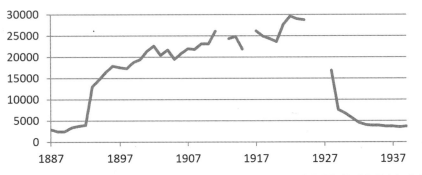

Figure 15.3 Diffusion of registered rickshaws in Singapore ([15-5], 61–62, Table 6.2).

Figure 15.4 Typical street scene in the mid-1930s in Singapore: pullers did not walk, but ran ([15-5], p. 203, Photo 32) (*courtesy James Warren*].

Figure 15.5 Cartoon showing the function of the rickshaw as last resource ([15-7], p. 120).

Mobility and Poverty

By looking at these pictures, reflecting the harsh sides of mobility in the East, the issue of pure existence pops up: poor farmers, driven from their land, migrate to the cities where the rickshaw was and is often the only means for an unskilled (in the industrial sense) illiterate of staying alive. Chinese rickshaw pullers started a violent uprising against the emerging tram system in 1929 in Beijing, destroying 60 of the 90 tram cars. Counting 60,000 in Beijing (and 70,000 in Shanghai) alone, rickshaw pullers took half a million fares per day, in a city of hardly 1 million people. They were the eyes and ears of the city (and of the Communist Party): while in the West the car became the object of poets and writers to sing the song of "the freedom of mobility," in China one of the famous modern novels (written by Lao She) showed how the poorest of the poor, peasant migrants from the countryside, used the rickshaw as a last resort to stay alive. Such stories reveal how it felt to be a human motor, especially when executing the *feipao* (the flying run). The income of one-fifth of the population of China's capital somehow depended on the rickshaw. The Chinese author Hu Shi (1891–1962) saw a "line of demarcation between Eastern and Western civilizations (which) is precisely the line of demarcation between rickshaw and automobile civilizations" ([15-8], p. 41), (15-7], p. 95]), [15-9]. Remarkably, a recent book on Indian car culture presents a similar thesis:

the difference between rich and poor coincides with the difference between pedestrianism and car culture ([15-10], p. 1).

Whereas the rickshaw was abandoned out of humane motives by the Chinese Communist Party when it came to power in 1949, it is still a crucial element in the urban modal split of many Asian cities, mostly in the form of the cycle rickshaw or pedicab (also in China, at least until 1972). The cycle rickshaw substituted for the hand-pulled version by the end of the inter-war years and the immediate post-war years, except in current Kolkata (former Calcutta, in West Bengal, India), now the only city in the world with hand-pulled rickshaws [15-11], [15-12]. The first cycle rickshaws appeared in Singapore, in 1929 (there were 60,000 in 1936 ([15-5], p. 77)). Worldwide, by the end of the last century, an estimated four to five million roam the roads, with India by far the most (1.7 million), followed by Bangladesh (1 million), but also the Philippines (15,000), Colombia (100,000), and Singapore (500). In Dhaka, capital of Bangladesh, the rickshaw accounts for more than one-third of the local modal split.

With the emergence of the tuk-tuk or auto rickshaw in the mid-1950s, the rickshaw also got a motorized version. But whether bicycle or auto, their construction has been often criticized as "not a good example of appropriate technology," either because of a lack of funds or because of a near-monopoly, as is the case with the Indian Bajaj Auto, which occupies nearly half of the auto rickshaw market of 800,000 per year. Electric versions have been introduced, and many local authorities have implemented regulation forcing such cars to drive on LPG (liquefied petroleum gas) or CNG (compressed natural gas) [15-13].

But also the existing technology is considered to be ready for improvement. The front-wheel fork, for instance, is directly derived (and bought) from bicycle technology. Whereas this wheel's camber (see Chapter 7) is beneficial for the bicycle's dynamic balance, for the rickshaw driver it means a constant readjustment of the steered wheel in its straight position, leading to extra strain on the driver's body ([15-6], p. 335-336). Rather than advocating radical innovation (projects to this effect initiated by good-willing Western engineers and development aid institutions in the 1980s all failed), current insights recommend a gradual technical adjustment, also in order to keep the vehicle affordable for the poor. In Dhaka, Bangladesh, a newly built rickshaw (Figure 15.6) nowadays cost between $40 and $50. Modernization here resides in the use of cell-phone-based call systems, showing that mobility also includes communication, or: the movement of information.[1] This also shows that innovation is not a privilege of the "new": also the "old" can benefit from it [15-15].

---

1. We cannot develop this issue any further, but think of the railway and the telegraph as an earlier example of such a mobility concept, conceived as a co-evolution between transport and communication. We have forgotten this, but the railways could not have functioned without the telegraph [15-14].

Figure 15.6 Embellishing and repairing rickshaws in
Dhaka, Bangladesh (*photos by the author*).

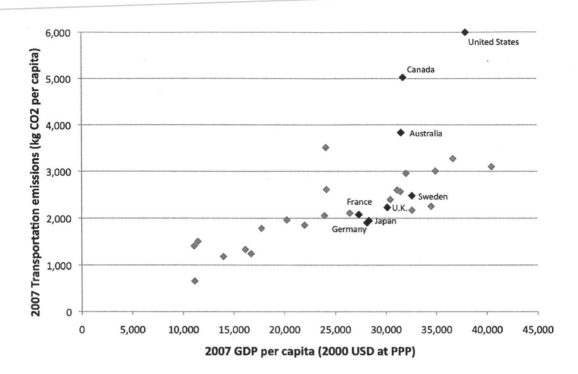

Figure 15.7 CO_2 emissions per capita as a function of per capita income for OECD countries (*reprinted from [15-26], p. 19, Fig. 1 with permission from Taylor & Francis*).

15.4 Conclusion: The Rest and the West

The rickshaw is only an example of the differences in the mobility cultures between the East and the West. Other examples are the two-wheeled bicycle (300 million in China in 1985; 50 million in India around the same time ([15-6], p. 76, 354)), but it is the hundreds of thousands of minibuses and "mosquito buses" that lead to a rethinking of the West's own mobility history, because they point to a special form of collective automobile use that has nearly been forgotten. And yet, in the United States, during the 1910s, more than 60,000 jitneys (owned by groups of blacks in the South for instance) roamed the streets until they were made illegal by concerted actions of railway interests and local authorities [15-16], [15-17], [15-18], but in several Western countries (Latin America, Turkey) there are still many semipublic forms of transport based on the shared use of automobiles or mini buses, such as the Turkish *dolmuş*.

The flow of transfers also goes in the other direction, as in the case of local initiatives to develop a non-Western alternative to the car. This is for instance the case with the Tata Nano in India, for sale for *Rs1-lak* as Indians say (meaning: 100,000 roupees) and originally conceived as a super-rickshaw. Then, under the influence of the Ford Model T, but also other popular cars like the Citroën 2CV, the Fiat 500, and the Volkswagen beetle, it was designed from scratch as a full-blown car, but affordable for the growing Indian middle class. Tata also claims that a special form of car engineering is involved. Whether this Ghandian engineering (as *The New York Times* coined it) included the famous *jugaad*, which has nowadays been identified by innovation theorists as a form of improvised, make-do engineering, is not clear ([15-19], p. 24).

China seems to steer another course, inviting foreign car companies to supply its growing middle classes with cars. Whether it is capable of leapfrogging generations of cars used in the West (as has been done with the wire-based telephone), remains to be seen, as the technology Western companies transfer are by far not the most sophisticated ([15-20], p. 141–143). In order to get its population in the car, the Chinese government will have to restructure its socialist market economy such that the enormous circular migratory flows (from the countryside to the cities and back) of tens of millions of Chinese are somehow channeled, because car culture, at least up to now, is thriving in a middle class sedentary society. The Chinese *hukou* (household registration) system tried to stem the devastating flows of poor people to the cities that haunt other developing countries, but still, one-third of the 17 million inhabitants of Shanghai are migrants ([15-21], p. 3, 12, 40). And although the Chinese, representing one-fifth of the world population, only own 1.5% of the world's cars (the United States, with only 5% of the world population, owns one quarter), world automotive conditions are shifting fast. For instance, recently China, with a CO_2 emission being the highest in the world, implemented fuel-efficiency standards 50% more stringent than the American CAFE standards, but it still struggles with out-dated emission control systems built into General Motors and Ford cars in local factories, such as Shanghai GM, Beijing Jeep, and Chang'an Ford ([15-20], p. 8, 17, 97), [15-22].

The Car as an Urban Problem

It is quite clear: the coming automotive problem is a city problem. If it is true that, above a certain threshold, a near-linear and above all universal relationship exists between national income and car ownership, then traffic engineering faces a daunting task indeed. According to this linear logic, which still has to be proven for non-Western countries, the only reason non-Western cities do not have as much casualties as the United States is that there are not yet so many cars. Some analysts have serious doubt about a correlation between income and car diffusion: they claim that the Western relation to the car is not much different from the Indian reverence towards the "sacred cow" [15-23], ([15-24], p. 105).

In other words, which elements of the much more variegated urban modal split in non-Western countries have to be neglected in order to maintain this linearity? Indeed, as long as one focuses upon the car (still a minority in many of these cities), a remarkable uniformity in its use results, such as the per-capita travel time budget, which oscillates around 1 to 1.5 hours, in Tianjin (China), Olomouc (Czechia), villages in Tanzania or U.S. cities ([15-25], p. 175). But will the same linearity exist when it comes to the production of greenhouse gases like CO_2? Figure 15.8 suggests the existence of such linearity for the OECD countries at least. Or will the new motorizing countries be below that lowest of the band width in Figure 15.8, Japan, the result of a conscious policy to ban automotive transport from very densely populated areas in cities?

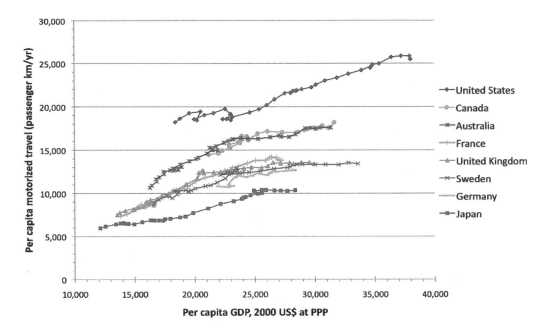

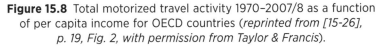

Figure 15.8 Total motorized travel activity 1970–2007/8 as a function of per capita income for OECD countries (*reprinted from [15-26], p. 19, Fig. 2, with permission from Taylor & Francis*).

Similar questions can be asked when it comes to road safety. The World Health Organization (WHO) commissioned research in worldwide leading causes of death and found that annually more than 1.2 million people are killed in road traffic (2.5% of all annual deaths). In earlier research (1994) the United Nations already found that more than half of all accidental deaths in developing countries are traffic-related ([15-24], p. 104).

Western versus Eastern Modal Splits

The problem is that such graphs are made with the Western modal split as a model. Figure 15.9 shows how such a modal split looks like for the country of Bangladesh if one would take all existing modalities into consideration, except walking and the bicycle, no doubt still the mostly used modalities. It is quite clear that a story based on the car's technology would not suffice here. Instead, and reminiscent of what we remarked earlier about the importance of innovation within the realm of the "old," a technical history of road mobility would have to take into account the (sporadic) proposals to improve the existing rickshaw technology into something which is less tiring for pullers (Figure 15.10).

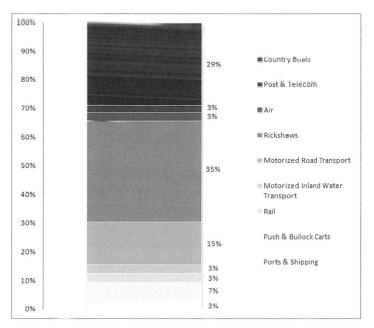

Figure 15.9 Modal split in Bangladesh in 1984/85 (*inspired by [15-6], p. 5, Fig. 1.1*).

Figure 15.10 Cycle rickshaw art ([15-27], p. 10) (*photo courtesy Kuntala Lahiri-Dutt*).

It would also have to take into account the efforts done by owners and pullers to embellish their cherished vehicle: rickshaw art represents a separate industry of artists and painters. These paintings express, it seems, a universal truth: the close connection between being mobile and feeling "the power," whether the power of God, the power of one's leg muscles, or the power of a powerful internal combustion engine.

Transport experts in the West now seem to observe deviations from this pattern as well for their own countries, observable only if one delves deeper in the user culture and acknowledges that different subgroups emerge with quite different mobility profiles, such as the elderly and the youth. Especially the latter seem to decouple their daily routines from car use as a matter of course, and if they do, they don't seem to follow the fascination for fully equipped automobiles like their parents did (and still do). Are we on the brink of a new mobility culture?

References

15-1. Gijs Mom, "Clashes of Cultures: Road vs. Rail in the North-Atlantic World during the Inter-War Coordination Crisis," in Christopher Kopper and Massimo Moraglio (eds.), *The Organization of Transport: A History of Users, Industry, and Public Policy* (London/New York: Routledge, forthcoming 2015).

15-2. W. J. de Graaff, "Groei van het verkeer en zijn problemen," *De Ingenieur* 60 No. 10 (5 March 1948) V.21–V.36.

15-3. Gijs Mom, *Atlantic Automobilism: The Emergence and Persistence of the Car, 1895–1940* (New York and Oxford: Berghahn Books, forthcoming).

15-4. Christopher I. Savage, *An Economic History of Transport* (London: Hutchinson University Library, 1966³) (revised ed.; first ed.: 1959).

15-5. James Francis Warren, *Rickshaw Coolie; A People's History of Singapore 1880–1940* (Singapore: Singapore University Press, 2003).

15-6. Rob Gallagher, *The Rickshaws of Bangladesh* (Dhaka: The University Press, 1992).

15-7. Fung Chi Ming, *Reluctant Heroes; Rickshaw Pullers in Hong Kong and Canton, 1874–1954* (Hong Kong: Hong Kong University Press, 2005).

15-8. David Strand, *Rickshaw Beijing; City People and Politics in the 1920s* (Berkeley/Los Angeles/London: University of California Press, 1989).

15-9. Lao She, *Rickshaw: The novel Lo-t'o Hsiang Tzu* (Honolulu: The University Press of Hawaii 1979) (first published in serial form in 1936/1937; transl. Jean M. James).

15-10. Vidyadhar Date, *Traffic in the Era of Climate Change; Walking, Cycling, Public Transport Need Priority* (Delhi: Kalpaz Publications, 2010).

15-11. Rajendra Ravi (ed.), *The Saga of Rickshaw; Identity, Struggle and Claims: A Study by Lokayan* (New Delhi: Lokayan, 2006).

15-12. Ziaush Shams, M. M. Haq, and Towhida Rashid, "Rickshaw-pullers of Dhaka City and Rickshaw Pulling as a Means of Livelihood," *Oriental Geographer* 49 No. 1 (January 2005). 33–46.

15-13. Akshay Mani and Pallavi Pant, *A Synthesis of Findings; Review of Literature in India's Urban Auto-rickshaw Sector* (n.p.: Embarq India, n.d.) (downloaded from www.embarqindia.org on 12 December 2012).

15-14. Heike Weber, "Mobile Electronic Media: Mobility History at the Intersection of Transport and Media History," *Transfers; Interdisciplinary Journal of Mobility Studies* 1 No. 1 (Spring 2011), 27–51.

15-15. David Edgerton, *The Shock of the Old; Technology and Global History since 1990* (Oxford: Oxford University Press, 2007).

15-16. Carlos A. Schwantes, "The West Adapts the Automobile: Technology, Unemployment, and the Jitney Phenomenon of 1914–1917," *The Western Historical Quarterly* 16 No. 3 (July 1985) 307–326.

15-17. Ross D. Eckert and George W. Hilton, "The Jitneys," *Journal of Law and Economics* 25 (October 1972), 293–325.

15-18. Walter Kudlick, "Carros Por Puesto–The 'Jitney' Taxi System of Caracas, Venezuela," *Highway Research Record* No. 283 (1969) (paper presented at the *Highway Research Board* annual meeting, January 1968).

15-19. Philip Chacko, Christabelle Noronha and Sujata Agrawal, *Small Wonder; The Making of the Nano* (Chennai/Mumbai/Bangalore/Hyderabad/New Delhi: Westland, 2010).

15-20. Kelly Sims Gallagher, *China Shifts Gears: Automakers, Oil, Pollution, and Development* (Cambridge, MA/London: The MIT Press, 2006) (Urban and Industrial Environments, ed.: Robert Gottlieb).

15-21. C. Cindy Fan, *China on the Move: Migration, the State, and the Household* (London/New York: Routledge, 2008) (Routledge studies in human geography).

15-22. Marlies ter Voorde, "CO_2-uitstoot hoger dan ooit na korte dip," *De Volkskrant* (13 December 2012) 27.

15-23. Peter Newman and Jeff Kenworthy, "The Ten Myths of Automobile Dependence," *World Transport Policy & Practice* 6 (2000) No. 1, 15–25.

15-24. Achim Saupe, "Human Security and the Challenge of Automobile and Road Traffic Safety: A Cultural Historical Perspective," *Historical Social Research* 35 (2010) No. 4, 102–121.

15-25. Andreas Schafer and David G. Victor, "The Future mobility of the World Population," *Transportation Research Part A*, 34 (2000) 171–205.

15-26. Adam Millard-Ball and Lee Schipper, "Are We Reaching Peak Travel? Trends in Passenger Transport in Eight Industrialized Countries," *Transport Reviews* 31 No. 3 (May 2011), 357–378 (downloaded from http://www.informaworld.com/smpp/content~db=all~content =a929755275~frm=titlelink on 12 December 2012).

15-27. Kuntala Lahiri-Dutt and David J. Williams, *Moving Pictures; Rickshaw Art of Bangladesh* (Ahmedabad: Mapin Publishing, 2010).

Chapter 16
Conclusions

The previous 15 chapters, the first half zooming in on the component and component assembly level, the second half zooming out toward, first, the car's inner structure and then, the broader car system in its societal and even world context, enable us to fine-tune our observations about the evolutionary development of automotive technology and the main trends that govern this development. On the basis of the results of the previous chapters we will also be able to answer our earlier question (see Chapter 14) about the frequency of innovations, most particularly whether it is true that the evolution of automotive technology is characterized by a surge of product innovations at the beginning of the product's life cycle superseded by *process innovations* at the production site, while the amount of *product innovations* decreases.

To start with the main trends: half-way in our analysis, in Chapter 9, we have distinguished between types of trends: evolutionary, automotive, and societal, which we are now able to characterize in more detail. In the chapters following Chapter 9 we have seen how, throughout the car's entire history, basic *evolutionary trends* were involved, most particularly, of course, the mechanism of *speciation* resulting in *variation* and *diversity*, but counteracted by another trend of *standardization* and *homogenization*, which is advantageous for economies of scale. In the course of our study of these trends, we have repeatedly emphasized that this evolution is not an autonomous process. Some students of technical change nowadays call such a naïve and utterly misleading opinion techno-fundamentalism ([16-1], p. 556). Engineers, nor users, should wish to be fundamentalists in this respect. Evolutionary trends are not specific to the automotive realm, but are nonetheless crucial to understand its development. Three of the main evolutionary trends that relate directly to the development of car technology are *functional drift*, *functional shift* and *functional split*, trends that describe the changes in (relational) *functions* of the car and its systems, enabled and constrained by the (technical) *properties* (see the dual nature of technology concept as explained in Chapter 1). *Drift* is the phenomenon that functions (as afforded by a specific set of properties) incrementally change because of changing habits and practices of users. This can result in a *shift* of functions from one set of properties to another and can, in the end, be so radical that functional profiles get *split* between two or more sets of properties. A nice example of this is the development of the driveline from a belt/chain system into a gearbox-cum-propshaft, where the two functions of distance and reduction began to shift and ultimate got split into two separate sets of properties (see Chapter 5). Another example is the changes in wheel suspension, where the damping function was split off the previously combined functions of damping and springing in the leaf spring, resulting in the introduction of separate dampers, and the elimination of the damping function in the (coil) spring (see Chapter 7).

Societal trends are likewise not specific to the car, but they are no less crucial for understanding what happened to the car during the past century. *Scientification* of automotive engineering, for instance, was embedded in a general scientification of engineering, as well as of much more sectors of society, such as urban planning, consumer research, and logistics, to give only three examples. The *safety* trend and the trend to a more outspoken *environmental consciousness* are other examples of societal trends that have meanwhile pervaded the engineering approach to the car. In the course of this book, we have often given examples and pointed at the importance of societal trends.

If we focus on the *automotive* trends, there is no doubt that the major trend consists in a cluster of sub-trends which all governed the shift of motoring skills within the human-machine cyborg to the technology. These crucial subtrends are *automation* and *scientification*, and the former appeared to be not very well possible without the latter, especially during the second half of the previous century when *electronization* came up. This shift from skills to technology (seen from a cyborg perspective: from software in the human to software in the car) coincided with a shift within automotive engineering from the engine to the car. More precisely, the engine stayed on as central focus but got company from the car as an entity worthy of its own engineering, and as a major object of scientification, as we saw in Chapter 12. Scientification also involved the emergence of new types of engineers, next to the classic chassis engineers: acoustical, chemical, biochemical, physics, and in general mathematically educated engineers. In other words: the car evolved from an engine on wheels to a dynamic vehicle propelled by an engine. Thus, it seems, propulsion was liberated, enabling engineers and users alike to rethink whether internal combustion, electric, or hybrid would be best.

Persuasive Technologies

In order to enable us to get a grip on such a long period of intertwining trends we have distinguished between five phases (Emergence, Persistence, Abundance, Doom and Confusion), derived from a cultural analysis of automobilism, but in Chapter 14 supported by a more or less parallel technical phasing. Will the next step be a car as an iPad on wheels as is so often suggested nowadays? It might be, if we assume that the connectedness of the car will become more important in the future than its dynamics in terms of riding and handling. It might well be, indeed, that we enter a phase where the car's media characteristics, its communicative aspects, its ability to be embedded in a "convoy" on the road, its development into an entertainment center on the long holiday trip, its cocoon-like protection in case of an accident, or in case of a sudden and wrongly executed lane-change, will become more important, now that the car has become as fast as the engine. Even if we will not all be driving an electric vehicle within a decade or so, we will certainly be handling an electrified, if not a digitized car, in which traffic rules, such as a speed limit, can be digitally monitored and, if need be, digitally enforced [16-2].

Modern-day cars now have (or are planned to have) a dozen features with "persuasive" characteristics (the next stage of what one student of the recent trends calls enabling technologies [16-3], such as forward collision warning, lane departure warning, parking aid, blind spot detection, driver alert, adaptive cruise control, adaptive high beam adjustment, eco mode, gear shift indicator (for "responsible" fuel saving behavior), and so on (Figure 16.1). National authorities hope through the introduction of such technologies, which even could be personalized and thus be made much more persuasive (if not coercive), to enhance altruistic behavior among motorists.

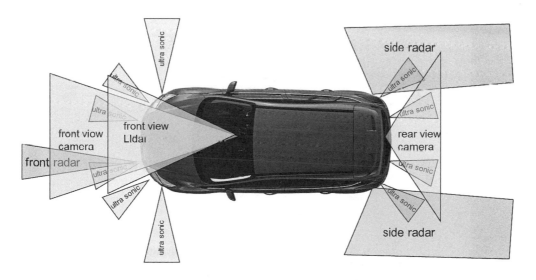

Figure 16.1 The digitized car: sensing devices on a modern-day Ford
(*courtesy of Ford Research and Advanced Engineering Europe*).

Understanding the Evolution of the Car

How can we understand this development and what is its relation to future, expected developments? The answer to these questions is twofold and can be gained by a view back and a gaze into the future.

As to the past: if we follow the periodization given in the first chapter, we can draw a picture of the consecutive steps that were taken in the evolution from the open-bodied *voiturette* of just before 1900 to the automated cocoon of today. In Table 16.1, these steps are formulated as research questions governing the periods.

Table 16.1 Steps in the evolution of automotive engineering research according to the periodization given in Chapter 1	
Period	Governing question
Emergence	does it work?
Persistence	does it move?
Exuberance	how does it move?
Doom	how does the other move?
Confusion	should we move (or be moved)?

From this stepped evolution of the car during the previous century one can conclude that engineering only in the third period started to understand the car's behavior as a whole, and that the car became "social" only in the fourth period, as it started to acquire properties with which it could be moved safely in traffic. Now that the car has acquired these properties we have entered a period in which we ask ourselves whether this way of moving the car must not be reconsidered, and whether we should not fully re-engineer it, in order to make it behave like an automaton, a monad in a flow in which we, for the first time in automotive history, will be able to delegate all our motoring skills to the technology, temporarily at least, and voluntarily (or, as some would have it: enforced).

We were able to acquire this general evolutionary perspective on the car by zooming out of the automotive ghetto and realizing that, at the beginning, a surprisingly large amount of innovations were derived from a pre-car period of industrial engines, bicycles, railways, and carriages. However, this cross-fertilization should not lead us to the conclusion that the automobile itself (whether propelled by electricity, steam, or internal-combustion engines) was not a major innovation. It was, because of its novel assemblage of existing technologies which quickly became the subject of continuous adjustments and adaptations to the ever changing context of the fledgling automobile culture. In other words: the innovation was at the systemic level, the level of the structure, and only then individual and intermingling paths of evolution at the separate components and component assemblies started.

We have seen in the previous chapters, that the user was often instrumental in this development, for instance in the case of the development of a more comfortable tire (Chapter 8). We have also witnessed a geographical cross-fertilization: the *divergence* of automotive technology and culture into two separate developments, a European and an American. Calling this development Americanization, as often happens, is not adequate, as we have seen many examples of transfers in opposite directions, at distinct levels within the car's structure, such as in the case of the brake system (see Chapter 7).[1]

We have also seen that many (perhaps a majority) of the proposals did not make it, such as electro-rheological fluids, memory metals, plastic leaf springs, but this does not mean that they might not, once in the future, pop up from the *state of the art* which will always be hanging, as it were, above the existing technology as a reservoir to choose from. We have also seen several solutions die out, such as the carburetor, or the hand-control of the ignition advance.

The second answer to our questions relates to the future and can be encompassed in one word: *expectations*. We should not forget that many innovations of the persuasive kind just suggested have not yet been implemented, let alone have become main-stream. Despite the attractiveness of the S curve (see Chapter 14), there is no law that this will happen, as we have seen on several occasions in the past. We should not forget either that many of the projections are expectations, science fiction if you will, despite the eagerness, especially among engineers, to deal with them as if they are reality. In a way they are real (as fantasies, and very powerfully so), as much as they are very necessary for engineers involved in design work. If a designer does not develop a fantasy about future use of her object-to-be-produced, she cannot work properly. This explains why the future has such a high reality force, especially for engineers. They must dream in order to innovate.

And yet, we should be careful about predicting the future, as the previous chapter testify. As the Austrian historian Peer Vries has shown, as late as the 1960s it was predicted that "the economic future belonged to the Soviet Union" which has now disappeared from the earth's surface. In the 1970s some scholars predicted that Japan would be number one, the same country that since more than a decade slipped into a severe crisis. In the 1990s experts predicted that the "Asian Miracle" would soon be over, whereas two decades later it absorbs nearly half of all cars produced worldwide. Around the turn of the new century it was predicted that Europe would be "an economic superpower rivalling the United States, with the potential to become a full world superpower," whereas the recent credit crisis raised doubt about the very survival of the European Union. But according to Vries there is no doubt about the decline of the West. He elucidates this conviction by pointing at the fact that per 2013 "over half of the global GDP is produced in countries that are not members of the OECD" ([16-5], p. 316, 318).

---

1. Quantitative analysis into general technology transfer between Europe and the United States suggests a parallel development "with little crossover" [16-4].

Interdisciplinarity

Another remarkable trend is the (often distant, but sometimes very close) influence of aeronautical engineering, a trend which deserves to be better researched. It seems as if the scientification of automotive engineering would not have been possible without the help of aeronautics: the two "adventure machines" (the car and the plane) have more in common than is often realized.

Perhaps in the current period, traditional automotive engineering needs the help of other disciplines as well, such as human-technology interaction design, which is working on persuasive technologies designed to invite drivers into more responsible behaviour. Are we moving into the direction of a "moral car", a car that doesn't allow us to do the "wrong" things? If so, which things are wrong and which are good? And who decides on this? However this may be, the ongoing process of driver deskilling may well pave the way for the development of new skills not geared to the mechanical part of the car, which possibly will be ever more *black-boxed* but geared toward its increasingly complex communicative parts. Might it be that the future of the car can only be a part of a general mobility concept, where collective aspects (never lost in the East) are reinvented, such as seems to be the case with car sharing, public bicycle use, and all those other practices that seem to be tied to the increasing urbanization trend of our planet?

The Super-Archetype of the Car

And how about innovation frequency?

Table 16.2 gives a "canon" of automotive archetypes including a trend analysis, executed by students of Eindhoven University of Technology, who followed courses that lie on the basis of this book. The set of 75 cars has been composed by experts in automotive history and is a mix of the most popular cars (measured in sales volume) and the technical importance of the car. On the basis of these archetypes, an average archetype per period could be constructed (taking the arithmetic average of the quantified properties, and, secondly, as Archetype 2.0, calculating the weighted average, correcting for production figures). Even a super average archetype could be constructed, a non-existing average of all archetypes over the entire century, resulting in a set of properties as given in Table 16.3. It appears that the Renault R4 comes nearest to the averaged values.[2]

---

2. This is a preliminary result, undertaken by the students of the course Automotive Trends II (0AU02) during the academic year 2012–2013, whom I thank for their efforts. It is planned to double the set in order to iron out the bias toward high-tech and innovations as well as toward Europe. For instance, the best-sold car ever (the Honda Civic in all its versions) is not yet part of the set.

	First Period	Second Period	Third Period	Fourth Period	Fifth Period
	Table 16.2 Canon of automotive archetypes, selected on the basis of a mix of sales volumes and technical interest				
1.	1886 Benz Patent Motorwagen	1922 Lancia Lambda	1948 Citroën 2CV	1978 Mazda RX7	1997 Alfa Romeo 156
2.	1899 Lohner Porsche	1924 Bugatti Type 35	1949 Saab 92	1979 Mercedes Benz W126	1998 Audi TT
3.	1902 Stanley Steamer	1925 Rumpler RU 4A 106	1950 Volkswagen T1	1984 Renault Espace	1998 Smart Fortwo
4.	1904 Rolls Royce 10 HP	1927 Volvo OV 4	1954 Mercedes 300SL	1987 Ferrari F40	2000 Volvo V70
5.	1904 Spyker 32 40	1930 Alfa Romeo 6C	1955 BMW Isetta	1987 Ford Bronco	2002 Mini Cooper
6.	1905 Peugeot Type 69 Bébé	1932 Fiat 508 Balilla	1955 Messerschmitt KR200	1990 Mazda MX-5	2005 Toyota Aygo
7.	1905 Renault Taxi de la Marne	1932 Mercedes 170	1957 Chevrolet Corvette C1	1992 Hummer H1	2005 Volkswagen Phaeton
8.	1906 Chenard & Walcker 14 16 CV	1933 BMW 303	1957 Lotus seven	1992 Jaguar XJ220	2007 Nissan GT-R R35
9.	1906 Krieger Landaulet	1934 Citroën Traction Avant	1959 Citroën DS	1992 McLaren F1	2008 Maserati Quattroporte
10.	1907 Detroit Electric	1935 Opel Olympia	1961 Renault R4	1992 Renault Twingo	2010 Nissan Leaf
11.	1908 Ford Model T	1936 Peugeot 302	1963 Porche 356C	1994 Audi A4	2011 Infiniti M35h
12.	1911 Audi C14 35	1938 Morris Eight	1964 Trabant P601	1995 Range Rover Vogue	2011 Opel Ampera
13.	1911 Lancia Epsilon Torpedo	1938 Renault Juvaquatre	1968 Ford Escort Mk1	1996 Ford Ka	2012 Fiat 500 Twinair
14.	1916 Mercedes Benz Knight 16 45 PS	1941 Willys Overland Jeep	1971 Volkswagen Käfer 1302	1996 General Motors EV1	2012 Renault Twizy
15	1922 Austin Seven	1942 Volkswagen Typ 166 Schwimmwagen	1974 Volkswagen Golf Mk1		2012 Tesla Model S

Table 16.3 The super average archetype, constructed on the basis of 74 archetypes (15 for each of the five periods) and representing *the* car of the 20th century.			
Specs	Archetype	Archetype 2.0	Average Archetype
General			
Brand	—	—	Renault
Model	—	—	R4
Current List Price	€ 134,639.7		€ 7,758
Start of Production	1958.4	1943.9	1961
End of Production	1967.2	1966.2	1993
Years of Production	9.9	22.3	32
Number of Production	891,092.1	891,092.1	8,000,000
Dimension			
Length	3851.6	3627.5	3670
Width	1665.8	1467.3	1490
Height	1565.3	980.5	1550
Weight	1083.9	681.4	670
Number of Doors	3.1	3.5	5
Number of Seats	4.0	4.2	4
Performance			
Power (kW)	83.2	24.7	19.5
Torque	214.5	86.0	59
Top Speed (km/h)	144.5	110.3	120
Acceleration 0-100 km/h(s)	15.3	9.9	22.4
Consumption (l/100km)	10.0	9.4	7
Number of Cylinders	4.3	3.6	4
Engine Displacement (l)	2.0	1.6	0.75
Number of Gears	3.9	3.4	4

The Pluto Effect, One Last Time

These results again (after what we said in Chapter 14) confirm our initial periodization, which was based on a *cultural* analysis of automobilism, but they also suggest that in terms of automotive technology, the energy crises of the 1970s are not such water sheds as an analysis of automotive culture suggests.

The results also show that Abernathy and Utterback formulated their conclusions of a succession of product and process innovations on the basis of a too limited collection of empirical material. Instead, the work of later innovation students such as Koen Frenken and Alessandro Nuvolari, especially their emphasis on an ever increasing variation by the founder of this later school, Pier Paolo Saviotti (see Chapter 14), cannot be confirmed by our results (Saviotti claimed that innovation diversity kept increasing over time) [16-6].

What perhaps hindered these previous researchers to see the true evolutionary, incrementally meandering changes in the course of the last automotive century is the fundamentally multilayered and hierarchical character of the car as an artifact: while the evolution of one element in this complex system of subsystems and component assemblies is interrupted for a while, at other levels innovative acceleration can be observed. By focusing on the Model T with its monolithic characteristics (as Abernathy and Utterback did) or by analyzing the artifact instead of its constantly changing internal structure (as many other innovation students do), the richness (and the ambiguity!) of evolutionary developments remains out of focus. It thus will require quite an amount of additional studies before we can truly start to understand the evolution of complex artifacts like the automobile. If our journey into one-and-a-quarter century of the automobile has taught us anything, it is the crucial importance of this fundamental hierarchical *layeredness of automotive technology*, the very basis of our evolutionary approach.

Perhaps these students of automotive technology evolution have underestimated the role of incremental change. Repeatedly in the previous chapters, we have emphasized the importance of the sub-artifact level and the tiny changes occurring at the micro-level. It seems as if at this level the real competition between alternatives takes place, and not at the level of the artifact and its hoped-for substitution. True substitution (instead of coexistence) seems to occur more often at these sub-levels than at the level of the entire artifact. At the sub-artifact level, alternative options are more easily adopted with the result that the artifact as a whole is capable of defending itself against the onslaught of alternatives. We have called this remarkable phenomenon, privileging the existing technology, the Pluto Effect, and we have shown that its opposite, the Sailing Ship Effect is an exception rather than a rule: only in very rare cases (which seem to occur more the deeper one descends in the artifact's structural hierarchy) true substitution occurs because the "old" technology cannot cope anymore with the requirements of the new functions. The Pluto Effect is only one of the several modalities of multimode interaction such as competition, symbiosis and predator-prey interaction ([16-7], p. 68), but it is the most underestimated one, probably because its working can only be understood by acknowledging the fundamental layeredness of technology.

In the previous chapters we have seen the extraordinary resilience of the existing automotive technology. If history can teach us anything it might be that there is no reason why this would be different in the future. The question, then, is whether the car's alternatives (such as the electric propulsion) will be strong enough to provoke a Sailing Ship Effect rather than a Pluto Effect. In other words, the electric vehicle has for a century played the role of a Car of Tomorrow; will it continue to play this role in the future, or will it become the Car of Today, in the future?

References

16-1. Siva Vaidhyanathan, "Introduction; Rewiring the 'Nation': The Place of Technology in American Studies," *American Quarterly* 58 No. 3 (September 2006) 555-567.

16-2. Andreas Spahn, "Moralizing Mobility? Persuasive Technologies and the Ethics of Mobility," *Transfers; Interdisciplinary Journal of Mobility Studies* 3 No. 2 (Summer 2013), 108–115.

16-3. Richard Bishop, *Intelligent Vehicle Technology and Trends* (Boston/London: Artech House, 2005).

16-4. Marvin J. Cetron, "Technology Transfer: Where We Stand Today (Chairman's Introduction)," in Harold F. Davidson, Marvin J. Cetron, and Joe D. Goldhar (eds.), *Technology Transfer* (Leiden: Noordhoff, 1974) (NATO Advanced Study Institute Series, Series E: Applied Science, Vol. 6 Technology Transfer), 3–28.

16-5. Peer Vries, "Decline of the West—Rise of the Rest?" *Journal of Modern European History* 11 (2013), 315–328.

16-6. Koen Frenken, *Innovation, Evolution and Complexity Theory* (Cheltenham/ Northampton, MA: Edward Elgar, n.y. [2005]).

16-7. C.W.I. Pistorius and J.M. Utterback, "Multi-Mode Interaction among Technologies," *Research Policy* 26 (1997), 67–84.

Bibliography

"10 S.A.E. Activities–Where They Started and How They Grew," *Journal of the Society of Automotive Engineers* 37 No. 1 (July 1935) 18–27.

75 Jahre Bosch; 1886–1961, ein geschichtliche Rückblick (Stuttgart: Robert Bosch GmbH, 1961).

"80% of Cars Reported Below Peak Efficiency," *Journal of the Society of Automotive Engineers* (January 1939) 21.

William J. Abernathy, *The Productivity Dilemma; Roadblock to Innovation in the Automobile Industry* (Baltimore/London: The Johns Hopkins University Press, 1978).

Dietmar Abt, *Die Erklärung der Technikgenese des Elektrouulomobils* (Frankfurt am Main/ Berlin/Bern/New York/Paris/Vienna, 1998).

W. Poynter Adams, *Motor-car Mechanism and Management; Part I: The Petrol Car* (London: Charles Griffin, 1907²).

Alan Alsthuler, Martin Anderson, Daniel Jones, Daniel Roos, James Womack, *The Future of the Automobile; The Report of MIT's International Automobile Program* (London/Sydney: George Allen & Unwin Ltd., 1984).

Horst Albach, *Culture and Technical Innovation; A Cross-Cultural Analysis and Policy Recommendations* (Berlin/New York: Walter de Gruyter, 1994) (Akademie der Wissenschaften zu Berlin, Research Report 9).

Daniel M. Albert, "Primitive Drivers: Racial Science and Citizenship in the Motor Age," *Science as Culture* 10 No. 3 (September 2001) 327–351.

Oliver E. Allen, "Kettering," *Invention & Technology* (Fall 1996) 52–63.

M. Arkenbosch, G. Mom and J. Nieuwland, *Het rijdend gedeelte; Band A: historie, theorie, banden en wielen, besturing* (Deventer, 1989) (part 4A of *De nieuwe Steinbuch; de automobiel; handboek voor autobezitters, monteurs en technici onder redactie en coördinate van drs.ing. G.P.A. Mom*).

M. Arkenbosch, G. Mom and J. Nieuwland, *Het rijdend gedeelte; Band B: veersysteem, wiel-geleiding, remsysteem, diagnose en uitlijnen* (Deventer, 1989) (part 4B of *De nieuwe Steinbuch;*

de automobiel; handboek voor autobezitters, monteurs en technici onder redactie en coördinatie van drs.ing. G.P.A. Mom).

"Automatische transmissie schakelt op," *Go! Mobility* 2 No. 2 (April 2012).

Robert U. Ayres and Ike Ezekoye, "Competition and Complementarity in Diffusion; The Case of Octane," *Technological Forecasting and Social Change* 39 (1991) 145–158.

Richard M. Bach, "Design and Style as Selling Factors," *Journal of the Society of Automotive Engineers* 22 No. 4 (April 1919) 468–474.

Sjoerd Bakker, *Competing Expectations; The case of the hydrogen car* (Oisterwijk: BoxPress, 2011).

Jean-Pierre Bardou, Jean-Jacques Chanaron, Patrick Fridenson and James M. Laux, *The Automobile Revolution; The Impact of an Industry* (Chapel Hill, 1982).

George Basalla, *The Evolution of Technology* (Cambridge: Cambridge University Press, 1989).

David Beasley, *Who Really Invented the Automobile; Skulduggery at the Crossroads* (Simcoe, ON: Davus Publishing, 1997).

W. Worby Beaumont, *Motor Vehicles and Motors; Their Design, Construction and Working by Steam, Oil and Electricity* (Westminster/Philadelphia: Archibald Constable & Company/J.B. Lippincott Company, 1906).

Ulrich Beck, "Risk Society and the Provident State," in Scott Lash, Bronislaw Szerszynski and Brian Wynne (eds.), *Risk, Environment and Modernity; Towards a New Ecology* (London/Thousand Oaks/New Delhi: Sage, 1996) 27–43.

Jörg Beckmann, *Risky Mobility; The Filtering of Automobility's Unintended Consequences* (Copenhagen: Copenhagen University, Sociological Institute: 2001).

David Beecroft, "Conditions in the Automotive Industry Abroad," *Journal of the Society of Automotive Engineers,* 4 No. 6 (June 1919) 521–525.

W. Beel, "De automatische gangwissel historisch gezien," *Auto- en Motortechniek* (1970) 514–526.

Max Bentele, "Learning from History: Fundamentals," in G.P.A. Mom, J.W. Möhlmann, J.C. Vorsterman van Oijen, H.J. Weegenaar and C. Wiers-Latooij, *Yearbook Autotechnical Trends 1993/Jaarboek Autotechnische Trends 1993* (Apeldoorn: HTS-Autotechniek, 1992) 4.4.1–4.4.18.

Michael L. Berger, *The Automobile in American History and Culture; A Reference Guide* (Westport, Connecticut/London: Greenwood Press, 2001).

Peter L. Bernstein, *Against the Gods; The Remarkable Story of Risk* (New York/Chichester/Brisbane/Toronto/Singapore: John Wiley & Sons, Inc., 1996).

Hal Bernton, William Kovarik, and Scott Sklar, *The Forbidden Fuel; Power Alcohol in The Twentieth Century* (New York: Boyd Griffin, 1982).

C. Besant, K.R. Pullen, M.R.S. Etemad, A. Fenocchi, M. Ristic, N.C. Baines, and W. Dunford, "Hybrid Traction—A Solution for a Not Too Distant Future," in G.P.A. Mom, J.W. Möhlmann, J.C. Vorsterman van Oijen, H.J. Weegenaar and C. Wiers-Latooij, *Yearbook Autotechnical Trends 1993/Jaarboek Autotechnische Trends 1993* (Apeldoorn: HTS-Autotechniek, 1992) 3.3.1–3.3.13.

Krish Bhaskar, *Innovation in the EC automotive industry—An Analysis from the Perspective of State Aid Policy* (Brussels/Luxembourg: Commission of the European Communities, April 1988).

Karin Bijsterveld, "Acoustic Cocooning; How the Car became a Place to Unwind," *Senses & Society* 5 (2010) No. 2, 189–211.

Karin Bijsterveld and Stefan Krebs, "Listening to the Sounding Objects of the Past: The Case of the Car," in Karmen Franinovic and Stefania Serafin (eds.), *Sonic Interaction Design* (Cambridge, MA/London: MIT Press, 2013) 3–38.

Karin Bijsterveld, *Mechanical Sound; Technology, Culture, and Public Problems of Noise in the Twentieth Century* (Cambridge. MA/London: The MIT Press, 2008).

Karin Bijsterveld, Eefje Cleophas, Stefan Krebs, and Gijs Mom, *Sound and Safe: A History of Listening Behind the Wheel* (Oxford/New York, etc: Oxford University Press, 2014).

Holger Bingmann, "Chapter 14: Competence; Case C: Antiblockiersystem und Benzineinspritzung (Anti-Blocking System and Fuel Injection)," in Horst Albach, *Culture and Technical Innovation; A Cross-Cultural Analysis and Policy Recommendations* (Berlin/New York: Walter de Gruyter, 1994) (Akademie der Wissenschaften zu Berlin, Research Report 9) 736–821.

Richard Bishop, *Intelligent Vehicle Technology and Trends* (Boston/London: Artech House, 2005).

E. Blaich e.a., *Internationales Automobil-Handbuch; Umfassendes Lehr- und Nachschlagewerk für alle Gebiete der Kraftfahrt* (Lugano: J. Kramer, 1954).

Herbert Blankesteijn, "Bewust bumperkleven; Auto's vormen trein op snelweg," *De Ingenieur* (30 November 2012) 19–20.

Günter Böcker, *Auf die Mischung kommt es an; Technik für die Mobilität: Erfinden–Entwicklen–Verwirklichen* (Meerbusch/Neuss: Lippert-Druck & Verlag/Pierburg, 1990).

Bosch und die Zündung (n.p., n.y. [Stuttgart: Robert Bosch GmbH, 1952]).

H. Bouvy, "De toekomstige onvermijdelijkheid van automatische transmissies" [The future inevitability of automatic transmissions], *Auto- en Motortechniek* (1969) 49–53.

H. Bouvy, "Over automatische transmissies en hun actualiteit" [On the timeliness of automatic transmissions], *Auto- en Motortechniek* (1974) 534–543.

T.A. Boyd, "The Self-Starter," *Technology and Culture* 9 (1968) 585–591.

W.F. Bradley, "Automobilism in America; How European Influence in Engine Design Is affecting American Designers," *Autocar* (31 July 1920) 180–182.

J.W. Brand, *De automobiel en haar* [sic] *behandeling* (Rotterdam: Nijgh & Van Ditmar, 1911[4]).

J.W. Brand, *De automobiel en haar* [sic] *behandeling* (Rotterdam: Nijgh & Van Ditmar, 1921[12]).

J.W. Brand, *De automobiel en haar behandeling* (Rotterdam: Nijgh & Van Ditmar, 1922[13]).

J.W. Brand, *De automobiel en haar behandeling* (Rotterdam: Nijgh & Van Ditmar, 1932[19]).

J.W. Brand, *De automobiel en zijn behandeling* (Rotterdam: Nijgh & Van Ditmar, 1950[25]).

Ernest Braun and Stuart MacDonald, *Revolution in Miniature; The History and Impact of Semiconductor Electronics Re-explored in an Updated and Revised Second Edition* (Cambridge/London/New York/New Rochelle/Melbourne/Sydney: Cambridge University Press, 1982[2]).

Peter Brimblecombe, "Arie Jan Haagen-Smit and the History of Smog," *Environmental Chemistry Group Bulletin* (January 2012) 15–17.

Joël Broustaille, "L' éternel retour de l'automatisme," *Culture Technique* No. 25 (1992) 49–50.

Mark B. Brown, "The Civic Shaping of Technology: California's Electric Vehicle Program," *Science, Technology, & Human Values* 26 No. 1 (Winter 2001) 56–81.

Hermann A. Brunn, "Body Comfort and Interior Appointments," *Journal of the Society of Automotive Engineers* 30 No. 1 (January 1932) 21–22.

Hermann A. Brunn, "Trends in Body Design," *Journal of the Society of Automotive Engineers* 22 No. 6 (June 1928) 679–683.

Lynwood Bryant, "Rudolf Diesel and His Rational Engine," *Scientific American* 221 (August 1969) 108–117.

Lynwood Bryant, "The Development of the Diesel Engine," *Technology and Culture* 17 No. 3 (July 1976) 432–446.

Lynwood Bryant, "The Role of Thermodynamics in the Evolution of Heat Engines," *Technology and Culture* 14 No. 2 (April 1973) 152–165.

Earle Buckingham, "Transmission Noise and Their Remedies," *Journal of the Society of Automotive Engineers* 17 No. 5 (November 1925) 460–462.

O.M. Burkhardt, "Wheel Shimmying; Its Causes and Cure," *Journal of the Society of Automotive Engineers*, 16 No. 2 (February 1925) 189–191.

David Burton, Amanda Delaney, Stuart Newstead, David Logan, and Brian Fildes, *Effectiveness of ABS and Vehicle Stability Control Systems* (Noble Park Noth, VI: Royal Automobile Club of Victoria (RACV), April 2004).

Colin Campbell, *Automobile Suspensions* (London: Chapman & Hall, 1981).

W. Bernard Carlson, "Invention and evolution: The Case of Edison's Sketches of the Telephone," in John Ziman (ed.), *Technological Innovation as an Evolutionary Process* (Cambridge: Cambridge University Press, 2000) 137–158.

J.R. Cautley and A.Y. Dodge, "Development of a Modern Four-Wheel Mechanical Braking-System," *Journal of the Society of Automotive Engineers*, 17 No. 1 (July 1925) 87–90.

Philip Chacko, Christabelle Noronha and Sujata Agrawal, *Small Wonder; The Making of the Nano* (Chennai/Mumbai/Bangalore/Hyderabad/New Delhi: westland, 2010).

R.E. Chamberlain, "This Body Business," *Journal of the Society of Automotive Engineers* 25 No. 2 (August 1929) 107–109.

Robin Cowan and Staffan Hultén, "Escaping Lock-In: The Case of the Electric Vehicle," *Technological Forecasting and Social Change* 53 (1996) 61–79.

F.F. Chandler, "Steering-Gear Analyses," *Journal of the Society of Automotive Engineers* (June 1924) 585–590.

Herbert Chase, "A Study of Modern Automotive-Vehicle Steering-Systems," *Journal of the Society of Automotive Engineers* (April 1923) 377–397.

Herbert Chase, "Practice and Theory in Clutch Design," *Journal of the Society of Automotive Engineers* (July 1921) 39–52.

Belinda Chen and Dan Sperling, "Analysis of Auto Industry and Consumer Response to Regulations and Technological Change, and Customization of Consumer Response Models in Support of AB 1493 Rulemaking; Case Study of Light-Duty Vehicles in Europe" (David, CA: Institute of Transportation Studies, June 2004) (report UCD-ITS-RR-04-14) (downloaded from http://www.its.ucdavis/edu on 9 December 2012).

Sally H. Clarke, *Trust and Power; Consumers, the Modern Corporation, and the Making of the United States Automobile Market* (Cambridge: Cambridge University Press, 2007).

M.R. Clements, *King of Stop and Go—The story of Bendix; A history, 1919–1963 in South Bend, Indiana* (n.p., n.y. [South Bend: Bendix Aviation Corporation, 1963]).

H.M. Crane, "How Versatile Engineering Meets Public Demand," *Journal of the Society of Automotive Engineers* 41 No. 2 (August 1937) 358–392.

Henry M. Crane, "The Car of the Future," *Journal of the Society of Automotive Engineers (Transactions)* 44 No. 4 (April 1939) 141–144.

C. Lyle Cummins, *Diesel's Engine; Volume One: From Conception to 1918* (Wilsonville, OR: Carnot Press, 1993).

C. Lyle Cummins, *Internal Fire* (Lake Oswego, Oregon: Carnot Press, 1976).

Tim Dant and Peter J. Martin, "By car: Carrying Modern Society," in Jukka Gronow and Alan Warde (eds.), *Ordinary Consumption* (London/New York: Routledge, 2001) 143–157.

Burgess Darrow, "Pneumatic Tires—Old and New," *Journal of the Society of Automotive Engineers (Transactions)*, 30 No. 5 (November 1932) 438–444.

Vidyadhar Date, *Traffic in the Era of Climate Change; Walking, Cycling, Public Transport Need Priority* (Delhi: Kalpaz Publications, 2010).

Daniel Dexter, "Case Study of the Innovation Process Characterizing the Development of the Three-Way Catalytic Converter System," (Final Report, prepared for the U.S. Department of Transportation National Highway Traffic Safety Administration, Office of Research and Development, November 1979) (Report No. DOT-TSC-NHTSA-79-36; HS-804-791).

Eugen Diesel, "Rudolf Diesel," in Eugen Diesel, Gustav Goldback and Friedrich Schildberger, *Vom Motor zum Auto; Fünf Männer und ihr Werk* (Stuttgart: Deutsche Verlags-Anstalt, 1958²) 205–255.

Eugen Diesel, *Diesel; Der Mensch, das Werk, das Schicksal* (Hamburg: Hanseatische Verlagsanstalt, n.y. [1939?]).

Eugen Diesel, *Die Geschichte des Diesel-Personenwagens* (Stuttgart: Deutsche Verlags-Anstalt, 1955).

Eugen Diesel, Gustav Goldback and Friedrich Schildberger, *Vom Motor zum Auto; Fünf Männer und ihr Werk* (Stuttgart: Deutsche Verlags-Anstalt, 1958²).

Eugen Diesel, "Robert Bosch," in Eugen Diesel, Gustav Goldback and Friedrich Schildberger, *Vom Motor zum Auto; Fünf Männer und ihr Werk* (Stuttgart: Deutsche Verlags-Anstalt, 1958²) 257–308.

J.H. Dodge, "Tripot Universal Joint (End Motion Type)," in *Universal Joint and Driveshaft Design Manual* (Warrendale, Pa: SAE, 1979) (Advances in Engineering Series No. 7) 131–140.

Erik Eckermann, "Fritz Ostwald—Protagonist des negativen Lenkrollradius; Sein Demonstrationsmodell: ein Technik-Füllhorn," *Automobil Revue* (1 August 1996) 19, 21.

Erik Eckermann, *Nathan S. Stern, Ingenieur aus der Frühzeit des Automobils* (Düsseldorf: VDI Verlag1985).

Erik Eckermann, "Nicht mehr wegzudenken; Zur Entwicklung der Benzineinspritzung," *Automobil Revue* (1993 No. 14).

Ross D. Eckert and George W. Hilton, "The Jitneys," *Journal of Law and Economics* 25 (October 1972) 293–325.

David Edgerton, *The Shock of the Old; Technology and Global History since 1990* (Oxford: Oxford University Press, 2007).

William Phelps Eno, *The Story of Highway Traffic Control 1899–1939* (n.p.: The Eno Foundation for Highway Traffic Control, Inc., 1939).

C. Cindy Fan, *China on the Move; Migration, the State, and the Household* (London/New York: Routledge, 2008).

Thomas L. Fawick, "Two Desirable Quiet Driving-Ranges for Automobiles," *Journal of the Society of Automotive Engineers* 21 No. 1 (July 1927) 99–106.

Alan R. Penn, "The English Light-Car and Why," *Journal of the Society of Automotive Engineers* 20 No. 4 (April 1927) 483–488.

Olaf von Fersen (ed.), *Ein Jahrhundert Automobiltechnik; Personenwagen* (Düsseldorf: VDI Verlag, 1986).

Olaf von Fersen, "'Negativer Lenkrollradius'; Ein geometrischer Trick entpuppt sich als wichtiger Sicherheitsfaktor," *Automobil Revue* (23 March 1973) 25, 27.

René Filderman, "Commentaires sur le phénomène du rejet de la boîte automatique en Europe," *Culture technique* No. 25 (1992) 49–50.

Kingston Forbes, "The Body Engineer and the Automotive Industry," *Journal of the Society of Automotive Engineers* 8 No. 5 (May 1921) 436–439.

"Four-wheel Brakes," *Journal of the Society of Automotive Engineers*, 13 No. 1 (July 1923) 70–72.

Kathleen Franz, *Tinkering; Consumers Reinvent the Early Automobile* (Philadelphia: University of Pennsylvania Press, 2005).

Tore Franzen, "Suspension Types Will Be Developed for Each Country," *Journal of the Society of Automotive Engineers (Transactions)* 33 No. 4 (October 1933) 347.

J.W. Frazer, "Bodies Considered from the Car Buyer's Viewpoint," *Journal of the Society of Automotive Engineers* 31 No. 1 (July 1932) 294, 299.

Koen Frenken, *Innovation, Evolution and Complexity Theory* (Cheltenham/Northampton, MA: Edward Elgar, n.y. [2005]).

Vincent Frigant and Damien Talbot, "Technological Determinism and Modularity: Lessons from a Comparison between Aircraft and Auto Industries in Europe," *Industry and Innovation* 12 No. 3 (September 2005) 337–355.

"Fuel Research Summarized," *Journal of the Society of Automotive Engineers* (February 1927) 193–195.

Kelly Sims Gallagher, *China Shifts Gears; Automakers, Oil, Pollution, and Development* (Cambridge, MA/London: The MIT Press, 2006).

Rob Gallagher, *The Rickshaws of Bangladesh* (Dhaka: The University Press, 1992).

Amy Gangloff, "Safety in Accidents; Hugh DeHaven and the Development of Crash Injury Studies," *Technology and Culture* 54 No. 1 (January 2013) 40–61.

David Gartman, *Auto Opium; A Social History of American Automobile Design* (New York/London: Routledge, 1994).

Giancarlo Genta and Lorenzo Morello, *The Automotive Chassis; Volume 1: Components Design* (n.p.: Springer Science+Business Media, 2009).

David Gerard and Lester B. Lave, "Implementing Technology-Forcing Policies: The 1970 Clean Air Act Amendments and the Introduction of Advanced Automotive Emissions Controls in the United States," *Technological Forecasting and Social Change* 72 (2005) 761–778.

Daan Gerrits, Harm Gijselhart, Tsvetan Balyovski, Johan van Uden and Jan van der Vleuten, "Anti-Blocking System," (student report for the course "Cars in Context," Eindhoven University of Technology, 2012).

Connie J.G. Gersick, "Revolutionary Change Theories: A Multilevel Exploration of the Punctuated Equilibrium Paradign," *Academy of Management Review* 16 (1991) Nr. 1, 10–36.

Heide Gjøen and Michael Hård, "Cultural Politics in Action: Developing User Scripts in Relation to the Electric Vehicle," *Science, Technology & Human Values* 27 No. 2 (Spring 2002) 262–281.

Glaenzer Spicer 1838–1988 (brochure Glaenzer Spicer, Poissy, n.y. [1988]).

H.L. Gleist, *Borg-Warner: The First 50 years* (Chicago: Borg-Warner Corporation, 1978).

E.W. Goodwin, "Automobile Body Design and Construction," *Journal of the Society of Automotive Engineers,* 2 No. 4 (April 1918) 271–278.

Philip G. Gott, *Changing Gears: The Development of the Automotive Transmission* (Warrendale: Society of Automotive Engineers, 1991).

Jonathan Gottschall en David Sloan Wilson (red.). *The Literary Animal; Evolution and the Nature of Narrative* (Evanston, IL: Northwestern University Press, 2005).

Jean-Albert Grégoire, *50 ans d'automobile; la traction avant* (Paris: Flammarion, 1974).

Jean-Albert Grégoire, *50 ans d'automobile; 2 la voiture* électrique (Paris: Flammarion, 1981).

Jean-Albert Grégoire, *Toutes mes automobiles; Texte présenté et annoté par Daniel Tard et Marc-Antoine Colin* (Paris: Ch. Massin, 1993).

Arnulf Grübler, *Technology and Global Change* (Cambridge/New York/Melbourne: Cambridge University Press, 1998).

Shane Gunster, "'You Belong Outside'; Advertising, Nature, and the SUV," *Ethics & the Environment* 9 (2004) No. 2, 4–32.

M. Hård and A. Jamison, "Alternative cars: The Contrasting Stories of Steam and Diesel Automotive Engines," *Technology in Society*, 19 no. 2 (1997) 145–160.

Michael Hård, *Machines Are Frozen Spirit; The Scientification of Refrigeration and Brewing in the 19th Century— A Weberian Interpretation* (Frankfurt am Main/Boulder, Col., 1994).

Horst Hardenberg, *Schiesspulvermotoren; Materialien zu ihrer Geschichte* (Dusseldorf: VDI Verlag, 1992).

Horst O. Hardenberg, *The Middle Ages of the Internal-Combustion Engine 1794–1886* (Warrendale, Pa.: Society of Automotive Engineers, 1999).

Charles K. Harley, "The shift from sailing ships to steamships, 1850–1890: A study in technological change and its diffusion," in Donald N. McCloskey (ed.), *Essays on a Mature Economy: Britain after 1840 (Papers and Proceedings of the Mathematical Social Science Board Conference on the New Economic History of Britain, 1840–1930, held at Eliot House, Harvard University, 1–3 September 1970)* (London: Methuen & Co Ltd., 1971) 215–237.

George R. Heaton and James Maxwell, "Patterns of Automobile Regulation: an International Comparison," *Zeitschrift für Umweltpolitik & Umwelrecht* 7 (1984) No. 1, 15–40.

Sander Heijne, "Autorijden, echt iets voor 40-plussers," *De Volkskrant* (24 November 2012) 24–25.

Gaston Heimeriks, Floortje Alkemade, Antoine Schoen, Lionel Villard and Patricia Laurens, "The Evolution of Technological Knowledge: A co-evolutionary analysis of patterns of Corporate Invention," (paper presented at the ECIS seminar, Eindhoven University of Technology, 8 November 2012).

P.M. Heldt, "Some Recent Work on Unconventional Transmissions," *Journal of the Society of Automotive Engineers* (July 1925) 127–141.

P.M. Heldt, "What is a European Light Car?" *Automotive Industries* 55 No. 7 (12 August 1926) 252–254.

"Helft auto's in 2010 voorzien van automaat," *rai voorrang* 5 No. 8 (14 November 2001), 1.

Arnold Heller, *Motorwagen und Fahrzeugmaschinen für flüssigen Brennstoff; Ein Lehrbuch für den Selbstunterricht und für den Unterricht an technischen Lehranstalten aus dem Jahre 1912* (Moers: Steiger Verlag, 1985) (reprint of 1922 edition, first ed.: 1912).

K.L. Herrmann, "Some Causes of Gear-Tooth Errors and Their Detection," *Journal of the Society of Automotive Engineers* 11 No. 5 (November 1922) 391–397.

Theodor Heuss, *Robert Bosch; Leben und Leistung* (München: Wilhelm Heine Verlag, 1981²).

Eric Higgs, "Revisiting Determinism," *Research in Philosophy and Technology* 16 (1997) 189–193.

J.B. Hill and T.G. Delbridge, "Seek Cracking Process for Production of Non-Detonating Fuel," *Journal of the Society of Automotive Engineers* (March 1926) 271–274.

Remigius Johannes Franciscus Hoogma, *Exploiting Technological Niches; Strategies for Experimental Introduction of Electric Vehicles* (Enschede: Twente University Press, 2000).

"How Fiat Designed Three Lightweight ESV's," *Automotive Engineering* 82 No. 2 (Feburary 1974) 46–47.

John Howells, "The Response of Old Technology Incumbents to Technological Competition–Does the Sailing Ship Effect Exist?" *Journal of Management Studies* 39 No. 7 (November 2002) 887–906.

Christian Huck and Stefan Bauernschmidt (eds.), *Travelling Goods, Travelling Moods; Varieties of Cultural Appropriation (1850–1950)*(Frankfurt am Main: Campus, 2012) 189–207.

Thomas P. Hughes, *Networks of Power; Electrification in Western society, 1880–1930* (Baltimore/London, 1983).

Peter J. Hugill, "Technology Diffusion in the World Automobile Industry, 1885–1985," in Peter J. Hugill and D. Bruce Dickson (eds.), *The Transfer and Transformation of Ideas and Material Culture* (College Station: Texas A&M University Press, 1988) 110–142.

Heinrich Huinink, "Die Entwicklung des Reifens; Der Reifen als Bindeglied zwischen Fahrbahn und Kraftfahrzeug," in Harry Niemann and Armin Hermann (eds.), *Geschichte der Strassenverkehrssicherheit im Wechselspiel zwischen Fahrzeug, Fahrbahn und Mensch* (Bielefeld: Delius & Klasing, 1999) (DaimlerChrysler Wissenschaftliche Schriftenreihe, Band 1) 169–183.

C.L. Humphrey, "Noise and Heat Control in the Automobile Body," *Journal of the Society of Automotive Engineers* 30 No. 5 (May 1932) 208–210.

Seth K. Humphrey, "Our Delightful Man-Killer," *Atlantic Monthly* 148 (1931) 724–730.

Lee Iacocca, *Iacocca: An Autobiography* (New York: Bantam, 1984).

R.K. Jack, "The Constantinesco Torque-Converter," *Journal of the Society of Automotive Engineers* (October 1927) 413–423.

Andrew Jamison, *The Steam-Powered Automobile; An Answer to Air Pollution* (Bloomington/London: Indiana University Press, 1970).

Wiel Janssen, Marcel Wierda and Richard van der Horst, "Automation and the Future of Driver Behavior," *Safety Science* 19 (1995) 237–244.

Thomas Janssoone, "The History of Navigation Systems" (student report, course History of Innovations, Eindhoven University of Technology, September 2009–January 2010).

Hans Jeekel, *The Car Dependent Society: A European Perspective* (Aldershot: Ashgate Publishers, 2013).

Thomas B. Johansson, Nebojsa Nakicenovic, Anand Patwardhan and Luis Gomes-Echeverri (eds.), *Global Energy Assessment (GEA)* (Vienna: International Institute for Applied Systems Analysys, 2012) (downloaded from www.iiasa.ac.at on 26 October 2012).

Ann Johnson, *Hitting the Brakes; Engineering Design and the Production of Knowledge* (Durham/London: Duke University Press, 2009).

Kevin Jost, "Mapping the Road to 54.5 mpg," *Automotive Engineering online* (16 October 2012) (www.sac.org/mags/aei/11461, retrieved on 1 April 2013).

Arthur W. Judge, *Modern Motor Cars; Their Construction, Maintenance, Management, Care, Driving, and Running Repairs, with Special Section on Cycle Cars. Commercial Cars, and Motor Cycles* (London: Clun House, n.y. [1923?]) 3 volumes.

Arthur W. Judge, *Modern Motor Cars; Their Construction, Maintenance, Management, Care, Driving, and Running Repairs, with Special Section on Cycle Cars. Commercial Cars, and Motor Cycles* (London: Clun House, n.y. [1927?]) (new and revised edition) 4 volumes.

Christian Kehrt, *Zwischen Evolution und Revolution; Der Werkstoffwandel im Flugzeugbau* (Karlsruhe: KIT Scientific Publishing, 2013).

René Kemp, Frank W. Geels and Geoff Dudley, "Introduction; Sustainable Transitions in the Automobility Regime and the Need for a New Perspective," in Frank W. Geels, René Kemp, Geoff Dudley and Glenn Lyons (eds.), *Automobility in Transition? A Socio-Technical Analysis of Sustainable Transport* (New York/London: Routledge, 2012) 3–28.

C.F. Kettering, "Cooperation of the Automotive and Oil Industries," *Journal of the Society of Automotive Engineers* (January 1921) 43–45.

Charles F. Kettering, "More Efficient Utilization of Fuel," *Journal of the Society of Automotive Engineers* (April 1919) 263–269.

Charlotte Kim, *Cats and Mice: The Politics of Setting EC Car Emission Standards* (Brussels: Centre for European Policy Studies, May 1992) (CEPS Working Document no. 64; CEPS Standards Programme: Paper no. 2).

David A. Kirsch, *The Electric Vehicle and the Burden of History* (New Brunswick,New Jersey/London: Rutgers University Press, 2000).

Billy Vaughn Koen, "The Engineering Method," in Paul T. Durbin (ed.), *Critical Perspectives on Nonacademic Science and Engineering* (Bethlehem–Lehigh University Press/ London and Toronto: Associated University Press, 1991) 33–59.

Wolfgang König, "Adolf Hitler vs. Henry Ford: The *Volkswagen*, the Role of America as a Model, and the Failure of a Nazi Consumer Society," *German Studies Review* 27 No. 2 (May 2004) 249–268.

William Kovarik, "Special Motives: Automotive Inventors and Alternative Fuels in the 1920s" (paper to the Society for the History of Technology, October 18, 2007).

M.A. Krebs, "Sur un carburateur automatique pour moteurs à explosions" (Paris: Gauthier-Villars, 1902) (paper presented at the Académie des Sciences, Paris, 24 November 1902).

Philippe Krebs, "Arthur Constantin Krebs (1850–1935); Autorité et stratégie à la direction de Panhard & Levassor (1913–1916)" (Master's thesis, Conservatoire Nationale des Arts et Métiers, 2009).

Philippe Krebs, "Arthur KREBS (1850–1935); Le successeur d'Émile LEVASSOR" (downloaded from http://rbmn.waika9.com on 11 September 2012).

Stefan Krebs, "The French Quest for the Silent Car Body; Technology, Comfort, and Distinction in the Interwar Period," *Tranfers* 1 No. 3 (Winter 2011) 64–89.

Peter Kroes and Anthonie Meijers, "Introduction; The dual nature of technical artefacts," *Studies in History and Philosophy of Science* 37 (2006) 1–4.

Walter Kudlick, "Carros Por Puesto–The 'Jitney' Taxi System of Caracas, Venezuela," *Hihgway Research Record* No. 283 (1969) (paper presented at the *Highway Research Board* annual meeting, January 1968).

K. Kühner, *Geschichtliches zum Fahrzeugantrieb* (Friedrichshafen, 1965).

Kuntala Lahiri-Dutt and David J. Williams, *Moving pictures, Rickshaw Art of Bangladesh* (Ahmedabad, Mapin Publishing, 2010).

W.E. Lay and L.C. Fisher, "Riding Comfort and Cushions," *Journal of the Society of Automotive Engineers* (*Transactions*) 47 No. 5 (November 1940) 482–496.

Edwin T. Layton, Jr., "A Historical Definition of Engineering," in Paul T. Durbin (ed.), *Critical Perspectives on Nonacademic Science and Engineering* (Bethlehem–Lehigh University Press/London and Toronto: Associated University Press, 1991) (Research In Technology Studies, Volume 4) 60–79.

B.J. Lemon, "Judging Super-Balloon Tires," *Journal of the Society of Automotive Engineers (Transactions)*, 31 No. 4 (October 1932) 403–411.

Harro van Lente, *Promising Technology; The Dynamics of Expectations in Technological Developments* (Delft: Eburon, 1993).

Stuart W. Leslie, *Boss Kettering* (New York: Columbia University Press, 1983).

Fabien Leurent and Elisabeth Windisch, "Triggering the Development of Electric Mobility: A Review of Public Policies," *European Transport Research Review* 3 (2011) 221–235.

A.D. Libby, "Advantages of Magneto Ignition," *Journal of the Society of Automotive Engineers* 7 No. 3 (September 1920) 277–290.

Benjamin Liebowitz, "The Measurement of Vehicle Vibrations," *Journal of the Society of Automotive Engineers* (January 1920) 17–25.

S.J. Liebowitz and Stephen E. Margolis, "Path Dependence, Lock-In, and History," *Journal of Law, Economics, & Organization* 11 No. 1 (Spring 1995) 205–226.

Alan P. Loeb, "Birth of the Kettering Doctrine: Fordism, Sloanism and the Discovery of Tetraethyl Lead" (unpublished presentation at the 41st Business History Conference, Fort Lauderdale, 19 March 1995).

Deborah Lupton, "Monsters in Metal Cocoons: 'Road Rage' and Cyborg Studies," *Body & Society* 5 (1999) No. 1, 57–72.

A. Maier, *Zur Geschichte des Getriebebaues* (n.p., n.y. [1962]).

Akshay Mani and Pallavi Pant, *A Synthesis of Findings; Review of Literature in India's Urban Auto-rickshaw Sector* (n.p.: Embarq India, n.d.) (downloaded from www.embarqindia.org on 12 December 2012).

Kanehire Maruo, "The Three-way 'Catalysis': How the Three-Way Catalyst Became the Ruling Technical Solution to the Automobile Emission Problem," in Mikael Hård (ed.), *Automobile Engineering in a Dead End: Mainstream and Alternative Developments in the 20th Century* (Gothenburg: Gothenburh University, 1992) 45–61.

Tom McCarthy, *Auto Mania; Cars, Consumers, and the Environment* (New Haven/London: Yale University Press, 2007).

F.F. Miller, D.W. Holzinger and E.R. Wagner, "Rzeppa Universal Joint," in *Universal Joint and Driveshaft Design Manual* (Warrendale, Pa: SAE, 1979) (Advances in Engineering Series No. 7) 145–150.

William F. Milliken and Douglas L. Milliken, *Chassis Design; Principles and Analysis; Based on previously unpublished technical notes by Maurice Olley* (Warrendale, PA: Society of Automotive Engineers, 2002).

Fung Chi Ming, *Reluctant Heroes; Rickshaw Pullers in Hong Kong and Canton, 1874–1954* (Hong Kong: Hong Kong University Press, 2005).

Carl Mitcham, "Do Artifacts Have Dual Natures? Two Points of Commentary on the Delft Project," *Technè* 6 No. 2 (Winter 2002) 9–12.

Carl Mitcham, "Engineering as Productive Activity: Philosophical Remarks," in Paul T. Durbin (ed.), *Critical Perspectives on Nonacademic Science and Engineering* (Bethlehem–Lehigh University Press/London and Toronto: Associated University Press, 1991) 80–117.

Gijs Mom, *Atlantic Automobilism: The Emergence and Persistence of the Car 1895–1940* (New York and Oxford: Berghahn Books, forthcoming).

Gijs Mom, "De auto-elektronica in historisch perspectief," in G.P.A. Mom and A.G. Visser, *Elektronica in de auto* (Deventer, 1986) (part 9 of *De nieuwe Steinbuch; de automobiel; handboek voor autobezitters, monteurs en technici onder redactie en coördinatie van drs.ing. G.P.A. Mom*) 17–46.

Gijs Mom, "Compétition et coexistence; La motorisation des transports terrestres et le lent processus de substitution de la traction équine", *Le Mouvement Social* No. 229 (October/December 2009) 13–39.

Gijs Mom, "Constructing the State of the Art: Innovation and the Evolution of Automotive Technology (1898–1940)," in Rolf-Jürgen Gleitsmann and Jürgen E. Wittmann (eds.), *Innovationskulturen um das Automobil; Von gestern bis morgen; Stuttgarter Tage zur Automobil- und Unternehmensgeschichte 2011* (Stuttgart: Mercedes-Benz Classic Archive, 2012) (Wissenschaftliche Schriftenreihe der Mercedes-Benz Classic Archive, Band 16) 51–75.

Gijs Mom, "The dual nature of technology: Archeology of Automotive Ignition and the Evolution of the Car," in Christian Huck and Stefan Bauernschmidt (eds.), *Travelling Goods, Travelling Moods; Varieties of Cultural Appropriation (1850–1950)* (Frankfurt am Main: Campus, 2012) 189–207.

Gijs Mom, *The Electric Vehicle; Technology and Expectations in the Automobile Age* (Baltimore: Johns Hopkins University Press, 2004).

Gijs Mom, "De elektrische installatie in historisch perspectief," in H. de Boer, Th. Dobbelaar and G. Mom, *De elektrische installatie* (Deventer, 1986) (part 8 of *De nieuwe*

Steinbuch; de automobiel; handboek voor autobezitters, monteurs en technici onder redactie en coördinate van drs.ing. G.P.A. Mom) 17–51.

Gijs Mom, "Fox Hunt: Materials Selection and Production Problems of the Edison Battery (1900–1910)," in Hans-Joachim Braun and Alexandre Herlea (eds.), *Materials: Research, Development and Applications (Proceedings of the XXth International Congress of History of Science (Liège, 20–26 July 1997)* (Turnhout: Brepols, 2002) 147–154.

Gijs Mom, "Frozen History: Limitations and Possibilities of Quantitative Diffusion Studies," in Ruth Oldenziel and Adri de la Bruhèze (eds.), *Manufacturing Technology: Manufacturing Consumers; The Making of Dutch Consumer Society* (Amsterdam; aksant, 2008) 73–94.

Gijs Mom, "'The future is a shifting panorama': The role of expectations in the history of mobility," in Weert Canzler and Gert Schmidt (eds.), *Zukünfte des Automobils; Aussichten und Grenzen der autotechnischen Globalisierung* (Berlin: edition sigma, 2008) 31–58.

Gijs Mom, *Geschiedenis van de auto van morgen; Cultuur en techniek van de elektrische auto* (Deventer: Kluwer Bedrijfsinformatie BV, 1997).

Gijs Mom, "Inleiding," in J. Kasedorf, *Automobielelektronica; principes en toepassingen* (Deventer: Kluwer Technische Boeken, 1987) 1–7.

Gijs Mom, "Mobility for Pleasure; A Look at the Underside of Dutch Diffusion Curves (1920–1940)," *TST Revista de Historia; Transportes, Servicios y Telecomunicaciones* No. 12 (June 2007) 30–68.

Gijs Mom, "Ongelode benzine: Een anti-klopjacht op wereldschaal," *Auto Service* 2 no. 11 (25 March 1988) 1, 3–4.

Gijs Mom, "Orchestrating Car Technology; Noise, Comfort, and the Construction of the American Closed Automobile, 1917–1940," *Technology and Culture* 55, no. 2 (April 2014) 299–325).

Gijs Mom, personal notes from the course "Automotive Technology"(Voertuigtechniek) at the HTS-Autotechniek, Apeldoorn, The Netherlands, by ing. G. Révèsz, 1978–1979.

Gijs Mom, "De vitale vonk (Techniekhistorie: de elektrische installatie–18)," *Auto en Motor Klassiek* 3 No. 11/12 (November/December 1987) 25–33.

Gijs Mom, "Wie Feuer und Wasser: Der Kampf um den Fahrzeugantrieb bei der deutschen Feuerwehr (1900–1940)," in Harry Niemann and Armin Hermann (eds.), *100 Jahre LKW; Geschichte und Zukunft des Nutzfahrzeuges* (Stuttgart: Franz Steiner Verlag, 1997) (Stuttgarter Tage zur Automobil- und Unternehmensgeschichte; Eine Veranstaltung des Daimler-Benz Archivs, Stuttgart, Bd. 3) 263–320.

Gijs Mom and Ruud Filarski, *Van transport naar mobiliteit; De mobiliteitsexplosie (1895–2005)* (Zutphen: Walburg Pers, 2008).

G. Mom and H. Scheffers, *De aandrijflijn* (Deventer, 1992) (part 3A of *De nieuwe Steinbuch; de automobiel; handboek voor autobezitters, monteurs en technici onder redactie en coördinatie van drs.ing. G.P.A. Mom*).

Gijs Mom and Laurent Tissot (eds.), *Road History; Planning, Building and Use* (Lausanne: Alphil, 2007).

Gijs Mom (met medewerking van Charley Werff en Ariejan Bos), *De auto; Van avonturenmachine naar gebruiksvoorwerp* (Deventer: Kluwer Bedrijfsinformatie, 1997).

Gijs Mom and Diekus van der Wey, "Ethanolverdamping en de dieselmotor; Literatuuronderzoek naar en ontwerp van een ethanolverdamper t.b.v. het gecombineerd gebruik van dieselbrandstof en ethanolgas in een DAF-dieselmotor" (BSc thesis, HTS-Autotechniek, Apeldoorn, The Netherlands, 7 June 1982).

J. Robert Mondt, *Cleaner Cars: The History and Technology of Emission Control Since the 1960s* (Warrendale, Pa.: Society of Automotive Engineers, Inc., 2000).

Kurt Möser, *Fahren und Fliegen in Frieden und Krieg; Kulturen individueller Mobilitätsmaschinen 1880–1930* (Heidelberg/Ubstadt-Weiher/Neustadt a.d.W./Basel: Verlag Regionalkultur, 2009).

Jürgen Mössinger, "Software in Automotive Systems," *IEEE Software* (March/April 2010) 92–94.

Tim Moran, "The Radial Revolution," *Invention & Technology* (Spring 2001) 28–39.

"More Flexible Springs Needed," *Journal of the Society of Automotive Engineers* 16 No. 1 (January 1925) 10–11.

The Motor Manual; The Original Practical Handbook, dealing with the Working Principles, Construction, Adjustment and Economical Maintenance of Modern Cars. Information on Touring and Motoring Law. The most comprehensive and up-to-date work on motoring. Embodying all the information a motorist should know of the subject (London: Temple Press, n.y. [1928][26]).

Richard R. Nelson and Sidney G. Winter, *An Evolutionary Theory of Economic Change* (Cambridge, Mass./London: The Belknap Press of Harvard University Press, 1982).

Christopher Neumaier, "Design Parallels, Differences and...a Disaster; American and German Diesel Cars in Comparison, 1968–1985," *ICON; Journal of the International Committee for the History of Technology* 16 (2010) 123–142.

Christopher Neumaier, *Dieselautos in Deutschland und den USA; Zum Verhältnis von Technologie, Konsum und Politik, 1949–2005* (Stuttgart: Franz Steiner Verlag, 2010).

Christopher Neumaier, "Eco-Friendly versus Cancer-Causing Perceptions of Diesel Cars in West Germany and the United States, 1970–1990," *Technology and Culture* 55 no. 2 (April 2014) 429–460.

"New Problems in Body Design; Low Cars and High Speed Necessitate Chassis Engineers' Aid, Buffalo Section Is Told," *Journal of the Society of Automotive Engineers* 21 No. 6 (December 1927) 629–630.

Peter Newman and Jeff Kenworthy, "The Ten Myths of Automobile Dependence," *World Transport Policy & Practice* 6 (2000) No. 1, 15–25.

Jan Nill and Jan Tiessen, "Policy, time and technological competition: lean-burn engine versus catalytic converter in Japan and Europe," in Christian Sartorius and Stefan Zundel (eds.), *Time Stategies, Innovation and Environmental Policy* (Cheltenham/Morthampton, MA: Edward Elgar, 2005) 102–132.

"Noise Studies Now Important in Design." *Journal of the Society of Automotive Engineers (Transactions)* 34 No. 2 (February 1934) 62.

Jan Norbye, *The Car and Its Wheels—A Guide to Modern Suspension Syustems* (Blue Ridge Summit: TAB Books, 1980).

Jan Norbye, "Straps, Springs and Swivels—The Story of Suspension," *Automobile Quarterly* 3 No. 4 (Winter 1966) 412–419.

Jan Norbye, *The Complete Handbook of Front Wheel Drive Cars* (Blue Ridge Summit: TAB Books, 1979).

Peter David Norton, *Fighting Traffic; The Dawn of the Motor Age in the American City* (Cambridge, Massachussetts/London: The MIT Press, 2008).

"Nos amis George Ageon…et Jean-Albert Grégoire nous ont quittés," *Journal de la Société des Ingénieurs de l'Automobile* (October 1992) 3.

M. Olley, "European and American Automobile Practice Compared," *Journal of the Society of Automotive Engineers*, 9 No. 2 (August 1921) 109–117.

Maurice Olley, "Why American, British and Continental Car Designs Differ," *Automotive Industries* 44 No. 10 (10 March 1921) 547–553.

Byron Olsen, "The Shift from Shift to Shiftless: Transmission Advances in U.S. Cars (1929–55)," *Automotive History Review* nr. 46 (Fall 2006) 25–40.

Renato J. Orsato, Marc Dijk, René Kemp and Masaru Yarime, "The Electrification of Automobility; The Bumpy Ride of Electric Vehicles Toward Regime Transition," in Frank W. Geels, René Kemp, Geoff Dudley and Glenn Lyons (eds.), *Automobility in Transition? A Socio-Technical Analysis of Sustainable Transport* (New York/London: Routledge, 2012) 205–228.

Wim Oude Weernink, "Europa blijft 'sportief' schakelen," *NRC Handelsblad* (3 February 2006) 10.

Nelly Oudshoorn and Trevor Pinch (eds.), *How Users Matter; The Co-Construction of Users and Technologies* (Cambridge, MA/London: The MIT Press, 2003).

Chris Paine (dir.), *Who Killed the Electric Car?* (Sony Pictures Classics, 2006 [DVD]).

Chris Paine (dir.), *Revenge of the Electric Car* (WestMidWest Productions and Area 23A, 2011 [DVD]).

"The Passenger Car of the Future," *Journal of the Society of Automotive Engineers* 5 No. 3 (September 1919) 236–241.

Diana de Pay, "Chapter 17: Commitment; Case G: Der Wankelmotor (The Rotary Engine)," in Horst Albach, *Culture and Technical Innovation; A Cross-Cultural Analysis and Policy Recommendations* (Berlin/New York: Walter de Gruyter, 1994) (Akademie der Wissenschaften zu Berlin, Research Report 9) 1013–1039.

Caetano C.R. Penna and Frank W. Geels, "The co-evolution of the climate change problem and car industry strategies (1979–2012): Replicating and elaborating the Dialectic Issue LifeCycle (DILC) model" (paper presented at the workshop "Electrification of the car: will the momentum last?", Technical University Delft, 29 November 2012).

Henri Perrot, "Four-Wheel Brakes," *Journal of the Society of Automotive Engineers*, 14 No. 2 (February 1924) 101–106.

Andrew Pickering, "Practice and posthumanism; Social Theory and a History of Agency," in Theodore R. Schatzki, Karin Knorr Cetina and Eike von Savigny (eds.), *The Practice Turn in Contemporary Theory* (London/New York: Routledge, 2001) 163–174.

Bruce Pietrykowski, "The Curious Popularity of the Toyota Prius in the United States," in Weert Canzler and Gert Schmidt (eds.), *Zukünfte des Automobils; Aussichten und Grenzen der autotechnischen Globalisierung* (Berlin: edition sigma, 2008) 199–211.

C.W.I. Pistorius and J.M. Utterback, "Multi-Mode Interaction Among Technologies," *Research Policy* 26 (1997) 67–84.

Jürgen Potthoff and Ulf Essers (with Helmut Maier and Barbara Guttmann), *75 Jahre FKFS–Ein Rückblick; Eine Chronik des Forschungsinstitut für Kraftfahrwesen und Fahrzeugmotoren Stuttgart–FKFS–aus Anlass seines 75-jährigen Bestehens, 1930–2005* (Stuttgart: Forschungsinstitut für Kraftfahrwesen und Fahrzeugmotoren Stuttgart FKFS, 2005).

Jürgen Potthoff and Ingobert C. Schmid, *Wunibald I.E. Kamm–Wegbereiter der modernen Kraftfahrtechnik* (Heidelberg/Dordrecht/London/New York: Springer, 2012).

Theodore M. Prudden, "Noise Treatment in the Automobile," *Journal of the Society of Automotive Engineers (Transactions)* 35 No. 1 (July 1934) 267–270.

John Quarles, *Cleaning Up America; An Insider's View of the Environmental Protection Agency* (Boston: Houghton Mifflin Company, 1976).

"Radio-Driven Auto Runs Down Escort," *New York Times* (28 July 1928) 28.

Rajendra Ravi (ed.), *The Saga of Rickshaw; Identity, Struggle and Claims; A Study by Lokayan* (New Delhi: Lokayan, 2006).

"Recent Progress in Automobile Design," *Journal of the Society of Automotive Engineers* 26 No. 2 (February 1930) 246–249.

Edward S. Reed, *James J. Gibson and the Psychology of Perception* (New Haven/London: Yale University Press, 1988).

K. Revers, "Kapers op de kust van D.A.F." [Competition for DAF], *Auto- en Motortechniek* (1962) 517.

Emilio Reyes, Alejandro Medina, Alexandru Dinu, Coen de Winter and Frank Rams, "World automotive congress FISITA" (student report of the course "Cars in Context", Eindhoven University of Technology, 2012).

H.E. Rice, "Problems in Ignition Development," *Journal of the Society of Automotive Engineers* 1 No. 2 (August 1917) 154–158.

Alois Riedler, *Wissenschaftliche Automobil-Wertung; Berichte I–V des Laboratoriums für Kraftfahrzeuge an den Königlich Technischen Hochschule zu Berlin* (Berlin/Munich, 1911).

Alois Riedler, *Wissenschaftliche Automobil-Wertung; Berichte VI–X des Laboratorium für Kraftfahrzeuge an der Königlich Technischen Hochschule, Teil II* (Berlin, 1912).

Søren Riis, "The Symmetry Between Bruno Latour and Martin Heidegger: The Technique of Turning a Police Officer into a Speed Bump," *Social Studies of Science* 38 No. 2 (April 2008) 285–301.

Joseph C. Robert, *Ethyl; A History of the Corporation and the People Who Made It* (Charlottesville: University Press of Virginia, 1983).

Everett M. Rogers, *Diffusion of innovations* (New York/London/etc: The Free Press, 1995[4]).

Arjan van Rooij, "Knowledge, money and data: an integrated account of the evolution of eight types of laboratory," *British Journal for the History of Science* 44 No. 3 (September 2011) 427–448.

Edmund Russell, *Evolutionary History; Uniting History and Biology to Understand Life on Earth* (Cambridge/New York etc.: Cambridge University Press, 2011).

Friedrich Sass, *Geschichte des deutschen Verbrennungsmotorenbaues von 1860 bis 1918* (Berlin/Göttingen/Heidelberg: Springer-Verlag, 1962).

Louis Baudry de Saunier, *Das Automobil in Theorie und Praxis; Band 2: Automobilwagen mit Benzin-Motoren; Mit einem Vorwort von Peter Kirchberg* (Leipzig: Reprintverlag, 1991) (first ed.: 1901).

Louis Baudry de Saunier, "Le carburateur à réglage automatique Panhard & Levassor" (Paris: Ch. Dunod, n.y] (reprint from *La Locomotion*, 27 December 1902).

Louis. Baudry de Saunier, "Le carburateur Krebs," *La Nature* 31, No. 1552 (21 February 1903) 177–179.

Achim Saupe, "Human Security and the Challenge of Automobile and Road Traffic Safety: A Cultural Historical Perspective," *Historical Social Research* 35 (2010) No. 4, 102–121.

Pier Paolo Saviotti, *Technological Evolution, Variety and the Economy* (Cheltenham/ Brookfield: Edward Elgar, 1996).

P. Saviotti, "The Measurement of Changes in Technological Output," in A.F.J. van Raan (ed.), *Handbook of Quantitative Studies of Science and Technology* (Amsterdam/New York/ Oxford/Tokyo: North-Holland, 1988) 555–610.

"Schalieolie zet Dakota op zijn kop," *De Ingenieur* (22 March 2013) 61.

Han Schaminée and Hans Aerts, "Short and Winding Road: Software in Car Navigation Systems," *IEEE Software* (July/August 2011) 19–21.

Michael Brian Schiffer (with Tamara C. Butts and Kimberly K. Grimm), *Taking Charge; The Electric Automobile in America* (Washington/London, 1994).

Friedrich Schildberger, *Bosch und der Dieselmotor* (Stuttgart: Robert Bosch GmbH, 1950).

J. Edward Schipper, "Passenger-Car Brakes," *Journal of the Society of Automotive Engineers* 10 No. 4 (April 1922) 273.

J. Edward Schipper, "Recent Trends in Automobile Design," *Journal of the Society of Automotive Engineers* (April 1918) 279–284.

Elske Schouten, "Indonesië ontdekt de welvaartsziektes," *NRC Handelsblad* (18–19 December 2012) 12.

F. Schmelz, H.-Ch. Graf von Seherr-Toss and E. Aucktor, *Gelenke und Gelenkwellen; Berechnung, Gestaltung, Anwendungen* (Berlin/Heidelberg/New York/Paris/Tokyo: Springer Verlag, 1988).

Hermann Scholl, "Elektronik im Kraftfahrzeug," in [VDI-Gesellschaft Fahrzeugtechnik], *100 Jahre Automobil; Tagung Fellbach, 17. und 18. April 1986* (Düsseldorf: VDI Verlag, 1986) (VDI Berichte 595) 307–312.

Carlos A. Schwantes, "The West Adapts the Automobile: Technology, Unemployment, and the Jitney Phenomenon of 1914–1917," *Western Historical Quarterly* 16 No. 3 (July 1985) 307–326.

James Scoltock, "Stephen Wallman," *Automotive Engineer* (July/August 2009) 7.

Bruce Seely, *Building the American Highway System; Engineers as Policy Makers* (Philadelphia: Temple University Press, 1987).

Ziaush Shams, M.M. Haq and Towhida Rashid, "Rickshaw-pullers of Dhaka City and rickshaw pulling as a means of livelihood," *Oriental Geographer* 49 No. 1 (January 2005) 33–46.

Lao She, *Rickshaw; the novel Lo-t'o Hsiang Tzu* (Honolulu: The University Press of Hawaii 1979).

G.R. Shearer, "The Rolling Wheel—The Development of the Pneumatic Tyre," *Proceedings of the Institution of Mechanical Engineers* 191 (1977) 75–87.

William Sierzchula, Sjoerd Bakker and Kees Maat, "The influence of financial incentives on the adoption of electric vehicles" (paper presented at the workshop "Electrification of the car: will the momentum last?", Technical University Delft, 29 November 2012).

R. Simons, *Geschichte der Automobile mit Frontantrieb* (München: Deutsches Museum, 1981).

John Kenly Smith, Jr., "The Scientific Tradition in American Industrial Research," *Technology and Culture* 31 No. 1 (January 1990) 121–131.

Andreas Spahn, "Moralizing Mobility? Persuasive Technologies and the Ethics of Mobility," *Transfers; Interdisciplinary Journal of Mobility Studies* 3 No. 2 (Summer 2013) 108–115.

Daniel Sperling (with contributions from Mark A. Delucchi, Patricia M. Davis, and A.F. Burke), *Future drive; Electric Vehicles and Sustainable Transportation* (Covelo, Cal., 1995).

Daniel Sperling and Deborah Gordon, *Two Billion Cars; Driving toward Sustainability* (Oxford, etc.: Oxford University Press, 2009).

F.C. Stanley, "Causes and Prevention of Squeaking Brakes," *Journal of the Society of Automotive Engineers* 18 No. 2 (February 1926) 160–162.

John M. Staudenmaier, S.J., *Technology's Storytellers; Reweaving the Human Fabric* (Cambdige, Ma/London: SHOT and MIT Press, 1989).

Roland Stephen, *Vehicle of Influence; Building a European Car Market* (Ann Arbor: The University of Michigan Press, 2000).

David Strand, *Rickshaw Beijing; City People and Politics in the 1920s* (Berkeley/Los Angeles/London: University of California Press, 1989) 41.

Edward B. Sturges, "A Mechanical Continuous-Torque Variable-Speed Transmission," *Journal of the Society of Automotive Engineers* (July 1924) 86–92.

Donald N. Sull, "The Dynamics of Standing Still: Firestone Tire & Rubber and the Radial Revolution," *Business History Review* 73 (Autumn 1999) 430–464.

Donald N. Sull, Richard S. Tedlow and Richard S. Rosenbloom, "Managerial Commitments and Technological Change in the US Tire Industry," *Industrial and Corporate Change* 6 (1997) 461–501.

Donald N. Sull, "Why Good Companies Go Bad," *Harvard Business Review* (July–August 1999) 42–52.

Bo Sundin, "From Waste to Opportunity: Ethanol in Sweden during the First Half of the 20th Century" (paper presented at the Society for the History of Technology conference, October 19, 2007).

Howard Taylor, "Forging the Job; A Crisis of 'Modernization' or Redundancy for the Police in England and Wales, 1900–39," *British Journal of Criminology* 39 No. 1 (Special Issue 1999) 113–135.

Alfred Teves GmbH; Eine Chronik im Zeichen des technischen Fortschritts (Frankfurt am Main/Stuttgart: Alfred Teves GmbH/Motorbuch Verlag, n.y. [1986]).

Eric Tompkins, *The history of the pneumatic tyre* (n.p. [Birmingham], 1981).

H. Ledyard Towle, "Projecting the Automobile Into the Future," *Journal of the Society of Automotive Engineers* 29 No. 1 (July 1931) 33–39.

Wilhelm Treue, *Achse, Rad und Wagen; Fünftausend Jahre Kultur- und Technikgeschichte* (Göttingen: Vandenhoeck & Ruprecht, 1986).

J. Trommelmans, *Het moderne auto ABC: Constructie en werking* (Deventer: Kluwer Technische Boeken, 1993).

James M. Utterback and William J. Abernathy, "A Dynamic Model of Process and Product Innovation," *Omega, The International Journal of Management Science* 3 (1975) No. 6, 639–656.

James M. Utterback and Fernando F. Suárez, "Innovation, competition, and industry structure," *Research Policy* 22 (1993) 1–21.

Pieter E. Vermaas and Wybo Houkes, "Technical functions: a drawbridge between the intentional and structural natures of technical artefacts," *Studies in History and Philosophy of Science* 37 (2006) 5–18.

J.G. Vincent and W.R. Griswold, "A Cure for Shimmy and Wheel Kick," *Journal of the Society of Automotive Engineers* 24 No. 4 (April 1929) 388–396.

J.G. Vincent, "The Internal-Combustion Engine as Developed by the Automotive Industry," *The Journal of the Society of Automotive Engineers* (September 1920) 235–256.

Walter G. Vincenti, "Real-world variation-selection in the evolution of technological form: historical examples," in John Ziman (ed.), *Technological Innovation as an Evolutionary Process* (Cambridge: Cambridge University Press, 2000) 174–189.

Walter G. Vincenti, *What Engineers Know and How They Know It, Analytical Studies of Aeronautical History* (Baltimore/London: Johns Hopkins University Press, 1990).

Marlies ter Voorde, "CO_2-uitstoot hoger dan ooit na korte dip," *De Volkskrant* (13 December 2012) 27.

Peer Vries, "Decline of the West—Rise of the Rest?" *Journal of Modern European History* 11 (2013) 315–328.

E.R. Wagner, "Driveline and Driveshaft Arrangements and Constructions," in *Universal Joint and Driveshaft Design Manual* (Warrendale, Pa: SAE, 1979) (Advances in Engineering Series No. 7) 3–10.

E.R. Wagner, "Tracta Universal Joint," in *Universal Joint and Driveshaft Design Manual* (Warrendale, Pa: SAE, 1979) (Advances in Engineering Series No. 7) 127–140.

James Francis Warren, *Rickshaw Coolie; A People's History of Singapore 1880–1940* (Singapore: Singapore University Press, 2003).

G. Weber and H. Zoeppritz, "Zur Entwicklungsgeschichte des Kraftfahrzeugreifens," *Automobiltechnische Zeitschrift* 64 No. 9 (September 1962) 257–265.

A. Wegener Sleeswijk, "Reconstruction of the south-pointing chariots of the Northern Sung dynasty; Escapement and differential gearing in 11th century China," *Chinese Science* 2 (January 1977) 4–36.

"Weights of 1921 Cars on Which Kansas Bases License Fee," *Automotive Industries* 45 No. 7 (18 August 1921) 330–331.

Heike Weishaupt, *Die Entwicklung der passiven Sicherheit im Automobilbau von den Anfängen bis 1980 unter besonderer Berücksichtigung der Daimler-Benz AG* (Bielefeld: Delius & Klasing, 1999).

Heike Weishaupt, "Die Entwicklung der passiven Sicherheit im Automobilbau von den Anfängen bis 1980," in Harry Niemann and Armin Hermann (eds.), *Geschichte der Strassenverkehrssicherheit im Wechselspiel zwischen Fahrzeug, Fahrbahn und Mensch* (Bielefeld: Delius & Klasing, 1999) 99–122.

E.E. Wemp, "History of Automotive-Clutch Development," *Journal of the Society of Automotive Engineers* (October 1925) 361–371.

Erich Werminghoff, "Die Einrichtungen zur Vergasung von Otto-Kraftstoffen," *Automobil Industrie* (1959) 48–60.

Jameson M. Wetmore, "Engineering with Uncertainty: Monitoring Air Bag Performance," *Science and Engineering Ethics* 14 (2008) 201–218.

"What Buyers Want on Their Cars in Foreign Countries," *Automotive Industries* 54 No. 18 (6 May 1926) 766–768.

"Where cars are better," *Autocar* 59 No. 1667 (14 October 1927) 735–736.

Roger B. Whitman, *Motor-Car Principles; The Gasoline Automobile* (New York: D. Appleton and Company, 1907).

[E.P. Willoughby], *The Motor Manual; A Practical Handbook, dealing with the Working Principles, Construction, Maintenance and Economical Running of the Motorcar, together with a Chapter on the Law as it affects the Motorist* (London: Temple Press/The English Universities Press, 1948[33]).

Roland Wolf, *Le véhicule électrique gagne le coeur de la ville* (n.p., n.d. [Paris, 1995]).

Theodore P. Wright, "Automotive Safety Research," in id., "Articles and Addresses of Theodore P. Wright; Volume IV: A Post-retirement Miscellany" (Buffalo, NY: Cornell Aeronautical Laboratory, Inc., Technical Services Department, of Cornell University, 1961) (typescript) 159–179.

A.Y., "De massa motorisering in Europa en de automatische transmissie (II)," *Auto- en Motortechniek* (1963) 150–154.

Jeffrey Robert Yost, "Components of the Past and Vehicles of Change: Parts Manufacturers and Supplier Relations in the U.S. Automobile Industry," (unpubl. diss., Case Western Reserve University, May, 1998).

Kenneth F. Zino, "Ist der Vergaser überholt?" *Automobil Revue* (April 1981) 19, 21.

"Zur Geschichte des Vergasers," *Kraftfahrzeugtechnik* 37 (1987) No. 1, 7–9.

Index

About the Author

Dr. ing. Gijs Mom teaches at Eindhoven University of Technology. A long-term SAE International member, he has been educated as both a literary historian and an automotive engineer (in that order). After having briefly worked at Renault, Paris (engine development), he turned to the history of technology and wrote a dissertation which was published in 2004, at Johns Hopkins University Press, *The Electric Vehicle: Technology and Expectations in the Automobile Age*. For this book, he received the ASME Engineer-Historian Award 2004 as well as the Best Book Award from the Society of Automotive Historians.

Dr. Mom is the founder of the Netherlands Centre for Automotive History Documentation (NCAD) and cofounder of the International Association for the History of Transport, Traffic and Mobility (T²M), of which he was the first president. He is editor of the journal *Transfers* and is finishing two monographs on the history of Atlantic Automobilism as well as World Mobility History.